Why Place Matters

Saburo Horikawa

Why Place Matters

A Sociological Study of the Historic Preservation Movement in Otaru, Japan, 1965–2017

Springer

Saburo Horikawa
Department of Sociology
Hosei University
Tokyo, Japan

Translated by
Kate Dunlop
Seattle, Washington, USA

ISBN 978-3-030-71599-1 ISBN 978-3-030-71600-4 (eBook)
https://doi.org/10.1007/978-3-030-71600-4

Translation from the Japanese language edition: *Machinami Hozon Undo no Ronri to Kiketsu: Otaru Unga Mondai no Shakaigakuteki Bunseki* by Saburo Horikawa (Tokyo: Tokyo Daigaku Shuppankai),

This Springer imprint is published by the registered company Springer Nature Switzerland AG
The registered company address is: Gewerbestrasse 11, 6330 Cham, Switzerland

Preface

March 27, 1984. 10 p.m. This is when I first stepped off the train onto the snowy platform of Otaru Station. Although I had no way of knowing at the time, this moment marked the beginning of thirty-six years of fieldwork in Otaru.

I had entered university the previous April and immediately developed an interest in urban sociology. My newfound interest was initially nothing more than a vague curiosity that I could not express with any degree of coherence. This in itself was frustrating, but the larger problem was that I had never personally experienced the various issues at the heart of my urban sociology lectures. The emergence of a "resident-led movement" was surely an event of major significance within a local community, but I could not grasp the reality behind the term. Other terms—"class struggle," "structural functionalism"—also belonged to this distant world, well beyond my comprehension. Day after day, I attended these lectures without any sense of personal connection or shared experience.

I made a plan to break this deadlock: I would travel to Kobe, and observe with my own eyes a residents' movement that had garnered national attention for challenging a municipal government so intent upon development that it was often referred to as "Kobe Co., Ltd." I wanted to personally experience every step in the process by which a sociological thesis is developed: I would first observe the field site with my own eyes and ears, and slowly discern how these experiences could be abstracted and woven together into a compelling thesis. I thought that if I could undergo this process just once, I would be able to glean the reality behind the term "residents' movement." The entire university career of this intensely earnest nineteen-year-old would hinge on the success of "Operation Kobe."

I happened to mention this plan to my eldest brother, who told me "if that's your goal, perhaps you should take a look at Otaru, in Hokkaido, where everyone is up in arms about the canal." I had visited Otaru briefly as a junior high school student, but had no real knowledge of the "canal controversy" that had divided the city. As I listened to my brother, I realized that Kobe was not the only place where I could achieve my goal of personally experiencing a local resident-led movement. In early spring I set out for Otaru. Sheer chance had led me to the city that would become my fieldwork site for the next three decades.

On that cold March night, Yamaguchi Tamotsu was waiting for me in front of Otaru Station. Yamaguchi, a key member of the campaign to save the Otaru canal, drove me through the snow-covered streets of the city to the coffeehouse he owned and operated. We passed through the shop to his narrow living quarters in the back. Some hours later, empty beer cans littered the room and I heard the sound of the morning newspaper delivery scooter. Yamaguchi had talked insatiably through the night about the state of the preservation movement.

I had listened intently to Yamaguchi, who was passionate but also extremely logical in his arguments. But I was puzzled. Why was Yamaguchi so insistent that the canal be saved? Why was the Otaru canal so important to him? What was so wrong with tearing down old structures and replacing them with new ones? Why did Yamaguchi continue to participate in this preservation movement, when there wasn't a single yen in it for him? And why was the Otaru city government so obstinate in refusing to make any change to the road construction plan that threatened the canal?

I didn't have the answers to these questions, but I wanted them—no, I think I *needed* to find them. Before my eyes was a man who was putting into practice, with enormous conviction, the principles of community development, preservation, and resident-led activism. Here was a man who was trying to do something for the entire city, with no regard for personal benefit. I sensed that if I could understand a man like Yamaguchi, who even as he clashed with his opponents was trying to collectively protect and create something for the city of Otaru, I would be able to finally discern the real significance of sociological inquiry. *Why preserve?* I had to find the answer to this question through every possible means. Slowly but surely, my research proceeded from that first nightlong conversation with Yamaguchi in 1984. Otaru was my new field site, and I was soon a frequent visitor.

Yamaguchi's coffee shop was a common meeting place for members of the preservation movement, and I was offered free lodging whenever I came to town. There I was able to focus on the subjective world of local activists, and interrogate their reasons for seeking to save the canal and their motivations for joining the movement. I gradually expanded the scope of my inquiry, from the people who visited Yamaguchi's coffee shop to their broader network, and from there, to the municipal government. In 1997, when I began teaching a social research practicum as a young faculty member at Hosei University I added a fixed point observation survey of Otaru's historic canal district to the scope of my fieldwork. I still return to Otaru to conduct this fixed point observation survey every year.

In the course of my three decades of research, I have interviewed countless people, and walked endlessly through the narrow streets of Otaru. My questions have deepened and become more complex, but I have never grown tired of my subject matter. This is because Japanese cities remain entirely indifferent to history, and enslaved to the new. In the face of the wholesale destruction of historic districts and the unchecked construction of new buildings, my inquiry has kept its relevance. *Why preserve?* The question that awaited me on that March night so long ago is the question I have attempted to answer in this volume. It is my hope that this case study of a small port city in northern Japan can nevertheless speak to the universal themes of cities and preservation, change, and control.

Tokyo, Japan Saburo Horikawa

Acknowledgements

Writing a book is a solitary act. Yet a book is never the creation of just one person. I could not have embarked on the project of writing this book without considerable support and encouragement—and the generous cooperation of countless people in Otaru, my fieldwork site. Even when I secluded myself in my office during the final stages of the writing process, I was never really alone. The long list of names that follows should thus come as no surprise, particularly given the fact that this book is the result of thirty-six years of fieldwork. Indeed, I am humbled and a little awed by the number of people I have the good fortune to acknowledge in these pages.

It was my great fortune to study under such excellent teachers as an undergraduate and graduate student. Shimazaki Minoru (formerly of Chuo University), Okuda Michihiro (formerly of Rikkyo University), Tomoeda Toshio (Osaka University), Kawai Takao (Keio University), and Fujita Hirō (formerly of Keio University) responded to my budding academic interests with rigorous introductions to urban sociological theory, fieldwork methodology, and research fundamentals. Kawai Takao, who never uttered a superfluous word and remained intently focused on the substance of my work, was particularly influential. His comments on my research reports were always brief and to the point. He would tell me, "You're not there yet. You won't be finished until the depth of your research can be appreciated not only by your fellow scholars, but even by the people at your fieldwork site." While his comments were unsparing, Prof. Kawai remained laser-focused on my research, and never raised more worldly concerns. I cannot recall a single instance in which he pushed me to finish my dissertation as soon as possible in order to secure employment, or to adopt a more fashionable research theme. Looking back, I realize that my ability to remained steadfastly focused on my field site, even when the going got tough, was because of Prof. Kawai, who patiently waited for my efforts to pay off. To Prof. Kawai, who has now retired to far-away Yamagata, I offer my heartfelt thanks. I would also like to acknowledge the support of my former classmates Takemura Hideki (Keio University) and Ōyane Jun (Senshu University), who studied with me under Prof. Kawai. Arisue Ken (Keio University), Terada Ryōichi (Meiji University), and Hama Hideo (Keio University) supervised my doctoral dissertation, upon which this book is based, and offered me endless encouragement along the way to its completion. I am deeply grateful for their advice and support.

I have been equally fortunate in the encouragement and guidance I received later in my career. In particular, I would like to acknowledge the support of Andrew Gordon (Harvard University), Timothy George (University of Rhode Island), Torigoe Hiroyuki (professor emeritus, Waseda University), Tamano Kazushi (Tokyo Metropolitan University), Nishimura Yukio (professor emeritus, the University of Tokyo), Diane Barthel-Bouchier (Stony Brook University, SUNY), Hasegawa Kōichi (professor emeritus, Tōhoku University), Machimura Takashi (Hitotsubashi University), Jinnai Hidenobu (professor emeritus, Hosei University), Hoshi Takashi (Kōgakuin University), Jordan Sand (Georgetown University), Christopher Gerteis (SOAS, London), and Peter Siegenthaler (Texas State University).

The late Funabashi Harutoshi (Hosei University) would shrewdly get to the very essence of the issue at hand, and always offered me the exact piece of advice that I needed. Indeed, Prof. Funabashi invariably brought a truly remarkable degree of clarity to my rather inchoate discussions, and helped me to organize my thoughts and clarify the underlying theoretical framework. Whenever I came to an impasse in my writing, I knew that a conversation with Prof. Funabashi would help me break the deadlock and press forward. Professor Funabashi passed away in 2014, before I could present him with a copy of my finished work. This is something I will always regret.

Such a prolonged fieldwork project would not have been possible without the financial support of a number of public and private organizations. This book is a product of the generous research support I have received from Japan's Ministry of Education, Culture, Sports, Science and Technology, the Japan Society for the Promotion of Science, Japan Economic Research Foundation, Dai'ichi Seimei Zaidan, and Hosei University. Harvard University, Stony Brook University SUNY, and the Sainsbury Institute for the Study of Japanese Arts and Culture generously provided me with the opportunity to share and refine portions of this work as an invited lecturer.

For their devoted support of their teacher, I would also like to extend my warm thanks to the young sociology students who have participated in the "Horikawa Seminar" and the Horikawa Social Research Practicum at Hosei University. These young men and women made me realize the truth of the old maxim, "to teach others is to learn deeply oneself."

The fact that the list of people to whom I am indebted is too long to continue with by name is a reminder of how this work came to completion. However, the advice and suggestions I received from my colleagues in the fields of environmental sociology, urban sociology, and regional sociology are incorporated throughout this work. I hope this will be understood as a reflection of my enduring appreciation.

The English-language publication of my work was supported by a grant from the Japan Society for the Promotion of Science (18HP6003) that supports the sharing of research findings. Fritz Schmul, an editor at Springer Nature, was the first to appreciate the value of publishing an English translation of the 504-page original, which was originally published by the University of Tokyo Press. Fritz and I first discussed the possibility of publishing an English language translation in July 2016, at an airy courtyard cloister on the University of Vienna campus. Since then Fritz has cheered me on in this challenging endeavor over countless cups of sake at his favorite tempura restaurant in Tokyo. This English edition would never have been possible without his warm encouragement. Fritz's editorial acumen was confirmed when my book went on to receive three major academic awards. Margaret Deignan took over editorial duties from Fritz in the final stages of bringing this book to publication. I thank Ms. Deignan and her team—Jill Ritchie, Catalina Sava, Ilona Isaeva, Ram Prasad Chandrasekar, Muruga Prashanth, Hermine Vloemans and Amudha Vijayarangan—for their splendid work and constant support.

My translator, Kate Dunlop, deserves special mention. During our frequent meetings to review the English manuscript, I was astonished by Kate's ability to retain the nuance of the Japanese original in her translation (although she would always dismiss my astonishment with an almost Japanese modesty). Our meeting in Kate's sunlit living room in Seattle is a particularly fond memory.

Of course, my greatest debt is to the citizens of Otaru. Due to privacy concerns, I cannot thank many of my informants by their real names. But I offer my heartfelt thanks to the citizens of Otaru and the Otaru city government—and most of all, to the former activists who led the campaign to save the Otaru canal. These preservation activists welcomed me into their home and talked with me late into the night about the history of their campaign. Many rummaged through their old storage spaces to dig out primary sources (often handwritten documents) from the preservation campaign. Meanwhile, city officials met with me in their offices to passionately explain the necessity of paving over the canal, and elaborate on their vision for community development in Otaru. Other city hall employees made endless trips down to the basement archive room on my behalf, and good-naturedly pushed their work aside whenever I came to ask more questions. I have been granted interviews from hospital beds. My requests for help with my survey of Otaru have been met with invitations to partake in delicious crab and sake. These countless acts of kindness are forever etched in my mind.

In some cases, this type of generous support can end up interfering with the pursuit of objective research. But this was not the case in Otaru. Every time I met with a new informant, I would remind him or her that I planned to write a report on my findings, and that my conclusions might not be to his or her liking. I warned my informants that, while I would of course correct any factual inaccuracies, I would not accommodate any other requests for revision. I would then ask if they still wished to proceed. I always received the same answer: "But of course! You should do all the research you can, and then write with conviction." I am profoundly grateful for the unparalleled tolerance and generosity displayed by the people of Otaru.

Notwithstanding privacy concerns, I would like to acknowledge the following individuals in Otaru. Yamaguchi Tamotsu always welcomed me cheerfully into his home. Ogawara Tadashi readily stayed up until dawn with me to continue our conversation. No matter how busy she was, Mineyama Fumi never refused to meet with me during my many visits to Otaru over the years. Former Otaru mayor Shimura Kazuo would prepare meticulous notes for me and always talked with real passion during our meetings. The cooperation of these individuals was absolutely essential to my research. Here I want to emphasize once more the enormous support I received from the Otaru municipal government. Even when they were at their busiest—preparing for questioning at the city assembly, for example—city officials readily provided me with the documents I requested, and patiently answered my questions.

The list goes on. During our many conversations, Sasaki Kyōjirō always spoke from his heart. Also deserving of special thanks are Yanagida Ryōzō, Ishizuka Masaaki, and Morishita Mitsuru—the so-called "Hokudai trio," who clearly and calmly guided me in our many conversations. I also owe these three men an apology: I was unable to adequately situate their work within my history of the canal preservation movement in Otaru. Their many accomplishments, including their advocacy of the "educational power of the environment" (*kankyō no kyōikuryoku*), cannot be contained within a single case study of Otaru; they demand further attention from Japan's architects and urban planners. I did not omit an analysis of their attitude to life and the specific terms they generated because they were insignificant, but because they were all *too* significant to be treated in anything less than comprehensive fashion. I have faith that their work will soon receive the attention it deserves.

Finally, I acknowledge with enormous gratitude the support of my family: my late parents, Atsuhiro and Hatsue; my elder brothers, Jun and Jōji; my sisters-in-law, Hikaru, Midori, and Yumiko; my father-in-law, Isamu; my mother-in-law, Keiko. Without their moral support, I could never have completed this book. But I have saved for last the two most important people of all. Parenting my son, Gen (pronounced as "Ghen"), has taught me the fundamental joy of learning. And it is no exaggeration to say that my wife, Yukiko, has supported and guided me every single step of the way. It is to Gen and to Yukiko that I offer my most profound and heartfelt thanks.

Saburo Horikawa

Introduction

Abstract The Great East Japan Earthquake of 2011 wrought enormous and diverse changes within Japanese society, and revealed that life cannot go on without a stable form. What mechanisms might allow us to control changes to the material "form" that is the basis of our daily existence—be that one's own house or the broader townscape? This is the question that guides this book. Bearing in mind the lessons of March 11, I address this question through an examination of the community development efforts practiced by residents of Otaru, a once prosperous city on Japan's northernmost island of Hokkaido. This book asks: "who is affected by changes to urban space, and how are they affected?" Local townscape preservation movements are one window into these issues. The preservation campaign launched by residents of a small port city in northern Japan offers a unique perspective on half a century of change within Japanese society. It also illuminates the relationship between urban society and the built environment, and suggests a sociological theorization of this relationship.

Keywords Townscape · Preservation · Social movement · Community development · Sociology

Ten years have passed since the devastating March 11 earthquake, tsunami, and Fukushima Daiichi nuclear disaster. Since then, the survivors of this triple disaster have fought to rebuild their lives. Yet they have also remained deeply attached to *photographs* of their old homes. Amid the struggle of their daily lives, they seek out old photographs that recall lost homes and communities. Photographs cannot fill empty stomachs or alleviate the discomfort of cramped temporary housing units. Why, then, are survivors so intent upon finding them? Perhaps they hope that these images might allow them to heal, or at least soothe their troubled spirits. However, this does not fully explain the intensity of their hunt for photographs of their family and hometowns. What is the real significance of their relentless search?

This is a question with no ready answer. What I would like to consider here is the fact that there is a shape to human activity. In other words, the society in which we live has a physical form, and changes to this physical form are capable of transforming society.

We tend to forget that while the images conjured up by the word "society" are abstract, *society itself is material.* Human activities comprise society, and these activities are not conducted in a vacuum, but rather through earthly materials. Houses, roads, clothing, food, tableware, musical instruments, buildings, railroads, harbors, seawalls—none of these things can exist without a material form. In a very literal sense, the March 11, 2011 earthquake and tsunami shook and destroyed the material basis of Japanese society.

At the same time, the *material components of society are social.* Just as the original apple was entirely different from the apples grown and consumed today, material substances have been transformed by society's tastes and inclinations, and by the significance society confers on them. The original apples were modified for the convenience of human society. Similarly, the forests that we often assume to be unchanged since time immemorial are, in most cases, second growth—the result of a decision made by society at a particular juncture in history. A beautiful forest is not necessarily a "natural" landscape. It would be more accurate to describe it as a "landscape with socially-assigned meaning."

Thus, society is material, but materials are equally social in nature. Consequently, when an earthquake, tsunami, or nuclear disaster fatally alters the material foundation, society can no longer survive in its old form and must submit to the inevitability of drastic change. Society and materials exert reciprocal influences upon each other: societal changes transform the material world, while the effects of these transformations return to reverberate through society.

It is because changes to physical space have such an enormous impact on people's daily lives and consciousness that people attempt to ward off these changes, or, when this proves impossible, to compensate for them. When a tsunami has swept away any trace of your home, or radioactive contamination has forever condemned your hometown, the sense of loss must be overwhelming. Imagine a feeling of helplessness so strong that it makes you uncertain of who you are, or where you belong. Or the desolation you would feel if the people who populated your daily life suddenly disappeared, without so much as a goodbye. Or a feeling of utter isolation akin to the sensation of walking alone through the soundless void of space. Japan's relatively quick resumption of nuclear power generation, meanwhile, must prompt a very different sense of alienation among survivors, who can never hope to resume their former lives. The hunt for old photographs is one manifestation of an effort to restore what was lost. For it is not enough for humans to remain living organisms—they must live embedded within the social relationships of their local community. Otherwise, simply surviving the earthquake and tsunami would have been enough, and there would be no need to sift through the rubble in search of images from the pre-3/11 world.

Ten years after the March 11 disasters, young volunteers continue to look for old photographs within the rubble of the disaster area. They clean whatever they find, and attempt to return the salvaged images to their owners. This is because people live and remember through certain material forms: it is the familiar landscape of our hometown that enables us to feel we are truly home. As social relationships are expressed intensively within a certain form, the loss or transformation of this form signals an immediate change to these relationships. Relationships were permanently severed when the earthquake and tsunami destroyed their material form. Yet photographs preserve these forms, and the social relationships that lie embedded within them. Having lost their homes and all other mementos of the deceased, "at least a photograph" has become the final plea of those left behind.

The Great East Japan Earthquake wrought enormous changes within Japanese society. These changes are diverse and wide-ranging. Given the inevitability of change, whether natural or man-made, people have always hoped to welcome it when it arrives. What was unexpectedly revealed by the 2011 earthquake was the fact that life cannot go on without a stable form. What mechanisms might allow us to control changes to the material "form" that represents the basis of our daily existence—be that one's own house or the broader townscape? This is the question that guides this book. Bearing in mind the lessons of March 11, I address this question through an examination of the community development efforts practiced by residents of the port city of Otaru. The guiding concern running through this book is "who is affected by changes to urban space, and how are they affected?"

It is generally accepted that urban spaces must be regulated and controlled, rather than abandoned to the chaos of unchecked change. However, the regulation of urban space is far from straightforward. We have learned that livable urban spaces cannot be achieved by the free market ("market failure"). Nor is there any guarantee that government intervention in city planning will achieve something more satisfactory ("government failure"). Despite all the indicators that attest to Japan's status as an economic superpower, the Japanese continue to live in cramped quarters frequently derided as "rabbit hutches," and city residents are often forced to walk along narrow roads with no sidewalks, dodging passing cars as they go. Verdant parks are but a distant dream for these city dwellers. As new apartment buildings and high-rise office buildings reach closer to the sky each day, the desire for something as simple as a sidewalk seems pathetically modest by comparison.

Given the reality of "market failure" and "government failure," it is time to ask what mechanisms might better control change within Japanese cities. It is easy to casually opine on the "lack of livability" in Japanese cities. However, this is hardly the sociological approach to the issue, to say the least. Social science demands that we ask *why* Japanese cities are so poorly equipped to meets the needs of residents. If Japan's economic strength and talented bureaucrats cannot create livable cities (or a sense of abundance and satisfaction among city residents), we can assume that certain decision-making processes and structures prevent Japan's ample resources from being channeled into community and urban development. To understand the current state of Japanese cities, we must begin by investigating the allocation of resources, existing decision-making processes and structures, and the consciousness and desires of local residents.

Local townscape preservation movements are one window into these issues. The preservation campaign launched by residents of a small port city in northern Japan offers a unique perspective on half a century of change within Japanese society. More importantly, however, it illuminates the relationship between urban society and the built environment, and suggests a sociological theorization of this relationship. The residents of Otaru rejected modern Japan's tendency to abandon history in its constant pursuit of the new, and raised their voices in defense of the historic Otaru townscape. An analysis of their philosophy and practices carries enormous implications for a rapidly globalizing and homogenizing world.

Contents

About the Author

Saburo Horikawa is a professor of sociology at Hosei University in Tokyo, Japan. He received his M.A. and Ph.D. in sociology from Keio University, and holds a B.A. *summa cum laude* from Chuo University. Dr. Horikawa teaches and conducts research in the fields of environmental and urban sociology. He was previously a Visiting Scholar at the Edwin O. Reischauer Institute of Japanese Studies at Harvard University, the editor of *Kankyō to Kōgai* (*Research on Environmental Disruption*), and guest lecturer at the University of Tokyo. He continues to teach graduate-level courses in sociology at Keio University as a visiting professor. Dr. Horikawa sits on both the Executive Council of the Research Committee on Environment and Society at the International Sociological Association and the International Exchange Committee of the Japanese Association for Environmental Sociology. He is also an Associate in Research at the Reischauer Institute of Japanese Studies. In 2018, his thirty-three years of fieldwork in Otaru, Japan culminated in the publication of *Machinami Hozon Undō no Ronri to Kiketsu: Otaru Unga Mondai no Shakaigakuteki Bunseki* from the University of Tokyo Press. This book has gone on to win three major academic awards: the Isomura Memorial Prize, from the Japan Association for Urban Sociology; the Ishikawa Prize, from the City Planning Institute of Japan; and the Okui Memorial Prize, from the Japan Society for Urbanology. All three prizes are the highest awards given by their respective associations. This volume, *Why Place Matters: A Sociological Study of the Historic Preservation Movement in Otaru, Japan, 1965–2017*, is the translated version of Dr. Horikawa's award-winning book.

Dr. Horikawa is co-editor of *A Chronology of the Fukushima Nuclear Plant Disaster and Evacuation* (2018, in Japanese), *A General World Environmental Chronology* (2014, in English), *The Environment and Dynamism of the Public Sphere* (2012, in Japanese), and *Urban Society and Risk* (2005, in Japanese). Articles by Dr. Horikawa appear in such publications as the *Japanese Sociological Review*, *Journal of Environmental Sociology*, *Annals of Japan Association for Urban Sociology*, *Kankyō to Kōgai*, *Advances in Social Research*, *Annual Review of the Kanto Society for Urbanology*, and *Ambiente & Sociedade*.

List of Figures

List of Tables

Chapter 1
Why Preserve?: Positioning the Issue and Methods of Analysis

Abstract Cities change. Old townscapes are destroyed and replaced with shiny new buildings. The cities of Japan are in a perpetual state of redevelopment. This process is so self-evident that no one gives it a second thought. Yet some people argue that old townscapes should be preserved. As cities change to meet the demands of a new age, those who would prefer to save existing townscapes appear resistant to change. Why would they attempt to protect old townscapes that are hardly convenient or the most comfortable? This is the central concern of this chapter. What does the act of preservation seek to achieve? What are the demands for preservation based upon? Are these demands simply a reflection of individual taste? What type of people participate in preservation movements? Are development and preservation always in conflict, or is it possible to conceive of a different relationship between the two? Most importantly, what do conflicts surrounding preservation reveal about society? In this chapter, I address these questions and, through the use of a detailed case study, undertake a sociological approach to the question of "why preserve?"

Keywords Townscape · Preservation · Otaru · Control of change

1.1 Why Preserve? Positioning the Issue

Cities change. Old townscapes are destroyed and replaced with shiny new buildings. The cities of Japan are in a perpetual state of redevelopment. This process is so self-evident that no one gives it a second thought.

Yet some people argue that old townscapes should be preserved. As cities change to meet the demands of a new age, those who would prefer to save existing townscapes appear resistant to change. Why would they attempt to protect old townscapes that are hardly convenient or the most comfortable? This is the central concern of this volume.

What does the act of preservation seek to achieve? What are the demands for preservation based upon? Are these demands simply a reflection of individual taste? What type of person participates in preservation movements? Are development and preservation always in conflict, or is it possible to conceive of a different relationship between the two? Most importantly, what do conflicts surrounding preservation reveal

S. Horikawa, *Why Place Matters*,
https://doi.org/10.1007/978-3-030-71600-4_1

about society? These are the questions that have shaped my research. Through the use of a detailed case study, I adopt a sociological approach to the question of "why preserve?"

In the spring of 1984, my interest in such questions led me to begin my research on the subject of urban "preservation." The core concern guiding my research can be summarized in a single question: How do society control and shape "changes" to the urban environment?

This question demands further explanation. When urban landscapes are transformed by redevelopment, the entire city does not change uniformly. Nor does all change elicit strong resistance. People who join preservation movements do not set out to prevent any and all change to their environment. They do not completely refute the necessity of change, but rather distinguish between "places that ought to be preserved" and "placed that ought to be torn down and redeveloped" (or "places where redevelopment is an acceptable option"). Their approach to the question of change, and the nature of that change, is governed by their perceptions. Sometimes distinctions are made according to clear standards, but in other cases they stem from non-explicit norms. In either case, the meaning ascribed to a particular urban environment by its residents or administrative authorities is not monolithic but rather replete with gradations: people may deem changes to certain places unacceptable, but tolerate sweeping changes in other areas.

How are such distinctions made? How is change controlled, and by whom? No single entity wields unambiguous control over change. Individual landscapes are produced or demolished through the conflict between governmental authorities, social movements, and other varied sectors and actors. Taking up the subject of cities and preservation can thus be distilled into the single question: "How does society control changes within cities?" (Horikawa 2010a). This book can be described as a sociological investigation into changes to the urban environment.[1]

This description must be qualified, however, through a discussion of four specific aspects of my approach to this subject. These four points could equally be described as the distinguishing characteristics of my research.

First, we must not take up the issue of "preservation" as a type of political indicator. Approaching changes to the urban landscape and the preservation movements opposing these changes as an occasional political phenomenon only leads to the labeling and exclusion of factors from analysis (for instance, preservationists could simply be equated with left-wing activists). While the social control of change is in itself a political phenomenon, using political party support (or affiliation) to subsume or exclude factors simply renders our original question impossible to answer. In this book, I ask how changes to the urban environment, which occur in response to market trends and may also involve party politics, are controlled by civil society. Posing the question in this way allows us to better comprehend the reality of preservation movements that do not conform to the conventional schema of political analysis,

[1] If we follow the use of the term "function" to describe the elucidation of certain mathematical principles from the course of change (rather than the simple observation of change), then my research might be described as a "study of cityscape functions."

such as "conservatives versus progressives," or "old residents versus new residents" (Horikawa 1998b).

Second, we must address the issue of "preservation" from the perspective of society. To be sure, matters of underlying system are absolutely essential, as the classification of urban space in Japan is determined by the legislative system in the form of the City Planning Act and the Building Standards Act. However, framing the issue of urban preservation in purely legalistic terms limits our capacity for understanding: problems can only be explained as the result of "imperfect laws," while "appropriate development" would be deemed "a legal act, and the exercise of legitimate authority prescribed by the law." We cannot ignore the impact of the legal system on real urban spaces—indeed, to do so would be exceedingly reckless. However, the legality of a particular form of urban development does not necessarily signify its relevance to residents. Indeed, *the real question here is why so many legal forms of urban development meet with repeated opposition from residents.* We must avoid stunting the question by adopting a narrow legal focus (Horikawa 2001). While the legal framework is a major factor in social decisions, it does not explain everything. The sites of conflict between the forces of preservation and redevelopment indicate the necessity of approaching the question from the level of society. Indeed, my refusal to compartmentalize the preservation issue as a legal one and my consideration of the issue within the context of civil society can be described as another distinctive feature of this research (Horikawa 2010a).

The third point pertains to the relationship between the individual and society. Many readers might question whether the issue of urban preservation is not really just a matter of personal preference. If you were to ask someone whether they preferred an old brick warehouse covered in ivy or a solid and lustrous black *minka* (a traditional Japanese-style residence), her answer would be a reflection of individual taste. However, during local conflicts surrounding the issue of preservation, one cannot easily persuade others of the significance of preservation by appealing to personal preference. This is even more true within bureaucratic organizations, where justifying a position in such terms is to be avoided at all costs. Indeed, I suspect that it is precisely because the issue of preservation has been presented not as a matter of individual preference, but rather as one in need of the consent or rejection of the local community, that movements for preservation have spread across Japan like wildfire[2] (Kankō Shigen Hogo Zaidan, ed. 1981; Zenkoku Machinami Hozon Renmei, ed. 1999).

Yet one frequently encounters a completely different discourse on the subject, expressed in such statements as: "scenery is subject to diverse assessments and there can be no unequivocal evaluation of its merit." There are many variations on this theme: "There is no uniform appraisal of a landscape," or, "as of yet there is no consensus on how to evaluate a particular landscape." From this point of view, treating movements seeking the preservation of old townscapes as a focus of the sociological inquiry is not only difficult but ultimately meaningless. The lack of consensus on the merit of a particular townscape confines the issue to "one of individual taste."

[2]See also the cases compiled in *Jurisuto*, ed. (1976, 1977).

But is this truly the case? Here I would like to draw attention to the background hypothesis informing this discourse. Underlying the argument that "scenery is subject to diverse assessments and there can be no unequivocal evaluation of its merit" is the assumption that "far from being objective, an assessment of a landscape is exceedingly subjective, and therefore a matter of individual taste." In this case, the real issue should be the validity of the underlying premise. Let us consider a real example of such an assertion:

> People's assessments of scenery, even with regard to the same place, will vary depending on the person. One person may perceive a *building* to be beautiful, while another person may deem the effect of the same *building* to be unpleasant.
>
> Nakano et al. 2011: 325. Emphasis added.

Now replace the italicized word *building* in the above passage with the word "policy," or "prime minister." One could also replace it with the word "fine art," or "Mr. A." In any case, the authors' meaning would still hold true. In other words, the possibility of differing assessments and the difficulty of establishing an unequivocal judgement is not specific to the question of the physical landscape. It is equally true of a wide range of social phenomena. Indeed, any attempt at the sociological analysis of a subject will confront the same difficulty pertaining to the diversity of potential assessments and the impossibility of an unambiguous conclusion; the issue of the physical townscape must thus not be dismissed on this basis. On the contrary, it confirms the fact that landscape preservation movements represent a viable focus for sociological inquiry. Before dismissing preservation movements as "nothing more than a matter of individual taste," *we must consider their significance at the societal level.* In fact, one could argue that it is precisely the structure of a society that dismisses an important *social* issue as a mere "matter of *individual* preference" that should be interrogated.

Finally, I focus on the relationship between society and the urban environment. We lead our lives within a physical environment. Changes to this environment inevitably produce changes to our way of living and our social relationships. This point—namely, the material nature of society and the sociality of the material world—is so self-evident as to be often overlooked, but we must consciously incorporate it into our field of vision (Horikawa 2011; Wakabayashi 2012). Social life never occurs in a vacuum. In the sense that it is led within the physical ordering of a specific urban environment, society is material in nature. Meanwhile, in that we transform nature according to our needs and thoughts, creating and deploying man-made objects, physical objects have a social existence.[3] Once constructed, buildings

[3]Those seeking a more complete explanation of the sociality of physical things may find helpful the following passage from Satō Kenji: "Above the fortieth parallel north, where evergreen oak and *madake* [timber] bamboo cannot grow, nor sweet potatoes, nor naked barley, forests are comprised entirely of camellia trees; in places, these trees grow in abundance. The rare sight of these camellias growing in such northern climes is a landscape that the Ministry of Home Affairs has designated as a 'natural monument' for preservation. This landscape, however, is more correctly described as a 'historic monument,' for it was created with what people transplanted to the area, or what they permitted to remain" (Satō 1994: 171). Even the natural forest before our eyes does not exist there

are incorporated as a new condition of our social existence, and our lifestyle changes as a consequence. Society does not simply produce new objects, it is remade by these objects in turn.[4] What kind of changes to cooperative relationships are brought about by changes to the physical urban environment? And conversely, how do these changes transform the urban environment? Our analysis must incorporate this "double-layered relationship" (Horikawa 1998a). Therefore, I situate the issue within a specific space and time, and attempt to grasp the interrelationship between the environment and society. For we cannot hope to gain a complete picture of the preservation issue unless we understand the particular environment that a preservation movement seeks to protect. Specifically, I adopt and apply the method of fixed point observation of architectural landscapes.[5] My use of both perception data obtained through traditional sociological fieldwork and this hard data on actual changes to the physical landscape is yet another distinguishing feature of this research. Repeated analysis of both survey and environmental data makes it possible to describe the relationship between the urban environment and society—but also illuminates the gaps within each approach. I do not treat these gaps as problems to be resolved; rather, I seek to identify what these very gaps might allow us to decode.[6]

As this overview should demonstrate, this book can be described as an inquiry into the ways in which society controls changes to the urban environment, and into townscape preservation as a means of "making cities what cities should be," in which I offer a sociological explication of how such movements seek to legitimize the involvement of city residents in public spaces. My application of the tools and perspectives of urban and environmental sociology signifies a fresh approach to the issues of "townscape preservation" and "historic environment preservation," which are more commonly addressed within the fields of architecture and urban planning.

---

naturally, but as a result of the human "history of practice" (Satō 1994: 171). In this sense, even a forest has an exceedingly social existence.

[4]I have previously discussed this as "the physical basis of the public sphere" (Horikawa 2011: 54–57). Wakabayashi Mikio offers a more refined version of the same argument: "The living bodies of humans, all forms of consciousness or principles, and all acts, relationships, and groups exist within, and have been held, formed, and molded by, a material environment, from the [first] natural environment to the huge and complex artificial environment of the modern city…Cities do not exist only as the actions, relationships, groupings or consciousness of its inhabitants, but also as such physical "things" as castle walls, roads, buildings, modes of transportation, communication media, which enable people to use these things and to be supported by them in turn….the material nature of society and the sociality of physical objects is one indispensable perspective for *Shakai(gaku) o Yomu*" (Wakabayashi 2012: 135).

[5]An overview of the fixed point observation method can be found in Chap. 3.

[6]Changes to the environment are not immediately reflected in human perception. Perceptions may change after a delay, or prove extremely resistant to change.

1.2 A Case Study of the Otaru Canal Preservation Issue

This book takes as its case study the preservation movement that unfolded in the city of Otaru, on Japan's northern island of Hokkaido. A detailed discussion of the Otaru case begins in Chap. 3; here a brief summary of the city's history will suffice to guide our discussion. Otaru is a commercial port city that underwent rapid development during the Meiji period (Fig. 1.1).

Located thirty kilometers northwest of Sapporo, Hokkaido's administrative and political center, Otaru was for many years an important distribution hub and economic center. In fact, the export of coal from the Otaru port led to such prosperity that

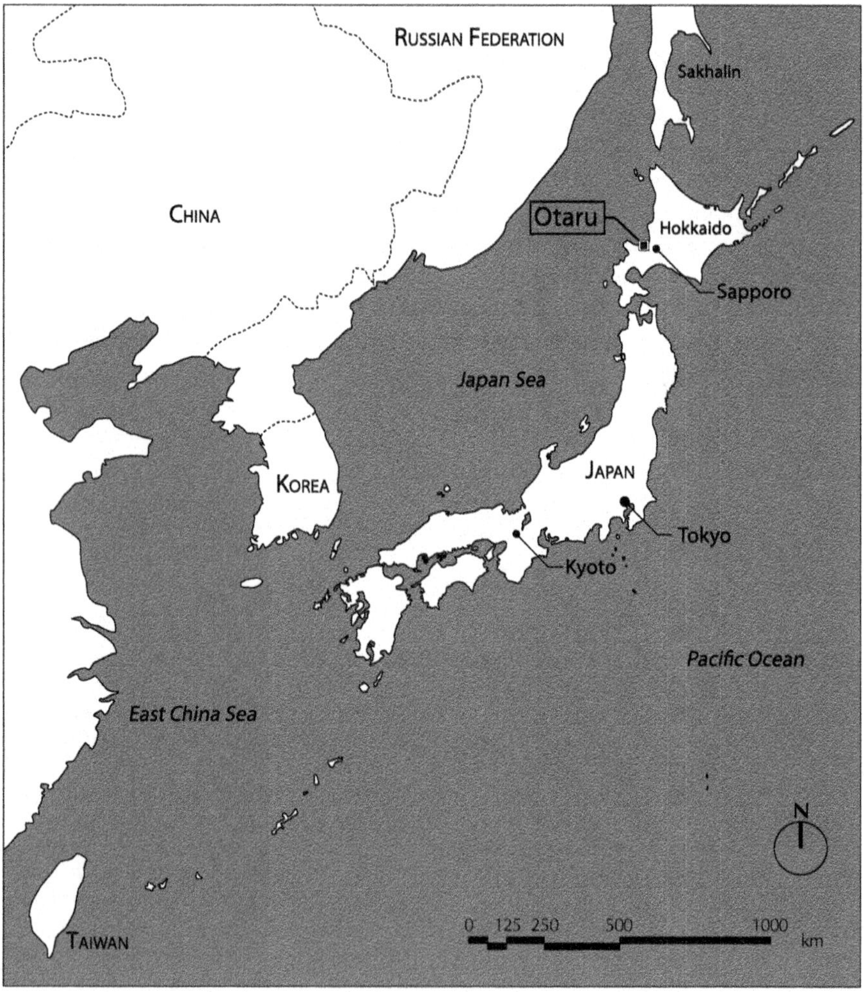

Fig. 1.1 The geographic position of Otaru (Author's illustration)

nineteen different banks, including the Bank of Japan, operated branches in the city, giving Otaru its nickname as "the Wall Street of the North." This prosperity fueled the construction of a canal, as well as the third railway in all of Japan. It was this vibrant and prosperous city of Otaru that the pioneering "modernologist" Kon Wajirō visited to discover the very latest fashions. Indeed, for a time Otaru was both the economic heart of Hokkaido and at the cutting edge of Japanese culture and customs.

The Japanese government's wartime concentration of economic functions in Sapporo, the transition from coal to oil as a primary fuel source, and the rise of the rival Tomakomai Port along Hokkaido's Pacific Coast put an end to the city's prosperity and sent postwar Otaru into decline. The canal that had once been at the heart of a thriving port city was deserted; from the 1960s onward only a few barges could be seen floating on its sludge-filled surface. In any discussion of Otaru, the phrase *shayō no* (meaning "declining," or more literally, "sunset") would inevitably precede the city's name. *Shayō no Otaru* was a dying city.

Enter a plan to transform the canal into a six-lane road known as the Rinkōsen—and a fierce debate over the relative merits of preservation and road construction, which divided the city against itself. Variously referred to as the "canal war" and the "canal debate," the question of whether the canal and surrounding townscape should be reclaimed for the construction of a major thoroughfare pitted the city government against preservation activists in a protracted conflict that spanned more than a decade, from 1973 to 1984.

This "canal war" was understood in very different ways. Viewed from the perspective of city officials, reclaiming the canal and constructing a new thoroughfare in its place represented the optimal means of pursuing Otaru's economic revitalization. The emergence of a campaign to save the canal prevented the city from proceeding with this road project as planned. Preservation activists, on the other hand, believed that preserving and restoring Otaru's distinctive canal and surrounding townscape was a far better way of pursuing regional revitalization. They demanded modifications to the city's road construction plan that would leave the historic townscape intact. In this sense, the "Otaru canal issue" could be described as a "dispute over a new road."

Yet how could a simple "road issue" bring about such a fierce and protracted confrontation between city officials and local activists? Framing the question in this way exposes the limitations of approaching the issue as a straightforward conflict over a road construction project. Beneath the debate surrounding the pros and cons of constructing a new road was a deeper disagreement over the correct strategy for urban redevelopment. It was this deeper conflict that shaped and sustained the decade-long confrontation in Otaru.

The Otaru city government's preferred strategy for redevelopment involved replacing an outdated form of cargo transport (barge—canal—warehouse) with a new one (truck-based transport along a major road). In other words, the city took a "scrap and build" approach to redevelopment. Pursuing redevelopment by demolishing something functionally obsolete in order to introduce new functionalities and technologies seemed an entirely reasonable approach.

The preservation movement opposed the city's plan with an oft-repeated catchphrase: "if the canal is destroyed, Otaru will no longer be Otaru." This naïve refrain aptly summarized the guiding ideal of the movement: urban rehabilitation through the re-use of historic buildings and structures as "stock." Activists espoused an ideology of "urban regeneration," in which the city's identity could be safeguarded through the preservation of historic structures—even as these structures were gently updated and put to new uses.

Thus, while the Otaru canal issue took the form of a dispute over the construction of a new road, it is better understood as a passionate debate regarding the direction of change, and control over change, in Otaru. Should the direction of change be entrusted to development policies spearheaded by political and bureaucratic authorities? Or should change be controlled by a social consensus that incorporates the wishes of local residents? The Otaru preservation movement garnered attention because it moved beyond the immediate issues of townscape preservation and community development to grapple with the broader themes of urban redevelopment and urban governance.

The preservation movement managed to extract a revised road construction plan from the Otaru city government. The modified blueprint did not necessitate the total reclamation of the canal; instead, the canal was narrowed to accommodate the construction of the new road alongside it. Although it won this concession from the city government, the preservation movement was ultimately destroyed from within, as intense disagreements between activists over the future course of the campaign led to the breakdown of the preservation effort. This marked the end of Otaru's "canal war." The canal and surrounding landscape that one can see today is the result of the revised blueprint produced by the city government.

The "canal war" had unintended consequences. Otaru's newfound national notoriety led to a tourism boom; at one point Otaru welcomed more than nine million visitors in a single year. Not only had Otaru come to represent a pioneering example of urban townscape preservation, it was also seen as a model case of "tourism development."

Yet Otaru's abrupt transformation into a major tourist destination has had a huge impact on the local landscape and the social life of the city. Regular fixed point observation surveys of Otaru confirm the dramatic transformation of the local landscape, as many of the historic buildings that residents tried to preserve have either been torn down and replaced with parking lots or drastically renovated and repurposed as souvenir stores or restaurants catering to tourists. Contrary to Otaru's reputation as a city of "historic tourism," the city is actually losing its distinctive townscape. This state of affairs has inspired new initiatives by former preservation activists. One former activist is now involved in third sector community development efforts, another has dedicated himself to tree-planting and other straightforward community activities, while a third has brought the spirit of the preservation movement to the city council as an elected official. Their efforts suggest it may be too soon to discuss the "canal war" in the past tense.

Otaru's experience with the preservation movement has thus not been entirely positive; many of the problems associated with tourism development are readily apparent, even in this brief overview. Residents' control over growth and urban governance were the real issues being contested during the long "canal war." Can officials and local residents create a new venue for discussing the city as a public good, or a new system for regulating change? Finally, what are the theoretical implications for the field of sociology? Otaru offers us a worthy case study for exploring such questions.

1.3 Methods, Concepts, and Survey Data

What are the distinguishing features of my analysis of the Otaru canal issue? What are the specific concepts and methodologies employed? Let us move from an overview of the subject matter to a discussion of research methods and concepts.

1.3.1 History Theories Through Case Studies

I have been guided in this investigation by sociological "history theory" involving the continued case study of a specific issue (Kōsaka 1998). To borrow the language of Funabashi Harutoshi, this is aimed at the "detection and explanation of regularity" with regard to a specific social phenomenon, and "the discovery of meaning" (Funabashi 2006: 6). In the movement toward general, foundational theory, this could also be described in terms of Robert Merton's "middle range theory" (Merton 1968: 39–72; Sztompka 1986; Funabashi 2006: 8). How can we explain the emergence of "preservation" as a social phenomenon? What does it say about society? What kind of significance can we identify in the act of preservation? This is the series of questions I seek to address.

My research is based on successive surveys, conducted in Otaru since 1984. Although the length of an investigation does not necessarily guarantee its quality, there is real significance to studying the process of local activism over an extended period of time. Taking a mass media approach and reporting on a local movement only at its zenith can mistake the position of the issue within the regional society. "Protest," "activism," and "litigation" are extremely rare events in the life of a local resident. Much like war itself, "activism" and "litigation" are tools of last resort. Just as one can only understand the significance of "war" in relation to the long stretches of peacetime that precede it, an exclusive focus on the "canal war" blinds us to the question of how Otaru residents lived in relationship to the familiar historic townscape during times of "peace." We must understand residents' "peacetime" environmental awareness in order to grasp the full significance of the fight to save the canal.

1.3.2 Overcoming the Schema of Binary Opposition and Oppositional Complementarity

Second, while drawing on the schematic of development versus preservation, I have focused on how the issue was shared and deepened in the process of opposition. The schematic of binary opposition may allow for only "simplistic" understanding, but it remains an excellent way of clearly delineating differences. This strength explains its survival in the face of the frequent criticism of oversimplification. While I employ the schematic of binary opposition, my strategy for analysis ensures we can move beyond a "simplistic" framework. Specifically, I focus on how different actors viewed the "issue" and the "canal," and how the city government and activists defined the district targeted for redevelopment in very different terms. My aim is to incorporate the significance within different actors' "gaze" upon the outside world. One could also describe this as an attempt to depict the diversity of "gazes" upon a single issue. Specifically, I focus on how even a major public works installation, such as a canal, can be assigned many different meanings; I refuse to reduce this diversity of understanding into a single "objective reality" or render it invisible through "flat description."

There is a simple explanation for this decision. It is precisely this discrepancy in understanding that gave rise to the gaps and fissures in the redevelopment strategies and the underlying logic of the preservation movement. "Why do they care so much about saving an antiquated canal?" "Why is it so hard for city officials to understand why we want to save the canal?" Both questions incorporate only the field of vision permitted by a specific "gaze," precluding the possibility of mutual understanding. This is why I seek to identify three things through my research: the "discourse" itself, "the subject responsible for the discourse," and "the location of that subject's gaze."

This triple aim is reflected in the book's structure. In Chap. 3, I offer a deliberately dispassionate, chronological account of the Otaru canal issue. Next, in order to understand how city officials and the preservation movement came to define the "canal redevelopment zone" in such different terms, the subsequent two chapters recount the same developments from different vantage points: Chap. 4 assumes the position of the "forces of change" in Otaru (primarily city officials), while Chap. 5 recounts events from the perspective of local preservation activists.[7] I make no attempt at an

[7] This concept and structure is my own re-working of an approach suggested by the work of Mike Pedler (Pedler et al. 1990). Pedler's account of the Manor Employment Project in the English city of Sheffield has a number of sociological implications. The book is divided into three sections. In the first, Pedler and his colleagues present an almost mechanical record of events ("Diary of Events") constructed from public records and the minutes of official proceedings. This is followed by a section entitled "The Tales," in which the interviews conducted by the fifteen members of Pedler's team are presented with almost no modification (Pedler et al. 1990: 55–128). The third section is entitled "Commentary," which is provided by Pedler and four fellow researchers from Sheffield College. Significantly, Pedler makes a deliberate distinction between the multiple "tales" presented in the book's second section and the singular "story" told by the book as a whole; this conscious distinction is clearly reflected in his terminology throughout (Pedler et al. 1990: 5). The aim of Pedler's approach is to illuminate how the same project was understood in entirely

omniscient, "godlike" narrative of developments in Otaru. The history presented in Chap. 3 (which is neither omniscient nor "objective" in the natural scientific sense, but may still function as a standard reference for analysis) is followed by accounts of the lived experiences of two sets of subjects. This means the same events are revisited three times: the events recorded in Chap. 3's standard "city history" reappear from the perspective of the pro-development city government in Chap. 4, and from the perspective of preservation activists in Chap. 5. This is a theoretical device that risks a certain redundancy in order to illuminate the overall structure of the "canal issue." The juxtaposition of these three different "gazes" allows us to maintain a critical distance from the subject matter, and helps us to comprehend what was lost and gained through the collision of two opposing logics of change and preservation. It also avoids the risk of relying uncritically upon attributes or positions that support misguided discourses involving "the ego of local residents," or "an unsympathetic local government."[8]

1.3.3 The Target of Preservation: A Two-Level Theory

A third characteristic of my methodology is the rigorous distinction between the real built environment and the mechanism of its production. This can also be described as a two-level approach to the target of preservation, which disentangles the physical reality of a building from the social relationships that produce and maintain it. If we were to parse the various physical structures which the Otaru preservation movement sought to save, we would find only aged wood and metal fittings rusted by years of exposure to wind and rain. It is the field of architecture that determines the overall value of a building created through the complex intertwining of physical elements such as wood and metal; the demand for preservation stems from this architectural value or the "rare" status of the structure. In that the basis of a building's preservation

different ways by the various actors involved. Collecting these different perspectives permits a pluralistic depiction and assessment of the project. Pedler's decision does not reflect a "naïve" attempt at "objective observation." On the contrary, it is rooted in his pragmatic desire to record a diversity of interpretations in order to derive lessons for future projects (Horikawa 1994). The overall flow of events is "the story," yet it is comprised of fifteen individual "tales" that do not arrive at a single, harmonious conclusion but rather continue to contradict and dispute each other. From a similar perspective, McPhee (1971) conducted exhaustive interviews with people holding different environmental philosophies and used these "narratives" to suggest a structure of conflict surrounding environmental issues. From the field of social movement theories, I am indebted to the work of Kajita (1988) and Tamano (2005).

[8]Put differently, I further my analysis with reference to the concept of "oppositional complementarity" (Funabashi and Funabashi, 1976; Kajita 1976, 1988), which seeks to understand the overall structure of the relevant social issue, including the question of what is produced by the structure of opposition between two sides. This is a necessary measure for avoiding fruitless ideological debates and focusing instead upon the refinement of sociological theory and the extraction of valuable lessons for urban policy.

stems from the characteristics of the building itself, this can be described as an essentialist definition.

It is human society, however, and the social relationships that comprise this society, which determine whether or not a particular building has value. Behind a "splendid" architectural specimen are the cooperative relationships that produced it: the techniques required to extract, transport, and process local stone or other materials, but also the prosperity and social structures that enable construction, and the local residents and residents' organizations upon which both this prosperity and social structures depend. Distinguishing the physical structure from the cooperative social relationships that produce it makes a non-essentialist definition possible. The value of a certain building or townscape is determined by human belief in its value—in other words, by a collective and subjective acknowledgement of value. I should hasten to add that I do not reject the value of architectural (history) research. Rather, my point is that something new may be brought into view by making a rigorous distinction between the architectural axiology of physical structures and the social consciousness and cooperation of the people who find significance in the act of preservation and enable the development of such theories of value in the first place.

1.3.4 The Analytical Tools of "Space" and "Place"

A fourth characteristic of my methodology is the distinction I make between the concepts of "space" and "place." This could also be described as the dualism of object consciousness. As this conceptual distinction is a principal component of my analysis, two hypothetical scenarios can be used to clarify its importance.

What would happen if a city official were to present preservation activists with the following proposal: "We need to pave over the entire canal in order to build a new road, but we will provide you with the same amount of land in a different location if you agree to our road plan." There is only one conceivable reaction to such a proposal: an emphatic "no." Indeed, there would have been no preservation movement, nor debate, if activists could have even conceived of answering in the affirmative. Activists did not seek to secure a particular length or breadth of land; what they wanted was the specific physical presence of the canal itself, and the scenery created by the canal's presence (Horikawa 2000, 2006).

Let us consider yet another scenario, in which the same official makes a different proposal to local activists: "We need to pave over the entire canal in order to build a road. If you can agree to this road plan, we will create an exact replica of the canal in a different part of the city."[9] The answer would be the same as before: "no." Even without a preservation movement, it is hard to imagine how a replica of the canal

[9]This hypothetical proposal is presented here as nothing more than a useful thought experiment, but a similar proposal actually emerged in the course of Otaru's "canal war." Inoue Takashi, the author of the city's blueprint for the Rinkōsen road plan, wrote: "with regard to the preservation of the warehouses that line the canal, my proposal is to create a canal park at a different location" (Inoue 1981: 44). Elsewhere, Inoue argued that "the best way to save the Otaru canal is to create an

could win social acceptance. Indeed, even if such a replica were constructed, it would lack the authenticity of the original canal, which played such a central role in Otaru's economic development. The disappearance of the *real* location and structures that comprised the historical cornerstone of the city meant that "Otaru would no longer be Otaru." The men and women who joined the movement to preserve the canal could never accept such a proposal.

These hypothetical scenarios suggest that local activists understood the canal as something more than a particular parcel of land. This is because there are two different ways to understand land. The first is to understand "land as *space*"—in other words, in terms of area and volume, with no reference to individual emotional attachments or history. The language of city planning laws reflects this approach to land. Land as space is an odorless, transparent "cube," devoid of the "background" bestowed by context and history (Kuwako 1999). Consequently, it is compatible with any number of purposes. The cube of space that presently functions as "my" family home could be sold tomorrow, and become the home of a stranger. The dictates of city planning could also transform it into a road, or a factory. This is what it means to understand a certain piece of land as "space."

By contrast, approaching land as something that incorporates memories, emotional attachments, and history—or in relation to someone's daily life—is to understand "land as *place*." Discussing "my family home" in terms of "a place with a meaning," rather than as "94 square meters of land," is to understand it as a "place." This place is imprinted with "my" life and family history, and represents a fusion of local environmental conditions and customs; it is always discussed in connection to "me." This "place" has a "background" that cannot be easily erased. As such, it cannot be readily converted to a different purpose. This is what it means to understand land as "place."

One can transfer the ownership rights to a certain piece of land (a "space"), but the physical land itself remains immobile. In this sense, land has a distinctiveness that cannot be separated from meaning, history, or specific geographic features (it is equally a "place"). We can thus conclude that land has a dual nature: it exists simultaneously as "space" and "place." I believe that we can utilize the concepts of "space" and "place" to explain the emergence of social conflicts surrounding preservation.

Let me explain. When a social conflict over the preservation of a certain place erupts, we should seek to explain the factors that led to this conflict, rather than attempt to ascertain which side is "in the right." In the case of Otaru, for example, *both* the development and preservation camps recognized the necessity of building a new road, yet became enmeshed in a decade-long conflict over how to deal with the canal district. The same canal was viewed in two different ways. The pro-development faction saw the canal as "unused land that could be used to build a road"—in other words, as a transparent and colorless "space." Preservation activists, by contrast, saw the canal as a deeply meaningful place essential to their very identity. Introducing

Otaru Canal Park *somewhere in Japan*" (Horikawa, Morishita, Yanagida & PRAHA Machizukuri Jōhō Center, ed., 1995: 70; emphasis added).

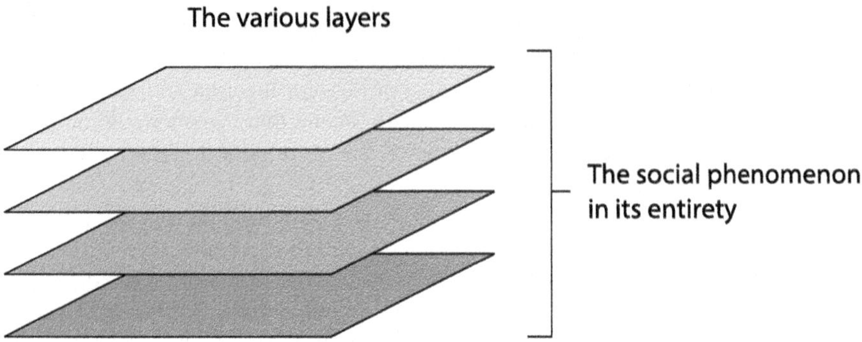

Fig. 1.2 The layer concept (Author's illustration)

the concept of "space versus place" underscores the discrepancy between different actors' "gazes" upon the canal.[10] This in turn can explain the emergence of vastly different emphases and mobilization strategies (Horikawa 1998a, b, 2000). Comprehending and describing the various subjective meanings ascribed to the target of preservation by different actors is not only necessary but suggests the potential for one form of sociological analysis.

1.3.5 The Articulation of "Layers"

The concept of "layers" is a fifth and final distinctive aspect of my approach (Horikawa 2010a; 2011). The word "layer" expresses the various components and registers of debate within a particular social conflict. As shown in Fig. 1.2, social conflicts typically involve diverse and complex disputes between multiple actors, and emerge as the composite of various actors, registers, and situations.

The "layer" is a conceptual apparatus for understanding the shape of multiple conflicts and the substance of debates in terms of the "register" in which they are discussed. If we articulate a debate into multiple layers, we can discern the social level at which a particular component (the pros and cons of road construction, for example) was addressed—we can identify the register of debate. A layer can be likened to a transparent sheet of plastic; each sheet represents a different level of the

[10]The concepts of "space" and "place" explain how two people living in Otaru for the same amount of time, looking out at the same canal, might nevertheless understand the significance of the canal in entirely different ways. Similarly, it illuminates the difference between looking at the canal "as a resident" and "as a government official." Although working as a civil servant at Otaru City Hall would clearly influence one's perception of the issue, we should note that certain city employees chose to join the preservation movement, while a number of Otaru residents supported the construction of a new road. Utilizing the concepts of "space" and "place" should prevent overly simplistic assumptions ("residents see place" and "city officials see space") and allow for a more nuanced recognition of the fact that some residents saw the canal as a "space" while certain city employees valued the canal as a "place."

social conflict. It is when we stack all of these transparent sheets into a single sheaf that a social phenomenon can be understood in its entirety.

The layer concept is simple, and yet it allows us to understand the "canal war" on multiple planes, and as an aggregation of different registers of debate (in other words, as an aggregation of layers). This layered understanding of the canal war explains why Otaru residents rejected certain discourses ("Isn't it enough that half of the canal was preserved?" "The preservation movement was still successful because it brought tourists to Otaru"). Articulating the different registers at which the issue was discussed offers a high-resolution image of each actor's narrative, and of the conflict's overarching structure. If we adhere to a single-layer understanding of the Otaru "road dispute," we will never understand the entirety of the "Otaru canal issue," which also incorporated the layers of "urban governance" and "urban ideology." It is precisely in order to detect these complex and subtle nuances that I have chosen to employ the layer concept in my analysis.[11]

These five aspects of my approach are the inevitable consequence of my quest to explore the value consciousness of preservation activists. They are equally a strategy for using the "issue or dispute" to uncover an invisible "coexistence."

1.3.6 Survey Methods and Data

Let us turn to the specific methodologies and data employed in my analysis of the "Otaru canal issue." Naturally enough, the target region and analytical approach to an issue will always inform the type of data that is collected. In this volume, my analytical approach has determined my survey design and the type of data I collected and reviewed. The following discussion of methods and data is organized into sociological and non-sociological approaches.

1.3.6.1 Sociological Methods and Data

In order to explore the themes of this volume, we must focus upon the "subjective world" of each actor (Arisue 1992, 1999; Horikawa 1998a; Tamano 2005). This is because we cannot understand why a resident might wish to preserve one thing but not another from the physical structure itself.[12] Without entering "the world of

---

[11] A similar concept is that of the "arena." The arena concept focuses upon the venue and time-space (singular or plural) in which a debate unfolds. In the case of multiple arenas, it is tacitly assumed that each arena is independent and unrelated to the others. The concept of layers, on the other hand, presupposes the existence of multiple components to a dispute. Unlike the arena concept, each layer represents a particular component; layers interact and operate as a sheaf of overlapping, transparent sheets. The layer concept successfully identifies subtle variations and demonstrates how different social groups can "talk past each other" in disputes.

[12] As we have already seen, no matter how closely the material components of a structure (roof tiles, pillars, beams, bricks) are examined, they will not tell us whether or not the structure should be

subjective meaning," we cannot explain why someone might wish to save a building that others dismiss as worthless. Within this world, moreover, we must understand time as something that accumulates in drift and piles, rather than something that progresses in a linear fashion (Hama 2000, 2010; Horikawa 2010a).[13]

Since the spring of 1984, I have conducted forty-seven on-site surveys in Otaru. I have spent a total of 275 days in the city. I have also conducted interviews with individuals related to the Otaru case in both Sapporo and Tokyo. In addition to these exhaustive interviews aimed at the generation of qualitative data,[14] I examined official documents, statistics, records of government proceedings, and petitions submitted to the city government, created a database of news coverage[15] of the Otaru canal issue for analysis, and conducted regular fixed point observation surveys of the Otaru landscape (Horikawa 1994, 1998a, 2003). Otaru-inspired essays, novels, travel guides, movies, and music were yet another source of data.[16]

"Semi-structured qualitative surveys" are designed to elicit a comprehensive and heartfelt narrative from informants. Centered upon an informant's own words, the method allows the interviewer great flexibility in questioning the informant about his or her life history, or other background information.[17] In that the informant is

preserved. The question is whether or not, upon examination, someone will acknowledge that "this is an important building that should be saved."

[13]Hama Hideo draws upon the 2009 Katō Kunio animation film *Tsukimi no Ie* (La Maison en Petits Cubes) to describe this as "a state of time and space that differs from absolute time or absolute space, in which time is linked to place, and memory to space" (Hama 2010: 469). In his discussion, Hama distinguishes between "time that flows along a horizontal plane" and "time that accumulates vertically" (Hama 2010: 469). See also Hama (2000, 2002, 2007).

[14]While this is roughly equivalent to an "interview" or "hearing," it can be described in technical terms as a "semi-structured qualitative survey."

[15]Creating a chronological table of events was essential to the development of a complete overview of the canal preservation movement. To do so, I used the newspaper clipping index of the "Otaru canal issue collection" at the Otaru University of Commerce economics research center to supplement information pre-dating the beginning of my fixed point observation surveys of Otaru. At the request of the Otaru University of Commerce research team, I created a database of over 4,000 newspaper articles spanning the seven-year period between 1979 and 1985; seminar students of Otaru University of Commerce Prof. Shinozaki Tsuneo assisted with the data input. The news clippings were obtained from four different collections, belonging to the reference room of the Otaru branch office of the *Hokkaido Shinbun*, the Otaru newsroom of the *Asahi Shinbun*, the preservation activist Morishita Mitsuru, and the Otaru Chamber of Commerce. The range of sources is sufficiently diverse so as to negate the risk of a pro-movement or pro-administration slant. I am deeply indebted to Prof. Shinozaki and the Otaru University of Commerce reference room assistant Konno Shigeyo for readily assenting to my request to use the university database. See Horikawa (2015) for a discussion of the methodology of the chronological table.

[16]Why did I choose not to adopt quantitative research methods, such as a questionnaire survey? Not only had internal divisions already brought down the canal preservation movement by the time I began my graduate studies, but these methods simply did not conform to the purpose of my investigation.

[17]Since the 1980s, the sudden rise of life history studies within sociology and the development of new research methods has expanded the range of materials that can be treated as sources. In other words, the expansion of sociological methodology has expanded the concept of primary sources. In the process, the use of life histories and oral histories has attracted new attention and won widespread

free to discuss anything he or she chooses, the conversation is unstructured. Yet it is structured in the sense that every informant is asked to supply a life history and other types of background details as supplementary information, which enables the interviewer to situate the informant's narrative or identify meanings. During the interview itself, a list of semi-structured questions often sparks a rich and wide-ranging narrative.

These semi-structured interviews centered around four key informants: Yamaguchi Tamotsu, Ogawara Tadashi, Sasaki Kyōjirō, and the former Association to Protect the Otaru Canal chairwoman Mineyama Fumi (now deceased). These four led me to other members of the Otaru preservation movement, with whom I also conducted extensive interviews (Rossi and Dentler 1961; Satō 1992, 2006). To obtain pro-development perspectives on the "canal war" I interviewed local officials from various departments of Otaru City Hall. I also conducted a series of extensive and highly detailed interviews with former Otaru mayor Shimura Kazuo (now deceased) (Horikawa, ed. 1999, 2000), whose terms in office "began and ended with the canal." My interviews with informants such as Yamaguchi, Ogawara, and Mineyama span a quarter century; other informants, including Shimura, agreed to participate in multiple panel interviews. If relatively short interviews of just one to two hours are included, my total number informants is quite remarkable.[18]

---

acceptance. (Nakano, ed. 1977; Arisue 1984, 2012; Mizuno, 1986; Nakano and Sakurai, eds. 1995; Seisaku Kenkyūin Seisaku Jōhō Project ed. 1998; Mikuriya 2002; Nakano 2003; Hama et al., ed. 2013.) The expansion of the very concept of an investigation or survey has also given rise to new research that draws upon an expanded idea of the history of sociological inquiry (Kawai, ed. 1989, 1991, 1994). I have also chosen to actively incorporate interview data as source material for tracing the changing meaning ascribed to the canal by preservation activists.

[18] I interviewed a number of people who did not participate in the preservation movement but were somehow involved with the canal issue from differing positions and at varying levels of intensity. It was impossible to secure interviews with an even number of road and preservation supporters; in the end, the number of informants who had supported the canal's preservation was larger. Within the preservation movement, moreover, I was unable to secure an equal number of interviews from each different "type" of activist. There were many reasons for this: certain former participants cited poor health, privacy concerns, bereavement, or their current distance from Otaru as reasons for declining interviews; other participants are now deceased, while others flatly rejected my interview request. The total number of informants, however, does not automatically suggest bias or a lack objectivity. The chief concern of this investigation is the underlying logic of the Otaru preservation movement; it is thus entirely natural that the majority of my informants are former preservation activists. We should also take note of a more fundamental disparity. The "institutional memory" (Akashi 2002: 139) of administrations such as the pro-development Otaru city government tends to weaken over time due to frequent personnel reshuffles. City officials do leave written records of their work and so, while I interviewed a number of city officials, I was forced to rely heavily upon official city documents. Members of the preservation movement, by contrast, lacked the time, manpower, organizational structure, and physical facilities to produce and preserve written records. This left me little choice but to track down former activists and ask for an interview. In this sense, one could argue that it is entirely natural to have more preservation activists as informants in order to compensate for the relative lack of a written archive. The study conducted by Funabashi et al. (1985) on bullet train-related pollution includes sixty interviews with local residents and ten interviews with employees of the former Japan National Railways for the same reasons.

These interviews were typically conducted in an informal fashion, with the aim of allowing informants to speak as freely as they liked (Pedler et al. 1990; Whyte 1955).[19] Interviews were conducted at informants' homes or workplaces, but also at coffee shops or bars—or in one case, in the informant's hospital room. The location was always determined by the convenience of the informant. When I was permitted to record the interview, I would later transcribe it from the audiotape[20] and encode it for analysis.

In some cases, an interviewer's prolonged contact with informants can lead to "emotional immersion in the site" (Okuda 1983: 209) or "over-identification" with informants, threatening the credibility of the data obtained. In order to avoid these risks, I combined multiple types of data to correctly position each individual source, and turned to written documents and interviews with informants unrelated to the preservation movement to ensure the reliability of information obtained from my chief informants. In an effort to relativize my own position within the research process, after synthesizing my findings I returned to the "site" of Otaru for question and answer sessions with my informants, city government representatives, and other former preservation activists, where I presented my findings for scrutiny and discussion. After reading all of my findings, participants joined me in a long discussion that extended over multiple days (Horikawa 1989, 1994). Taken together, these measures should ensure the general reliability of my data (Satō 2006).

1.3.6.2 Non-sociological Methods and Data

If we use the term "soft data" to describe the data obtained through the sociological methods discussed above, then the data obtained through architectural methodologies

[19]Methodologically, interview content was derived through three approaches: the "reputation" method, the "thematic" method, and the "observation" method (Yasuda and Hara 1982: 1–26). In the "reputation" method, people who come up in the course of an interview with the first informant were subsequently approached and interviewed in turn; I gradually expanded my interview scope as each informant introduced me to one or more additional informants (Rossi and Dentler 1961: 1–9). Through the repetition of this process, I was able to identify and secure interviews with key people within the preservation movement. There is another merit to this approach: it allowed me to identify informal sub-groups within the movement and the movement and characteristics of these groups. When adopting the "thematic" approach, I asked informants to discuss a particular topic or theme (for example, their motivation for joining the preservation movement). By comparing the answers of all informants, I could begin to ascertain the position of individual informants within the movement, and identify differences in their individual perception of all aspects of the movement. The observation method involved attending and observing meetings of preservation activists to witness their interactions. There I could ascertain whether or not the men and women previously identified as "key" figures within the movement through the "reputation" interview method were as important as informants had asserted. I was also able observe the decision-making process and methods (through votes, etc.) of the preservation movement.

[20]Over more than thirty years of research, my recording device has changed from a cassette tape to a microcassette, then to a MiniDisc, and finally to an IC memory card. Regardless of the specific device used, the fundamentals remain the same: the recording was faithfully transcribed for analysis.

could be described as "hard data." As I have already noted, this fusion of "soft" and "hard" data is one of the distinguishing features of my research.

This hard data was obtained through fixed point observation surveys of the Otaru canal and surrounding districts. Panel surveys of buildings in the port district of Otaru, conducted continuously throughout the 1980s and 1990s by members of the Hokkaido University engineering department, provided the foundation for my surveys. The Hokkaido University surveys targeted the exterior appearance and use of 272 buildings (as of 1992) in the designated survey district. I took over this survey in 1997. Since then, my sociology students at Hosei University and I have regularly conducted the same survey (dubbed the *tatemono chōsa* or "building survey") of the very same buildings originally targeted by the Hokkaido University team. By restoring the remaining Hokkaido University survey forms and replicating their format exactly in subsequent surveys, we have developed a data set that captures more than thirty years of architectural changes within a particular city district. The data vividly represents the landscape targeted for preservation by local activists. It offers an equally eloquent depiction of the sweeping changes to the district following Otaru's transformation into a tourist destination. Every year, the survey is conducted over roughly the same period in September; our team also photographs each building and conducts interviews with the building's residents. In 1998 I initiated a separate fixed point observation survey of 119 shops in an Otaru shopping district[21] (the *shōtengai chōsa*, or commercial district survey) in order to ascertain architectural changes within a central commercial district as well.

I am not an architectural planner by training, but I have chosen to incorporate these non-sociological forms of research as a means of "triangulating" my social research.[22] This has produced a hybrid approach. Nevertheless, data obtained through sociological fieldwork and sociological analysis remain the backbone of this investigation. The data collected through secondary methods such as the fixed point observation surveys simply supplements and enriches the sociological approach.

[21]The national supermarket chain MyCal opened an Otaru branch on March 11, 1999. This was preceded by a major controversy surrounding the question of granting of an operation license to MyCal. In an effort to capture the changes between "pre MyCal" and "post MyCal" Otaru, I initiated the commercial district survey in September 1998—six months before MyCal Otaru opened—and have continued with fixed point observation surveys of the district ever since. The basic survey format is identical to that of the "building survey," with some modifications and additions to address the features of a commercial district.

[22]I approached the Hokkaido University research team for advice and guidance before undertaking the fixed point observation surveys; team members accompanied me on the first survey to give practical advice on its execution. When developing the specific survey items for the commercial district survey I solicited the guidance of two commercial district associations, the Otaru Miyakodōri Shōtengai Shinkō Kumiai and the Otaru Hanazono Ginza Shōtengai Shinkō Kumiai.

1.4 An Overview of Previous Research

How has "preservation" been addressed as a sociological phenomenon? In this section, I offer a brief overview of existing work on this subject. After a brief review of previous research on Otaru (Sect. 1.4.1), I turn to the literature on the preservation of historic environments (Sect. 1.4.2), and conclude with the "academic genealogy" of my own research (Sect. 1.4.3).

1.4.1 Previous Research on Otaru

The men and women who dedicate their daily lives to a preservation movement or community development efforts have little time to keep a written record of their activities. This was certainly true in the case of Otaru. The few documents that do exist can be divided into two categories: written documents produced by activists in order to elicit sympathy or support from the outside world, or surveys and research conducted by outside researchers who were either commissioned or motivated by a personal interest in the subject matter.

This first category includes publications compiled by the Association to Protect the Otaru Canal (Otaru Unga o Mamoru Kai) (1977, 1981), the Kankō Shigen Hogo Zaidan (now the Japan National Trust) (1979), Otaru Unga Kenkyū Kōza Jikkō Iinkai (1979, 1981, 1983), Masuda (1980a, b, 1982), and Sasaki (1982), as well as individual publications by Fujimoto (1985), Kawabata (1986), and Shinozaki (ed. 1989).

The Kankō Shigen Hogo Zaidan 1979 survey team was led by Kyoto University Prof. Nishiyama Uzō, a leading authority on city planning. However, it was three young Hokkaido University architecture students (known in Otaru as the *Hokudai sanningumi*, or "Hokudai trio") and preservation activists who undertook the actual survey, research, and writing of the report. In this sense, the report is very much the creation of local activists. The report, which includes a review of Otaru's historical development, an exhaustive survey of the condition of buildings in Otaru's port district, an analysis of the current state of city functions, and a list of final recommendations, is of high quality—even by today's standards. During the "canal war," however, the survey report was viewed as the product of the preservation camp, and its academic significance was never fully considered. Other works in architectural planning by the "Hokudai trio" have also stood the test of time, and remain important references today (Ishizuka 1980, 2004; Morishita et al. 1983a, b; Yanagida et al. 1983).

The writers Natsubori (1980, 1992, 1997) and Ogasawara (1986) have both taken up the theme of Otaru in their work. As writers, both keenly understood the importance of the historic environment as a venue for expression; what is interesting, however, is their shared interest in the organizational theory of activism. As Otaru natives, Natsubori and Ogasawara wanted to preserve the landscape that had inspired much of their creative work. Perhaps this is why they did not convey their feelings with a literary flourish, but instead devoted themselves to developing theories of activism that could achieve their goal of preservation.

After the "canal war" came to an end, Sapporo-based supporters of the Otaru preservation movement collaborated with the Otaru Unga Mondai o Kangaeru Kai to produce what remains to this day the most comprehensive collection of documents pertaining to the Otaru canal issue (Otaru Unga Mondai o Kangaeru Kai 1986a, b). Entitled "Documents," the first of this two-volume compilation reproduces a vast number of important original documents from the APOC and the subsequent mayoral recall movement, among other events, and a number of news clippings from the period. The second volume, "History," was compiled and partially authored by Ogasawara Masaru, who had also lead the Otaru Unga o Kangaeru Kai. Ogasawara records activists' own description of events, from the establishment of the APOC through the breakdown of the preservation effort. This makes the second volume an extremely valuable resource.

The most famous work by a member of the preservation movement is the thorough and earnest memoir of former APOC chairwoman Mineyama (1995). The canal's intimate connection with Mineyama's life history is of particular significance. Mineyama describes major life events—finding work, her marriage, and move away from (and eventual return to) Otaru—against the backdrop of the canal and Otaru cityscape. For Mineyama, this cityscape was inseparable from her daily life; her description of its decisive influence on her life is candid and entirely without artifice. Mineyama eloquently describes how the Otaru canal, ostensibly nothing more than a port facility, was actually a physical presence intimately interwoven into the lives of Otaru residents.

The observations of a university researcher who participated in the preservation movement are also noteworthy. A grant-in-aid for a scientific research team led by Shinozaki Tsuneo published a report that directly addressed the Otaru canal issue (Shinozaki, ed. 1989). A number of topics critical to the canal issue—the structure of the Otaru economy (and particularly the transformation of the local wholesale industry), and the results of a questionnaire on the canal's preservation distributed among Otaru residents—are discussed by the report's authors. As a whole, however, the report lacks a clear focus. The real significance of Shinozaki's work is as an early interdisciplinary attempt to address the canal issue, and as the source of a number of foundational secondary sources for more full-fledged research on the subject.

Local officials and other members of the pro-development camp, who opposed the preservation movement's efforts to save the Otaru canal, also wrote on the subject: Kumashiro (1981), Inoue (1981), Hokkaido Jūtaku Toshibu Seibika (ed. 1989), Shibuya (1990), Satō (1990), Nishio (1992), Iida (1993), Otarushi-shi Hensan Iinkai (ed. 1995), and Tanada (1996).

Writing as an administrative official involved in the city's road construction project, Kumashiro (1981) discusses the importance of developing the Otaru port, and maintains the city's decision to reclaim land from the canal to construct the new road was justified from an economic perspective. Shibuya (1990), Nishio (1992), and Tanada (1996) are all city officials who were forced to grapple with the Otaru canal issue over the course of many years; all three authors review measures undertaken at the time and reconfirm the necessity of the road construction project. As the authors of the city's original and revised blueprints for the new road, Inoue (1981) and Iida (1993) elaborate the logic informing their respective proposals.

Hokkaido Jūtaku Toshibu Seibika (1989) incorporates the recollections of officials at the Hokkaido prefectural government involved in the Otaru Rinkōsen road project, while Satō (1990) recalls the same events from a city planning and civil engineering perspective. Both works reiterate the necessity of the Rinkōsen, and touch upon technological innovations in civil engineering, officials' efforts to persuade all parties of the value of the new road, the many hardships that accompanied the execution of the project, and the teamwork that finally led to the completion of the Rinkōsen.

The nine-volume *Otarushi-shi* (History of Otaru City) compiled by the Otarushi-shi Hensan Iinkai (1995) dedicates a significant number of pages to the canal issue, reversing the city government's longstanding stance of barely alluding to the issue.[23] This official city history offers no comment or assessment of the canal issue, but the narrative, which is based on a collection of internal city documents, makes it an extremely valuable resource.[24] While the treatment of city documents does not meet the standards of true historical research—indeed, at times it fails to meet even conventional standards for handling historical sources—this city history offers an indispensable overview of the city government's response to the canal issue.

There are also more impartial accounts of the Otaru "canal war."

The decade-long Otaru canal preservation movement naturally attracted the attention of many journalists, including Tosaki (1979), Honma (1980), Odagiri (1982), Kihara, ed. (1983), Shin'chi (1983, 1983–1985), Miyamaru (1983, 1984, 1985), Harada (1986), Ōwada (1986, 1987a, b, c, d, e, 1988a, b, c, d), Andō (1992), Hoffman (1995, 1996), and Tamura (2009). The work of Shin'chi (1983–1985), an Otaru-based reporter for the *Asahi Shinbun*, is particularly notable for the thoroughness of his interviews with preservation activists, Shin'chi did not cover the issue as a "current event," but rather, over the course of a two-year series, delved into the life historie s and world views of his sources to vividly depict "the canal issue as a way of life." While Shin'chi has been accused of bias by members of the city administration, his detailed life histories of activists remain a precious secondary source for the present investigation. Former journalist and non-fiction writer Tamura (2009) focuses on the pro-road construction camp. Tamura fills a significant gap in the literature by incorporating the recollections of Iida Katsuyuki, an assistant professor at Hokkaido University during the Otaru canal dispute. Iida revised the Rinkōsen blueprint at the

---

[23]The educational text *Watashitachi no Otaru* ("Our Otaru"), compiled and edited by the Otaru Board of Education, makes no mention of the canal issue or preservation movement (Otarushi Shakaika Fukudokuhon Henshū Iinkai, ed. 1982, 1992). The only Otaru city publication to mention the canal is the retrospective on the Shimura administration by the Shisei no Ayumi Henshū Iinkai (1988).

[24]In the course of my research I was permitted to review to a massive collection of historical documents collected from various city government departments for use in the official history of Otaru City (Otarushi-shi Hensan Iinkai ed., c1994a, c.1994b).

request of the Otaru city government, and gives Tamura an account of how the new road plan was formulated.

There are also a number of academic treatments of the Otaru canal issue. The Otaru case has been taken up within the field of environmental economics by Miyamoto (1989), regional economic theory by Kanekura (1986) and Wakita (1993), urban administration theory by Tamura (1999), geography by Aramaki (1987), local political history by Shimizu (1990), and social education theory by Satō (1998). Okamoto and the Nippon no Minatomachi Kenkyū Kai (2008) address the Otaru case from the perspective of modern port cities and the question of land ownership, while Itō (2000) discusses the case in relation to Japan's modern civil engineering heritage. The majority of these works date from the late 1980s and 1990s, after the preservation movement had come to an end. As such, they are notable for the conclusions they draw. One shared theme is the idea that: "while activists were unable to save the entire canal, we should not forget that resurrecting the city as a tourism destination was a major achievement." Yet another motif running through these various treatments is the idea that, when viewed comprehensively, the preservation movement did not "lose" the canal war, and actually accomplished a great deal by transforming Otaru into a successful example of tourism development.

Are such conclusions really correct? How do the men and women who actually participated in the preservation movement evaluate the results of their efforts? What is the basis for deeming their campaign a "success?" The remainder of this book is dedicated to answering these questions. First, however, let us examine how other sociologists have approached the issue.

There are exceedingly few sociological (or any other social science) treatments of the Otaru preservation issue. In addition to my own work on the subject, one can cite Nishiyama (1993), Ōyama (2001), the Shakai Idō Kenkyūkai (2005), Naitō (2015), and Naitō and Sakuma (2017).

The work of Nishiyama (1993) represents one of the first sociological works to address the Otaru case. However, Nishiyama makes only passing mention of the Otaru canal preservation movement to corroborate her theory of citizen participation. As such, it cannot be considered a sufficiently dense account or analysis of the Otaru case. The same is true of Ōyama (2001). Although the Otaru preservation campaign and subsequent community development efforts are not the primary focus of the Shakai Idō Kenkyūkai (2005), it is distinguished by its painstaking account of the mechanisms behind Otaru's development as a city. Naitō and Sakuma (2017) include an analysis of the Otaru case in an addendum. Given that the authors cite just one primary source, the reliability of their analysis is highly suspect.

This brief overview underscores the fact that the Otaru preservation movement and subsequent community development initiatives remain almost entirely unaddressed within the field of sociology. The bulk of existing work on Otaru, much like work on the topic of preservation itself, belongs to the fields of architectural and urban planning.

1.4.2 Research on the Preservation of Historic Environments

It is to this research on the preservation of historic environments that we now turn. In this section, "existing research" refers to research on the issue of historic preservation itself, or to research on the advocates of preservation. This research within the fields of architectural planning and urban planning is generally directed at one of two concerns: whether or not a building is of sufficient value to merit preservation, or to the achievement of a higher quality of urban life through the configuration of buildings and parks. Both lines of inquiry have yielded impressive results. Representative works include those by Ashiwara (1979, 1983), Inagaki (1984), Nishiyama (1990), Nishimura (1997, 2004), Nishimura et al. (ed. 2003), and Jinnai (ed. 2009).

Is it really possible to discuss the value of a particular structure, or an appropriate configuration of buildings, without considering the meanings ascribed to the urban environment, or residents' affection for (and dependence upon) a particular urban lifestyle borne of this environment—in other words, without entering the "world of subjective meaning?" To be sure, this type of question is the most primitive and naïve form of criticism that sociology can make of architectural planning. Nor is this perspective entirely ignored by the highest quality works in architectural and urban planning. Fukukawa Yūichi, for example, discusses "urban community conservation" in the following terms:

> The mutually enforcing relationship between the "physical environment" and the "urban community" is what matters. The question of what kind of building creates what kind of townscape, and how (the physical environment) is inextricably linked to the question of what type of social relationships emerge as a consequence (urban community).
>
> Fukukawa (2003: 122). Parentheses in original.

It is clear from this excerpt that Fukukawa, an expert in architectural planning, recognizes the indivisibility of the physical environment and the urban community, and acknowledges the necessity of preserving buildings in order to protect the community.

Nevertheless, we cannot fully dismiss sociology's critique of architectural planning, because architectural planners are never really free from one essential constraint: the fact that their craft is to "build something."[25] The project of architecture cannot exist unless the consent of the local community for the construction of a new building is somehow obtained. While the project of "building" can be undertaken in many different ways, something *must* be built. "Consensus building" is thus a constant theme—and a roadblock that frequently confronts planners (Ishizuka 2004).

---

[25]Of course, architects may present plans or concepts even in the absence of a client, or without regard to the feasibility of implementation—as demonstrated by the actions of such architects as Kurosawa Kishō, Isozaki Arata, and Andō Tadao. These architects believed that presenting blueprints were a way of critiquing the architectural and societal status quo; their plans had meaning even if they were never implemented. Even is these cases, however, architects likely remained intensely aware of the importance of "building something."

From the standpoint of local residents, however, the chief concern is not whether or not a certain project is executed, but whether or not an irreplaceable place will be protected.[26] There is no need to treat "the world of subjective meaning" as a golden rule, but we must enter this world if we are to explain the structure and significance of the stubborn local resistance to a road construction project in Otaru.

This demonstrates the need for a sociological approach to the issue of preservation, which, as we shall see, remains exceedingly rare.

1.4.2.1 Western Research

Let us begin our review outside of Japan. Apart from the urban theories of Lefebvre and Castells, which also have enormous implications for the issue of historic environment preservation and theories of collective memory (see Lefebvre (1972 = 1975; Castells 1983; Halbwachs 1950 → 1968 = 1989; Boyer 1994; Zerubavel 2003; Rosenfeld and Jaskot, eds. 2008), the findings of Gans (1962, 1968), Weinberg (1979) Barthel (1996), and Nolan and Buckman (1998) are relevant to our approach. More recent works of note include Cintron (2000), Siegenthaler (2004), Jordan (2006), Ren (2008), Zukin (2010), Barthel-Bouchier (2013), George (2013) and Page (2016).

The sociologist Herbert J. Gans, who had practical experience as an urban planner and consultant, employed the participant observation method to analyze how American urban renewal policies ended up dividing communities. In particular, Gans (1962) developed an essential critique of federal government-led redevelopment initiatives (disparagingly referred to as "the federal bulldozer"), which razed old buildings to the ground in an effort at "slum clearance," and erected new ones in their place. Gans created one prototype for how to sociologically engage with the subject of urban policy. At the same time, the concept of the historic environment remained latent in Gans' work, which focused on the broader theme of urban renewal.

Weinberg (1979) was the first sociologist to directly address the subject of preservation movements. Weinberg introduces important case studies of preservation efforts in Charleston, North Carolina, Boston's Beacon Hill neighborhood, and Savannah, Georgia and focuses on such factors as "historic district" systems and the transfer of development rights. In this sense, he covers most items on the preservation issue "menu." However, although Weinberg introduces the historic preservation movements that were emerging throughout the United States at the time, his account is not sufficiently analytical: instead of employing sociological concepts to his subject, he organizes different cases from across the U.S. according to theme, and concludes

> Preservation is concerned with both history and architecture, but, more than that, its task is the preservation of the remains of an earlier civic culture.
>
> Weinberg (1979: 217)

[26]Neighborhood movements are often criticized as examples of "regional egoism" or NIMBYism. However, if local residents do not express their own interests in the first place, other "interested parties" (Nakanishi and Ueno 2003) may also lose their "expression circuit." Japanese readers may refer to my translation of Dunlap & Mertig, eds. (1992 = 1993: 51–73).

While there is nothing wrong with this sentiment, it is disappointing that Weinberg makes no attempt to identify what the sudden rise in historic preservation movements might say about American society. Weinberg's work is significant only in that it was the first to address the issue of historic preservation.

It is the work of Barthel (1996), published seventeen years after Weinberg, that represents the truly pioneering study of the subject matter. This is because Barthel[27] adopts a comparative approach to her study of British and American preservation efforts (represented by the National Trust for Places of Historic Interest or Natural Beauty and the National Trust for Historic Preservation, respectively), and uses points of difference and similarity to develop a comparative analysis of British and American societies. How does society perceive and ascribe meaning to the physical preservation or destruction of architectural structures? In what ways are these acts received or deployed within different societies? It is Barthel's consistent attention to such questions that confirms her work as the first genuine sociological inquiry into the preservation of historic environments.

Barthel (1984) began her career with a study of the Amana, a German Pietist religious commune in rural Iowa, which includes a detailed historical account of the sect's evolution. Members of the Amana commune originally sought to live by God's teachings alone, and rejected American materialism and commercialism. In 1932, however, the commune's members made the agonizing decision to undergo the "Great Change" and accept the reality of modern American life (Barthel 1984: 63–118). In the process, the Amana realized their commune's buildings and way of life could become a resource for tourism, and dedicated their efforts to preservation. This is what sparked Barthel's interest in the social act of preservation, which culminated in the 1996 publication of *Historic Preservation: Collective Memory and Historical Identity*. It is thus no exaggeration to conclude that Chap. 15 of Barthel's 1984 work represents the very origin of the sociological study of historic environment preservation. This is an excellent example of how the final chapter of a debut work often signals the future direction of an author's research (Barthel 1984: 159–169).

The central concept of Barthel's later work (1996) is that of "staged symbolic communities." These are not natural or spontaneous communities, but rather communities that have been shaped with intention, and staged and performed to convey a symbolic meaning. Barthel's examples include Colonial Williamsburg [28] in Virginia and the Plimouth Plantation,[29] which replicates the original settlement of Plymouth

[27]Readers interested in referring to Diane Barthel's work should note that her surname has since changed to Barthel-Bouchier.

[28]The Colonial Williamsburg is an open-air museum owned and operated by the Colonial Williamsburg Foundation, an NPO based in Williamsburg, Virginia. The entire village was purchased and restored to reflect life in 18th century (pre-American independence) Williamsburg. The museum was the 1926 brainchild of the Reverend Dr. W. A. R. Goodwin, and funded by John D. Rockefeller, Jr. The various structures the village were restored or reconstructed according to meticulous historical research, and all museum staff wear period garments and interact with visitors using the language of the period. See Coffman (1998) and Greenspan (2002).

[29]The Plimouth Plantation is an NPO-operated open-air history museum in Plymouth, Massachusetts, founded in 1949 by Henry Hornblower II. Centered around a reproduction of the

Colony in Massachusetts. Of course, the Amana commune where Barthel began her research is yet another example of a staged symbolic community (SSC).

The concept of staged symbolic communities is informed by the community and utopia theories of Gerald Suttles, Albert Hunter and Ralf Dahrendorf (Dahrendorf 1968; Suttles 1972; Hunter 1974; Barthel 1996: 35ff). Barthel employs Dahrendorf's definition of utopia to enumerate the following four characteristics of staged symbolic communities (Barthel 1996: 37–48):

1. ahistoricity,
2. social consensus,
3. repetitive activity, and
4. isolation.

Barthel asserts that SSCs with these four characteristics are, like Colonial Williamsburg, comprised of reconstructions based on a generic imaginary community, and do not refer to real historic communities, and are engaged in the act of historic preservation. Day after day, the lives of colonists living in pre-independence Virginia are recreated at Colonial Williamsburg within a discrete time frame that will never connect to the present day. Visitors to Colonial Williamsburg will travel back in time before returning to the "present" at the end of the day. In Williamsburg they leave behind, by contrast, all the same activities will be repeated the next morning, with no forward movement through time.

Barthel does not dismiss SSCs as "artificial" tourist attractions. On the contrary, Barthel offers a critique of modern society by identifying the underlying commonalities between these preserved communities (the SSCs) and the principles informing the construction of modern-day suburban residential communities. Barthel also points out a significant irony, or paradox: although SSCs are not "communities" in the sociological sense, visitors to places like Colonial Williamsburg often see these places as more "authentic" than the actual communities in which they live. In other words, Barthel argues that SCCs are becoming an important aspect of the modern cultural landscape, not only for their representation of an "utopian image," but also for the ideological objectives they offer to modern-day society. We can thus understand Barthel's aim as the development of an ideological critique of preserved communities (SSCs).

It is not sufficient, however, to simply identify something as "an ideology." Barthel's findings are important, and her concept of staged symbolic communities is a valid one. Yet the concept's origins in agricultural communities is a cause for concern. It is worth repeating that Barthel's concept is derived from her early work on the Amana commune in rural Iowa. As such, it offers a persuasive approach to her other cases: theme parks, open-air museums, and other religious communities, such as Shaker villages.

Mayflower, the ship that brought English colonists to Plymouth, Massachusetts, the museum offers visitors a number of "educational recreations," including reproductions of 17th century buildings, vessels, and ways of life. See http://www.plimouth.org.

Urban historic environments, by contrast, can never be characterized by ahistoricity, repetitive activity, or isolation. The Otaru case alone is an ample demonstration of the need for a different theoretical framework for urban preservation efforts. Not only would it be difficult to apply Barthel's analytical approach to an urban area, but her concept could actually obscure the issue of preservation in an urban setting. While I remain indebted to Barthel's pioneering work on the subject, my own research is distinguished by a focus on *urban* case studies of historic preservation.

Nolan and Buckman (1998) undertake a comparative study of the preservation of two different historic structures: Monticello and Montpelier, the respective homes of former U.S. presidents Thomas Jefferson and James Madison. Nolan and Buckman interview the people involved in the preservation and restoration of these two historic residences to identify the underlying differences to their "logic of preservation." While the authors deserve credit for their approach to the intrinsic issue of historic environment preservation, their analytical concept of "modern/postmodern" requires further refinement to be effective.

Although Leslie Gwen Cintron's 2000 dissertation remains unpublished, her study of the British National Trust is a noteworthy addition to the literature. Cintron painstakingly reviews the National Trust's asset list over the past century in order to identify changes to the organizational goals of the Trust. She demonstrates how the properties acquired by the Trust ended up re-defining the organization's goals and mission—and ultimately prescribed the type of British heritage to be preserved. Cintron thus outlines theories of organizational change and "British heritage." My own research is indebted to Cintron's focus on the organizational culture of preservation movements.

Jordan (2006) undertakes a detailed case study of preservation in postwar Berlin as a means of exploring questions of collective memory. Jordan interrogates not only how Nazi atrocities are remembered, but also the mechanisms that determine whether a particular event is commemorated or consigned to oblivion. Jordan thus identifies the type of place that is selected as a site of collective memory and preservation, and the places that are simply returned to anonymity. This is an excellent sociological study of preservation and memory. Jordan's focus on how society determines the "places to be changed" and the "places to be preserved without change" underlines the fact that urban change is far from monolithic, and carries enormous resonance for my own research.

We can thus conclude that Western sociologists have gradually come to address the meaning behind the act of preservation. While there have long been theories of cities and the production of space, a growing number of sociologists are now consciously addressing the subject of preservation movements in cities, where change is assumed to be a natural and constant part of life.

1.4.2.2 Research in Japan

Sociological approaches to the subject of preservation have been equally sporadic in Japan.

If we exclude my prior research on Otaru (Horikawa 1989, 1990), the first sociological work on historic environment preservation was Fujita (1998) *Toshi no Kyōdōsei to Machinami* (Urban Cooperation and Townscape).[30] As the title suggests, Fujita's approach is rooted in urban sociology, rather than the environmental sociology approach to historic environment preservation, which became more commonplace in the 2000s. Fujita compares townscape regulations in Japan and the West in an attempt to explain why Japanese townscapes are so unattractive. Fujita finds the answer in the differing societal positionings of "public" and "private." In the West, where "private" is more powerful than "public," preservation-oriented regulations protect "public" land from the risk of exploitation by "private" interests. In Japan, where the "public" is far stronger than "private," "individuals" cannot hope to protect their own homes and land. It is thus up to the "public" sphere to prevent any regulatory incursions into "private" ownership. As a result, the Japanese system is utterly uninterested in maintaining a certain level of coherence to the physical townscape. In Fujita's assessment, the ugly townscapes of Japanese cities are the perfect manifestation of this system.

Fujita's (1998) study is a comparative social theory of townscapes, and Fujita's attempt to explain the disconnected nature of Japanese townscapes makes this a pioneering sociological approach to the issue. There is no question that Fujita's decision to go beyond case studies of individual movements to expose the underlying principles of societal composition makes for a persuasive argument. Yet Fujita's analysis is extremely static. Fujita's argument is characterized by an attention to underlying structures: Fujita explains words like "city" and "region" with reference to their Latin roots, or discusses them as a product of culture. However, an explanation that reduces everything to "structure" cannot take into account the people who are dissatisfied with the present condition of their city and seek to change it. While I am deeply indebted to Fujita's work, we need a more dynamic perspective, which can pose such questions as "why preserve?" and "how can society control changes to the urban environment?"

Katagiri edited volume (2000) was the first to include the title "the sociology of historic environment preservation." This volume, to which I contributed, included case studies of preservation efforts in the villages and cities of Asuka, Kyoto, Tomonoura, Otaru, and Gujō Hachiman, but also discussed such themes as the preservation of "negative legacies" and the burial and mourning of the deceased. This was an ambitious work that set out to discover the significance of the social act of preserving historic environments through the use of multiple case studies. The discrepancy among authors, and variations in subject and approach, is not uncommon during the emergence of a new field of research, and is hardly an essential flaw. In the words of Katagiri himself, the simple fact that the volume embarked upon a new "sociology of the historic environment" was of enormous significance at a time when

[30]Fujita first introduced this argument in a 1994 essay in the magazine *Sōbun* (Fujita 1994). He revised and expanded this essay in 1995 (Fujita 1995), and further developed it into his 1998 dissertation. A final version of the argument appears as a chapter in Fujita (2003).

"one can only conclude that sociology has failed to address the subject sufficiently" (Katagiri 2000: 17).[31]

The critical reappraisal of museum studies in Ogino (ed. 2002) represents another contribution to the sociology of historic environment preservation. After defining the modern age as one "in which anything can become part of our cultural heritage," Ogino states his objective in straightforward language: "to explain the significance of preservation and the collective desire to preserve all sorts of things in modern life" (Ogino, ed. 2002: ii). To do so, Ogino employs the concept of "museological desire." As this term suggests, Ogino sets about interrogating the present wisdom of museums that seek to preserve and exhibit. This is an avenue of research that carries many implications for both museology and the sociology of historic environment preservation.

Let us stop to briefly summarize this overview of existing research.

As we have seen in Sect. 1.4.1, research into the Otaru preservation movement cannot even be described as "research," given the dearth of foundational primary sources. In addition to the sheer significance of addressing the Otaru case, there is a need to excavate and organize essential sources in order to develop a dense case study of Otaru. This immediately confirms the value of the current volume, which is based on thirty-three years of intensive investigation.

In Sect. 1.4.2, we confirmed the fact that historic environment preservation has not been adequately addressed as a theme in the existing literature. Moreover, alluding to the theme with reference to other social issues is not the same as asking the question: "what kind of changes to society are represented by the emergence of people who resist development and attempt to preserve their environment?" What we need now is more of the latter: research that directly addresses the issue of "cities and preservation."

1.4.3 Situating the Present Research Within the Literature

Having confirmed the paucity of prior research on the subject, how can we trace the intellectual lineages of the current investigation?

1.4.3.1 The First Lineage: Urban Community Theory

The primary subject of my research is the movement to preserve a historic urban environment. This was a movement instigated by members of an urban community to save their living environment; it was also a movement to oppose development. The urban community theory within the field of urban sociology can thus be described as the first "lineage" of my work. As I have argued elsewhere (Horikawa 1998b), postwar Japanese urban sociology has used the base concept of "to live" (*sumu koto*)

[31] A more recent work is that of Morihisa (2016).

to develop theories of resident-led movements in opposition to postwar urban development. Nitagai Kamon, a representative theorist, argues that urban issues emerge from the conflict between the city residents' logic of "to live" (utility value) and the logic of capital's private ownership of the city (surplus value). Nitagai envisioned the collective recapture of urban space based upon the logic of "to live" (Matsubara and Nitagai, eds. 1976: 331–396). Nitagai's concept of "to live" (*sumu koto*) is different from the everyday expression of "living" (*sumu*): "living" is the simple act of dwelling in a particular place, while the concept "to live" also incorporates the idea of cooperation within the local society. Nitagai paints a picture of modern urban life in which people take out loans to purchase a single-home or apartment in a suburban neighborhood, and are compelled to commute to work in the far-away city—leaving them without the time to become involved with, or contribute to, their local community. They have no ability "to live," and must dedicate their time and resources to mere "living." Nitagai's depiction holds true today, as the trend of concentration in urban areas continues.

Okuda Michihiro identifies a similar crisis of residence within a very different theoretical and conceptual framework.

First, Okuda identifies two categories of city. The first is the "city" (*toshi*) as work space. The second is the "region" (*chiiki*) as the foundation of residents' ability "to live." Okuda sees the modern city as the multilayered product of these two lineages, and defines the problem as the destruction of the "region" by the "city." Okuda saw the potential for a new form of community, borne of "regional" resident-led movements opposing the pressures of the "city" (Okuda 1993: 33–61). Put differently, Okuda identified how big cities have become "monoculturized" into a "no-man's land" of work spaces, and argued that "community" depended on the coexistence of the layers of "city" and "region." This theme runs throughout Okuda's subsequent work.

These arguments are clearly applicable to my analysis of the Otaru preservation movement. Preservation activists who demanded to live within their family homes and familiar landscape were not seeking to exist in a biological sense (in a "concrete box"); they aspired "to live" in Nitagai's sense of the phrase: cooperatively and embedded within the local society.

However, we must carefully examine how Okuda and Nitagai composed their theories. Nitagai and many other urban sociologists critiqued the postwar pro-development approach of Japanese urban planning, and saw as its counterpoint the resident-led opposition movements that emerged to defend livelihoods and "living space" (*seikatsu kūkan*). Their use of such abstract, non-place-specific terminology as "utility value" or "space" suggests that their vision of the "region" in which people wish "to live" was, ultimately, a modern one—in other words, an entirely replaceable living space. Behind this approach was their vision of the "universal citizen" posited by postwar modernization theory: someone who simply sought "to live somewhere." The emerging residential class in the new housing developments along the outskirts of big cities—people who had left their ancestral homes in rural villages, and therefore broken free of the feudalistic social order to live and conduct themselves as "citizens" in their new communities—were precisely what Nitagai and Okuda had in mind. When these "citizens" broke free of feudalistic ties to protest pollution or inadequate infrastructure, this took the form of a neighborhood movement based on "the power of resistance focused on 'to live' (*sumu koto*) (Satō 1993: 166). What we must appreciate is the abstracted concept of space (*kūkan*) in which these citizens will seek "to live" as they oppose the modern restructuring of the city. Such arguments carried enormous significance during Japan's period of rapid economic growth, as the national landscape underwent dramatic upheaval. They reveal both the issues that Japan's postwar sociologists felt they had to address and also what they posited as the "enemy."

But what if the power of resistance in order "to live" does not emerge from a generic space but a more specific one—a space with a name? In this case, we must also consider the specific environmental conditions of this space. We must examine the local topography, and the collective historical memory of the region. Living space, in other words, is not the homogenous and interchangeable "space" of urban planners, but a "place" (Horikawa 1998b, 2000, 2010a, b).

Japan's postwar sociology (particularly urban sociology) would dismiss as "feudalistic" the attachment of local residents to a specific piece of land, or place ("Otaru will no longer be Otaru without the canal"). This is because postwar urban sociology in Japan has relegated the obsession with "placeness"—for example, a deep affection for one's own hometown—to the background, and argued that the more crucial issues are the creation, by the modern "citizens" of Japan's postwar democracy, of "publicness" and "fundamental principles of citizenship." The "placeness" of urban space, increasingly under threat by the latter's abstraction into the concept of "space," was the central concern of historic preservation movements. Indeed, these movements were an attempt to bring "placeness" back to the foreground.

This could also be described as the difference between "an issue for wherever I may choose 'to live'" and "an issue that affects whether I can continue to live *here*." Protecting a specific, individual environment—in other words, arguing for the distinctiveness of space—was an essential question raised by the preservation movement. Today, as the second and third generations of the families who originally moved to suburban commuter belts surrounding major cities begin to involve themselves in regional movements, an attachment to place is seen as a strategy for environmental protection—indeed, it is this very "placeness" that residents use to oppose city planning measures.

This investigation builds upon and expands Japan's postwar urban sociology (and particularly urban community theory) and the work of Fujita Hirō. In doing so, I move away from the abstract language of "space" and "citizens;" my research is based upon social movements rooted in the direct "power of the environment" and "the immediacy of physical objects." In short, it can be considered a critical successor within the lineage of urban sociology.

1.4.3.2 A Second Lineage: Environmental Sociology

There is, however, a second lineage to my inquiry. The field of environmental sociology has been equally influential. As I noted in Sect. 1.1, my research addresses the relationship between the urban environment and society. Our urban lives cannot be separated from the material constraints imposed by the environment. Society is both social and material. Moreover, in the sense that the multitude of things we produce reflect the society of their production, material substances are social.

Sociology has traditionally neglected this point. Yet if social relations can be transformed by changes to the physical world, sociology must incorporate the environment itself as a variable. This was the point of departure for the newly emergent environmental sociology of the 1970s (Catton and Dunlap 1978a, b). The relevance of this perspective to the current investigation is only too obvious, and should explain why my approach is equally rooted in the field of environmental sociology.

Riley Dunlap and William Catton heralded the advent of environmental sociology by criticizing the dependence of conventional sociology on a "Human Exceptionalism (Exemptionalism) Paradigm (HEP)," characterized by a faith in technological progress and the exceptional nature of the human activity. They attacked the latent worldview of this sociology, which saw environmental destruction as the occasional, unavoidable byproduct of humanity's "exceptional" existence—and as something that could always be remedied by future human innovation. Dunlap and Catton argued that it was time for sociology to transition from this HEP to a "NEP"—a "New Environmental Paradigm" that incorporated the environment as a meaningful variable in the sociological analysis (Dunlap and Catton 1979). In other words, they heralded the arrival of a new environmental sociology.

The field of sociology, however, originally emerged as the systematization of the study of social relations amid the conflict between environmental determinism and instinct theory. While the latter would invariably explain human behavior in terms of "instinct," the former argued that most human behavior is acquired, and regulated by the surrounding environment. By contrast, the emerging field of sociology rejected these assumptions of human passivity, and contended that humans are capable of independent action. The fact that sociology did not incorporate the environment as a variable was thus not the result of neglect, but rather a fundamental strategy from the very beginning.

Not surprisingly, Dunlap and Catton's proposition was roundly criticized. A history of the HEP-NEP debate is outside the scope of the current review, but two points are worth noting: first, Dunlap and Catton's attempt at a sociological paradigm shift was highly abstract; as such, it failed to provide a model from which empirical research could be derived.

The HEP-NEP paradigm controversy was never decisively concluded. While the NEP proposal did warn traditional sociology of the need to actively address environmental issues, it was ultimately reduced to a single element within the existing paradigm. Dunlap himself has concluded that the HEP-NEP debate fulfilled its historical mission (Dunlap 1997, 2010); as a theory, it is now receding into the past (Mitsuda 1995; Fujimura 1996; Taniguchi 1998).

Yet there may be another way of understanding Dunlap and Cotton's proposition: as a simple argument for the sociality of the physical world and the material nature of society. Let us return once more to the case study of Otaru.

Shortly after I plunged into my case study of Otaru, a participant in the canal preservation movement told me an interesting story. Initially, preservation activists had little support from the broader public. Several young activists came up with the idea of holding a public event, which they called "the Port Festival," along the canal, as a way of convincing local residents of the canal's continued relevance and value. Without status or money, however, these young activists struggled to build sufficient support for their cause. One night, after a long discussion at the usual coffee house failed to produce any new ideas, one of their numbers suggested suspending the Port Festival: "no matter how hard we work at this, it's useless. It's time to give up." As members of the group began to assent to his proposal, another person spoke up: "even if we end up cancelling the event, let's go to the canal one last time, and talk about it there." And so, at around midnight, the group settled into a barge floating along the canal, and fell silent. At some point, this silence was broken: "The canal is a pretty great place, after all, isn't it?" "Wouldn't it be wonderful to restore it to the way it was in the past?" "Imagine drinking out here, under such a vast expanse of sky." The situation felt less hopeless now that they had left their corner of a smoke-filled coffee house, and sat out under the starry sky, gazing at the stone warehouses that had stood silently along the canal for so many years. On that night, this group of young activists vowed to redouble their efforts to ensure the success of the Port Festival.[32]

There is more to this story than the fact that a group of young activists reached a different decision out on the canal, with the salty sea breeze blowing in their faces, than they had in a windowless room. The real significance of this episode is the fact that the very people fighting to save the canal physically experienced anew the full extent of the environment's power, and reconfirmed the significance of preservation. According to my informant, even these activists were surprised by what a powerful effect the environment of the canal had upon them when they actually spent time in its presence.

Even more significant is the fact that these young people described their experience using the words "the educational power of the environment" (*kankyō no kyōiku ryoku*).[33] The environment does not regulate everything (the young activists made an independent decision to carry on with the Port Festival), but the environment of the Otaru canal was influential enough that one hesitates to ascribe everything to the will of human actors. By the 1980s, this delicate nuance had already crystallized into the expression "the educational power of the environment." Because these activists opposed a road construction plan that was entirely indifferent to "the educational

[32]The Port Festival was held in July, 1978 and attracted more than 100,000 visitors. The festival was held annually for seventeen years. The final Port Festival was held in August, 1994. For more details, see Chap. 5.

[33]This concept and term was developed by the "think tank" of the Otaru preservation movement, otherwise known as the "Hokudai trio," comprised of three talented graduate students in architecture and urban planning: Yanagida Ryōzō, Morishita Mitsuru, and Ishizuka Masaaki. Although a full discussion is outside the scope of this investigation, which remains focused upon the logic of preservation activists, the concept is of indisputable significance. See Ishizuka et al. (1982) and Yanagida et al. (1984).

power of the environment," the term aptly summarizes the preservation movement's main argument.

How should sociology approach that midnight visit to the canal, or "the environmental power of the environment?" The HEP-NEP debate has been consigned to the past, but have sociologists taken sufficient heed of the full weight and meaning of the environment's effect upon humans in their search for a new theory? Have they thoroughly discussed the disposition of townscape preservation activists, or the actual mechanics of preservation movements? How should sociology, a discipline that focuses on social relations, discuss the immediate and powerful impact of the environment? Academic history may have relegated this question to the past, but is that really where it belongs (Horikawa 2008, 2017)? This is both the concern that guides my investigation and the reason it remains rooted in the field of environmental sociology.

Thus, by way of such questions as "why do people preserve?" and "what is the material nature of society?" I arrive at a different vantage point from which to consider the question of "what is the urban environment?" A thorough analysis of the specific case of Otaru enables a "sociological examination of the very phenomenon of preservation." By focusing our gaze *on* Otaru we may identify problems with existing theories and policies, but it is *from* Otaru that we may explore new solutions.

1.5 An Overview of the Book

Let us turn to a brief overview of the book's structure (see Fig. 1.3).

In Chap. 2, "The Historical Environment as Object: The Rise and Significance of Townscape Preservation Movements," I review the formation and development of townscape preservation movements in postwar Japan, and consider their significance. A brief historical overview of these preservation movements and Japan's legal system underscore the nature of the former: preservation movements were not motivated by the nostalgic or "retro" tastes of their participants. On the contrary, they were developed by activists to oppose Japan's system of city planning.

After positioning the issue of the landscape and its preservation in Chap. 2, I move to the case study of Otaru in Chaps. 3, 4 and 6. In Chap. 3, "A City with a Grudge: A History of the Port City of Otaru," I offer a detailed history of Otaru to clarify the reasons behind the city's continued resentment of its neighbor, Sapporo, and the emergence of a "revitalization" plan predicated on the reclamation of the Otaru canal. This chapter traces the historical processes leading up to the outbreak of the "canal war" in the 1970s.

Having reviewed the city's history, I turn to the "canal war" itself. I examine how the various actors interacted around the subject of change, and the social consequences of their actions. There were two diametrically opposing approaches to change: development and preservation; I review the events of the "canal war" from the perspective of both the "development" and the "preservation" camps. The canal preservation movement, quite obviously, sought preservation. The agent of change

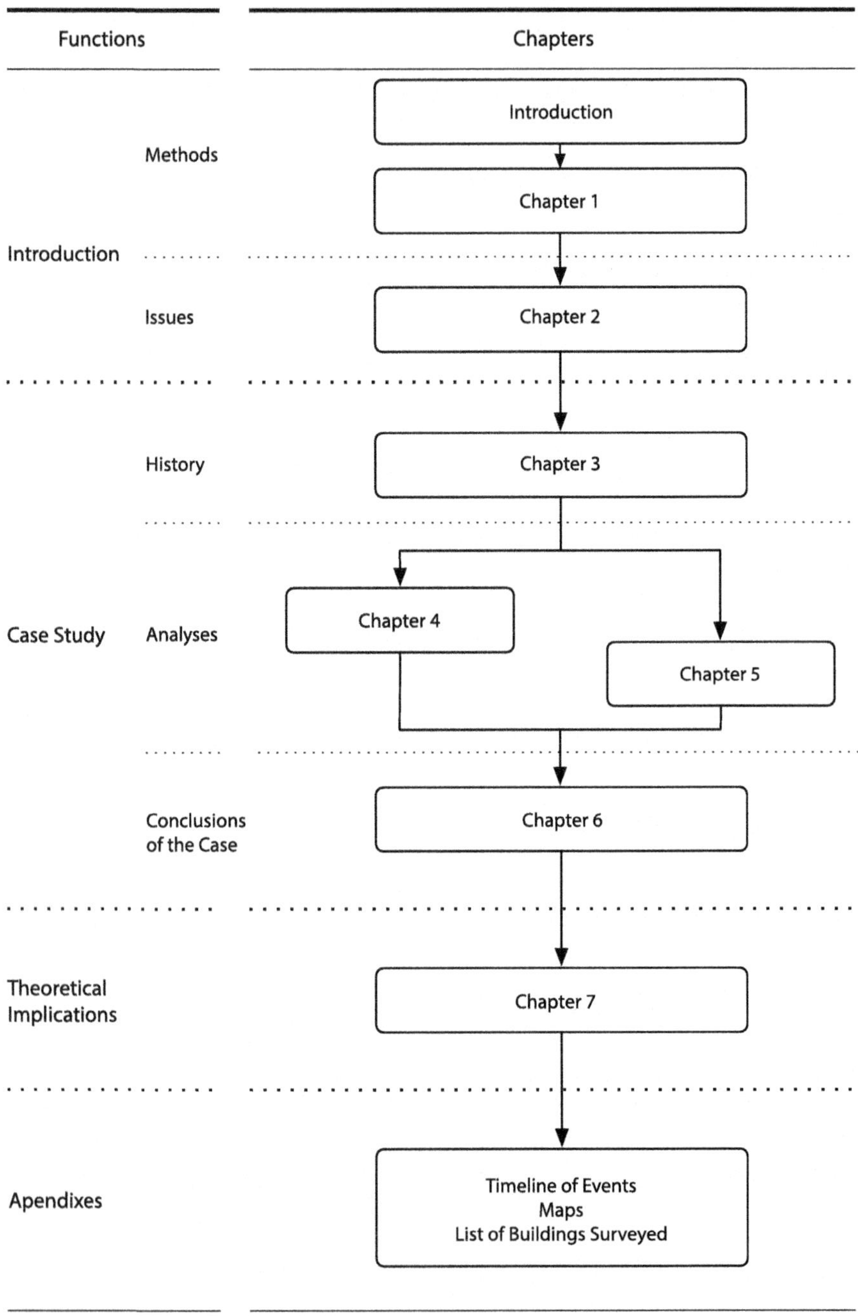

Fig. 1.3 An overview of the book (Author's illustration)

(and development) was the city government. In Chaps. 4 and 5 I offer two deep readings of the canal conflict by tracing the "lived time"[34] of city officials and preservation activists. The narrative in Chap. 3 remains the standard point of reference for both subsequent chapters.

In Chap. 4, "The Case for Change: The 'Canal Issue' from the Perspective of the Otaru City Government," I analyze the logic of development shared by the "forces of change" (i.e., local government officials) and explore the meaning they assigned to the canal and the canal war. In this undertaking, I avoid overly simplistic criticisms of the bureaucracy, and seek to clarify the inherent principles and logic of local government to explain why officials sought to reclaim the canal.

I assume the opposite vantage point in Chap. 5, "The Logic of Preservation: The Preservationists' View of the 'Canal Issue,'" in which I explore what the canal meant to preservation activists, and how it comprised the very identity of many local residents. I take care to avoid an overly sympathetic treatment of the movement's participants, and use their statements and actions to isolate different layers of activism, and to develop four different categories of the activist. I examine the characteristic logic of each type of activist, and the various conflicts and alliances between different factions of the preservation movement. The conclusion that emerges from this examination is a surprising one: notwithstanding the inherent conservatism of the word "preservation," the movement to save the Otaru canal was in fact a pioneering form of activism. Preservation, it turns out, can be an extremely radical act.

In Chap. 6, "What Was Won? What Was Lost? The Transformation of the Otaru Townscape in the Post 'Canal War' Period," I examine how the canal war changed the local landscape, and identify the implications of these changes for other urban environmental issues and related policies. Using previous architectural survey data and the results of my own fixed point observation surveys, I offer an empirical verification of tourism's impact on the Otaru landscape and clarify the structure behind the "paradox of tourism development."

Having completed my case study of Otaru, I turn to the theoretical implications of my findings in the concluding chapter, "To Preserve is to Change." I first unravel the preservation movement's "logic of preservation" to demonstrate how it identifies the problems with modern urban planning. This logic of preservation equally confounds the typical framework of "preservation or development:" what local residents really sought was autonomous control over changes to their city. Their desire can be expressed through the paradoxical assertion that "to preserve is to change"—a finding that may represent the first step toward the development of a new form of control over urban spaces. I conclude with a discussion of how this analysis may contribute to theories of urban and environmental sociology. Finally, a timeline of events and other reference materials have been appended for the reader's reference.

[34]I use the term "lived" because it incorporates the double significance of contrasting the present and the past (through the use of the past tense) and the abstract and the concrete (through the use of the passive tense). This anticipates an inductive argument based on the detailed account of the specific lived experiences of flesh-and-blood humans, not abstract, tense-less cities and times.

References

Akashi, T. (2002). *Jichitai Esunogurafī: Chihō ni Okeru Soshiki Henyō to Aratana Shokuinzō* (Law and Society Debut Series 3). Shinshinsha.

Andō, T. (1992). Otaru Unga Hozon Undō no Isan: Wakamono ga Hiraita Machizukuri no Mirai. *Gekkan Times, 16–9,* 100–105.

Aramaki, M. (1987). *Kanashiki wa Otaru no Machiyo: '86 Otaru e no Kugen/Chokugen, Soshite Teigen o Matomeru.* Chiri Konwakai (B5, 36 pp.).

Arisue, K. (1984). Seikatsu Kenkyū to Raifu Hisutorī: Seikatsushi Kenkyū Kara. In N. Kawazoe (Ed.), *Seikatsugaku e no Apurōchi* (pp. 49–68). Domes Shuppan.

Arisue, K. (1992). Shitamachi no Seikatsu Sekai: Jūsōteki Toshi Bunka e no Seikatsushiteki Apurōchi. In K. Morioka & Y. Matsumoto (Eds.), *Seikatsu, Kankei, Bunka* (Toshi Shakaigaku no Furontia 2) (pp. 197–222). Nippon Hyōronsha.

Arisue, K. (1999). *Gendai Daitoshi no Jūsōteki Kōzō: Toshika Shakai ni Okeru Dentō to Henyō.* Kyoto: Minerva Shobō.

Arisue, K. (2012). *Seikatsushi Sengen: Raifu Hisutorī no Shakaigaku.* Keio University Press.

Ashiwara, Y. (1979). *Machinami no Bigaku.* Iwanami Shoten.

Ashiwara, Y. (1983). *Zoku, Machinami no Bigaku.* Iwanami Shoten.

Barthel, D. (1984). *Amana: From pietist sect to American community.* Lincoln, NE: University of Nebraska Press.

Barthel, D. (1996). *Historic preservation: Collective memory and historical identity.* New Brunswick, NJ: Rutgers University Press.

Barthel-Bouchier, D. (2013). *Cultural heritage and the challenge of sustainability.* Walnut Creek, CA: Left Coast Press.

Boyer, M. C. (1994). *The city of collective memory: Its historical imagery and architectural entertainments.* Cambridge, MA: MIT Press.

Castells, M. (1983). *The city and the grassroots: A cross-cultural theory of urban social movements.* Berkeley, CA: University of California Press.

Catton, W. R., Jr., & Dunlap, R. E. (1978a). Environmental sociology: A new paradigm. *The American Sociologist, 13,* 41–49.

Catton, W. R., Jr., & Dunlap, R. E. (1978b). Paradigms, theories, and the primacy of HEP-NEP distinction. *The American Sociologist, 13,* 256–259.

Cintron, L. G. (2000). *Preserving national culture: The national trust and the framing of British national heritage: 1895–2000.* Unpublished Ph.D. dissertation, Harvard University, Cambridge, MA.

Coffman, S. (1998). *Official guide to Colonial Williamsburg* (New ed.). Williamsburg, VA: The Colonial Williamsburg Foundation.

Dahrendorf, R. (1968). Out of Utopia: Toward a reorientation of sociological analysis. In R. Dahrendorf (Ed.), *Essays in the theory of society* (pp. 107–128). Stanford, CA: Stanford University Press.

Dunlap, R. E., & Catton, W. R., Jr. (1979). Environmental sociology. *Annual Review of Sociology, 5,* 243–273.

Dunlap, R. E., & Mertig, A. G. (Eds.). (1992). *American environmentalism: The U.S. environmental movement, 1970–1990.* New York, NY: Taylor and Francis.

Dunlap, R. E. (1997). The evolution of environmental sociology: A brief history and assessment of the American experience. In M. Redclift & G. Woodgate (Eds.), *The international handbook of environmental sociology* (pp. 21–39). Cheltenham, UK: Edward Elgar.

Dunlap, R. E. (2010). The maturation and diversification of environmental sociology: From constructivism and realism to agnosticism and pragmatism. In M. Redclift & G. Woodgate (Eds.), *The international handbook of environmental sociology* (2nd ed., pp. 15–32). Cheltenham, UK: Edward Elgar.

Fujimoto, T. (1985). Rekishi, Bunka Katsuyōgata no Chiiki Shinkō: Otarushi. In *Chiiki Kaihatsu, 249,* 1–8. Nihon Chiiki Kaihatsu Sentā.

Fujimura, M. (1996). Shakaigaku to Ecorojī: R.E. Dunlap no Riron no Kentō. *Kankyō Shakaigaku Kenkyū, 2,* 77–90.

Fujita, H. (1994). Nihon Toshi no 'Keikaku' to 'Mukeikaku'. *Sōbun, 356,* 14–17. Sōbunsha.

Fujita, H. (1995). Nihon no Toshi to Machinami: 'Keikaku' to 'Mukeikaku' no Hazamade.' In K. Kasahara, K. Nishimura, & T. Miyauchi (Eds.), *Shakai Kōzō no Tankyū: Genjitsu to Ronri no Intāfueisu* (pp. 205–227). Shinsensha.

Fujita, H. (1998). Toshi no Kyōdōsei to Machinami: Toshi Keikaku no Hikaku Shakaigaku. In Rekishi to Hōhō Henshū Iinkai (Ed.), *Toshi to Gengo* (Rekishi to Hōhō 2) (pp. 119–146). Aoki Shoten.

Fujita, H. (2003). *Toshi to Bunmei no Hikaku Shaikaigaku: Kankyō, Risuku, Kōkyōsei.* University of Tokyo Press.

Fukukawa, Y. (2003). Toshi Komyuniti no Hozen. In Y. Nishimura, T. Ōnishi, S. Ōgaki, T. Kishii, & K. Koide (Eds.), *(Toshi Kōgaku Kōza) Toshi o Hozen Suru* (Chapter 5). Kajima Shuppankai.

Funabashi, H., Hasegawa, K., Hatanaka, S., & Katsuta, H. (1985). *Shinkansen Kōgai: Kōsoku Bunmei no Shakai Mondai* (Yūhikaku Shosho 749). Yūhikaku.

Funabashi, H. (2006). Ronri Keisei wa Ika ni Shite Kanō ka? o Tō Shōshiten. *Shakaigakuhyōron, 57(1),* 4–24.

Funabashi, H., & Funabashi, K. (1976). Taikōteki Bungyō no Riron. *Gendai Shakaigaku,* 3-2, 114–129.

Gans, H. J. (1962). *The urban villagers: Group and class in the life of Italian-Americans.* New York, NY: Free Press.

Gans, H. J. (1968). *People and plans: Essays on urban problems and solutions.* New York, NY: Basic Books.

George, T. S. (2013). Furusato-zukuri: Saving home towns by reinventing them. In C. Gerteis & T. S. George (Eds.), *Japan since 1945: From postwar to post-bubble* (pp. 27–46). London: Bloomsbury.

Greenspan, A. (2002). *Creating Colonial Williamsburg: The restoration of Virginia's eighteenth-century capital* (2nd ed.). Chapel Hill, NC: The University of North Carolina Press.

Halbwachs, M. ([1950] 1968). *Le mémoire collective* (2e edition, Bibliothèque de sociologie contemporaine). Paris: Presses Universitaires [Koseki, T. trans. (1989). *Shūgōteki Kioku.* Kyoto: Kōrosha].

Hama, H. (2000). Kioku no Topogurafi. *Mita Shakaigaku, 5,* 4–16.

Hama, H. (2002). Rekishi to Shūgōteki Kioku: Hikōsen Graf Zeppelin-gō no Hirai. *Nenpō Shakaigaku Ronshū, 15,* 3–15.

Hama, H. (2007). Rekishi to Kioku. In K. Hasegawa, H. Hama, M. Fujimura, & T. Machimura (Eds.), *Shakaigaku* (New Liberal Arts Selection) (Chapter 6). Yūhikaku.

Hama, H. (2010). Kioku to Basho: Kindaiteki Jikan, Kūkan no Henyō. *Shakaigaku Hyōron, 60–4,* 465–480.

Hama, H., Arisue, K., & Takemura, H. (Eds.). (2013). *Hibakusha Chōsa o Yomu: Hiroshima.* Nagasaki no Keishō. Keio University Press.

Harada, H. (1986). Otaru Unga no Ue o Kuruma ga Hashiridashita. *Gijutsu to Ningen, 15–3,* 7–15.

Hoffman, M. (1995). Otaru: The river in the sands. *Japan Quarterly, 42–3,* 298–305. Asahi Shinbun Publishing.

Hoffman, M. (1996). Otaru: glass wires, stone circles, and all that jazz. *Japan Quarterly, 42–4,* 451–462. Asahi Shinbun Publishing.

Hokkaido Jūtaku Toshibu Toshi Seibika (Ed.). (1989). *Machi no Saisei o Mezashite: Otaru Rinkōsen Kensetsu no Kiroku.* Sapporo: Hokkaido Toboku Kyōkai.

Honma, Y. (1980, December 12). Otaru Unga o Umetatete Yoi no ka: Bunka o Kirisuteru An'ina Kaihatsu. *Economist, 58–6,* 76–79. Mainichi Shinbunsha.

Horikawa, S. (1989). Otaru Unga Hozon Undō no Bunretsu Katei: Undō Rinen no Hensen to Tenkai. *Hōgaku Kenkyūka Ronbunshū, 30,* 111–128. Keio University Press.

Horikawa, S. (1990). Sengo Nippon ni Okeru 'Machinami Hozon' Undō no Tenkai: Otaru Unga Hozon Undō o Jirei toshite. M.A. thesis, Keio University.

Horikawa, S. (1994). Chiiki Shakai no Saisei Toshite no Machinami Hozon: Otarushi Saikaihatsu Chiku o Meguru Undō to Gyōsei no Ronri Kōchiku Katei. In Shakai Undōron Kenkyūkai (Ed.), *Shakai Undō no Gendaiteki Isō* (pp. 95–143). Seibundō.
Horikawa, S. (1998a). Toshi Kūkan to Seikatsusha no Manazashi. In A. Ishikawa, K. Satō, & K. Yamada (Eds.), *Mienai Mono o Miru Chikara: Shakai Chōsa to Iu Ninshiki* (pp. 133–149). Yachiyo Shuppan.
Horikawa, S. (1998b). Rekishiteki Kankyō to Chiiki Saisei: Machinami Hozon ni Okeru 'Bashosei' no Sōtenka. In H. Funabashi & N. Iijima (Eds.), *Kankyō* (Kōza Shakaigaku 12) (pp. 103–132). University of Tokyo Press.
Horikawa, S. (2000). Unga Hozon to Kankō Kaihatsu: Otaru ni Okeru Toshi no Shisō. In S. Katagiri (Ed.), *Rekishiteki Kankyō no Shakaigaku* (Shirīzu Kankyō Shakaigaku 3) (pp. 107–129). Shinyōsha.
Horikawa, S. (Ed.). (1999). *Otarushi ni Okeru Rekishiteki Kankyō Hozon to Kankō Kaihatsu (2): 1998-nendo Hosei Daigaku Shakaigakubu Shakai Chōsa Jisshū Hōkokusho.* Hosei Daigaku Shakaigakubu Shakai Chōsa Jisshūshitsu.
Horikawa, S. (Ed.). (2000). *Otarushi ni Okeru Rekishiteki Kankyō Hozon to Kankō Kaihatsu (3): 1999-nendo Hosei Daigaku Shakaigakubu Shakai Chōsa Jisshū Hōkokusho.* Hosei Daigaku Shakaigakubu Shakai Chōsa Jisshūshitsu.
Horikawa, S. (2001). Keikan to Nashonaru Torasuto: Keikan wa Shoyū Dekiru ka'. In H. Torigoe (Ed.), *Shizen Kankyō to Kankyō Bunka* (Kōza Kankyō Shakaigaku 3) (pp. 159–189). Yūhikaku.
Horikawa, S. (2003). Toshi to Hozon no Fīrudowāku: 'Komyunikēshon Toshite no Chōsa' e no Sukecchi. In I. Watado, Y. Hirota, & J. Tajima (Eds.), *Toshiteki Sekkai/Komyuniti/Esunishiti: Posuto Metoroporisu no Toshi Esunogurafi Shūsei* (pp. 189–211). Akashi Shoten.
Horikawa, S. (2006, January 31). *'It wasn't a matter of length': Historic preservation movements in Japan.* Paper read at Department Colloquium, Department of Sociology, State University of New York at Stony Brook, Stony Brook, NY.
Horikawa, S. (2008). Kantō Essei: 'Kotoba' to Kankyō Shakaigaku no Kakushin. *Kankyō Shakaigaku Kenkyū, 14,* 1.
Horikawa, S. (2010a). Basho to Kūkan no Shakaigaku: Toshi Kūkan no Hozon Undō wa Nani o Imi Suru no ka. *Shakaigaku Hyōron, 60–4,* 517–534.
Horikawa, S. (2010b). Kankō Kaihatsu. In Nihon Shakaigakkai Shakaigaku Jiten Kankō Iinkai, (Ed.), *Shakaigaku Jiten* (pp. 614–615). Maruzen.
Horikawa, S. (2011). Kindai Toshi no Mizubeto Kōkyōken: Toshin no Mizube wa Dare no Mono ka. *Kantō Toshi Gakkai Nenpō, 13,* 50–59.
Horikawa, S. (2015). Nenpyō ni yoru Mondai Kōzō no Haaku: *Kankyō Sōgō Nenpyō* no Kokoromi. In H. Satō, A. Hamamoto, M. Sano, & T. Takimura (Eds.), *Kaihatsu Shakaigaku o Manabu Tame no 60-satsu* (pp. 119–120). Akashi Shoten.
Horikawa, S. (2017). Nihon ni Okeru Kankyō Shakaigaku no Bokkō to 'Seidoka:' Hitotsu no Shiron. *Hōgaku Kenkyū, 90–1,* 379–406.
Hunter, A. (1974). *Symbolic communities: The persistence and change of Chicago's local communities.* Chicago, IL: University of Chicago Press.
Iida, K. (1993). Gekijō Toshi—Otaru: Koseitekina Keikan Shigen o Yomigaeraseru Machi e. *Tomorrow's, 30,* 13–14. Dai Nippon Toboku Kabushikigaisha Kikakushitsu Kōhōka.
Inagaki, E. (1984). *Bunka Isan o Dō Uketsugu ka* (Toshi no Jyanarizumu). Sanseidō.
Inoue, T. (1981). Tōkō, Otaru Unga Hozon ni Kan Suru Iken. *Yusō Tenbō, 178,* 44. Nittsu Sōgō Kenkyūjo.
Ishizuka, M. (1980). (Otaru Hōkokukai) Otaru, Unga, Soshite Watashitachi no Ayumi. *Kankyō Bunka, 47,* 16–18. Kankyō Bunka Kenkyūjo.
Ishizuka, M. (2004). *Sanka no "Ba" o Dezain Suru: Machizukuri no Gōi Keisei, Kabe e no Chōsen.* Kyoto: Gakugei Shuppansha.
Ishizuka, M., Yanagida, R., & Morishita, M. (1982). Otaru no Rekishiteki Kankyō no Saisei ni Kan Suru Kenkyū: Sono 8, Shimin Ishiki to Hozon Undō (2) ("Kankyō Gakushūgata" Ibento no

Kokoromi o Tsūjite. *Nippon Kenchiku Gakkai Taikai Gakujutsu Kōen Kōgaishū* (Toshi Keikaku) (pp. 1861–1862).
Itō, T. (2000). *Nihon no Kindaika Isan: Atarashii Bunkazai to Chiiki no Kasseika* (Iwanami Shinsho Shin Akaban 695). Iwanami Shoten.
Jinnai, H. (Ed.). (2009). *Rekishiteki Kankyō Hozon Ron* (Inagaki Eizō Chosakushū 7). Chūō Kōron Bijutsu Shuppan.
Jordan, J. A. (2006). *Structures of memory: Understanding urban change in Berlin and beyond* (Cultural memory in the present). Stanford, CA: Stanford University Press.
Juristo. (Ed.). (1976). *Kaihatsu to Hozen: Shizen, Bunkazai, Reshiteki Kankyō*. Yūhikaku.
Juristo. (Ed.). (1977). *Zenkoku Machizukuri Shūran*. Yūhikaku.
Kajita, T. (1976). Shakai Mondai no Atarashii Tokushitsu to Tekunokurashī. *Gendai Shakaigaku, 6*, 99–113.
Kajita, T. (1988). *Tekunokurashī to Shakai Undō: Taikōteki Sōhosei no Shakaigaku* (Gendai Shakaigaku Sōsho 15). University of Tokyo Press.
Kanekura, T. (1986). Kōwan Keizai no Henka to Kadai. *Toshi Mondai, 77–5*, 88–106.
Kankō Shigen Hogo Zaidan. (Ed.). (1979). *Otaru Unga to Sekizō Sōkogun* (Kankō Shigen Chōsa Hōkoku 7). Kankō Shigen Hogo Zaidan.
Kankō Shigen Hogo Zaidan. (Ed.). (1981). *Rekishiteki Machinami Jiten: Furusato no Machi, Sono Hozon to Saisei no Tame ni*. Kashiwa Shobō.
Katagiri, S. (Ed.). (2000). *Rekishiteki Kankyō no Shakaigaku* (Shirīzu Kankyō Shakaigaku 3). Shinyōsha.
Kawabata, M. (1986). Furui Machi ni Atarashii Bunka o: Otaru Unga no Saisei o Megutte. In T. Himeda (Ed.), *Tsutaeru (2): Konton no Naka Kara* (pp. 121–144). Miraisha.
Kawai, T. (Ed.). (1989). *Kindai Nihon Shakai Chōsashi (1)*. Keio Tsūshin (Keio University Press).
Kawai, T. (Ed.). (1991). *Kindai Nihon Shakai Chōsashi (2)*. Keio Tsūshin (Keio University Press).
Kawai, T. (Ed.). (1994). *Kindai Nihon Shakai Chōsashi (3)*. Keio Tsūshin (Keio University Press).
Kihara, K. (Ed.). (1983). *Rekishiteki Kankyō* (Jirei Chihō Jichi 7). Porupu Shuppan.
Kōsaka, K. (1998). Shakaigaku Riron no Riron Kōzō. In K. Kōsaka & Y. Kōtō (Eds.), *Riron to Hōhō* (Kōza Shakaigaku 1) (pp. 42–64). University of Tokyo Press.
Kumashiro, M. (1981). Otaru Unga to Chihō Jichi: Otarushinai Kansendōro Keikaku no Suishin ni Tsuite. *Yusō Tenbō, 178*, 38–43. Nittsu Sōgō Kenkyūjyo.
Kuwako, T. (1999). *Kankyō no Tetsugaku: Nihon Shisō o Gendai ni Ikasu* (Kōdansha Gakujutsu Bunko 1410). Kōdansha.
Lefebvre, H. (1972). *Espace et Politique* (le droit à la ville). Paris: Éditions Anthropos [Imai, Shigemi, trans. (1975). *Kūkan to Seiji*. Shōbunsha.].
Masuda, Y. (1980a). Otaru Unga Hozon Mondai: Hozon no Igi. *Jurisuto* (Tokushū: Bunkazai no Hozon to Saisei), *710*, 73–76. Yūhikaku.
Masuda, Y. (1980b). Yomigaeru Otaru Unga: Atarashii Machizukuri o Mezashite. *Chihōshi Kenkyū, 30–4*, 57–62. Chihōshi Kenkyū Kyōgikai.
Masuda, Y. (1982). Kosei Aru Machizukuri no Tame ni: Otaru Unga to Sōkogun o Ikasō. In Chihōshi Kenkyū Kyōgikai (Ed.), *Nihon no Toshi to Machi: Sono Rekishi to Gendai* (pp. 311–328). Yūzankaku.
Matsubara, H., & Nitagai, K. (Eds.). (1976). *Jūmin Undō no Ronri: Undō no Tenkai Katei, Kadai to Tenbō*. Gakuyō Shobō.
McPhee, J. (1971). *Encounters with the Archdruid: Narratives about a conservationist and three of his natural enemies*. New York, NY: Farrar, Straus and Giroux.
Merton, R. K. (1968). *Social theory and social structure*. New York, NY: Free Press.
Mikuriya, T. (2002). *Ōraru Hisutorī: Gendaishi no Tame no Kōjutsu Kiroku* (Chūkō Shinsho 1636). Chūō Kōron Shinsha.
Mineyama, F. (1995). *Chiiki ni Ikiru: Otaru Unga to Tomo ni*. Sapporo: Hokkaido Shinbunsha Shuppankyoku.
Mitsuda, H. (1995). Kankyō Shakaigaku to wa Nani ka: Beikoku de no Paradaimu Ronsō Saikō. *Kankyō Shakaigaku Kenkyū, 1*, 53–71.

Miyamaru, Y. (1983). Kokoro ni Kui wa Utarenai, Otaru Unga no 10-nen. *Kankyō Bunka, 60*, 49–59. Kankyō Bunka Kenkyūjo.

Miyamaru, Y. (1984). Kōten, Soshite Hanten, Nami Sawagu Otaru Unga. *Kankyō Bunka, 61*, 63. Kankyō Bunka Kenkyūjo.

Miyamaru, Y. (1985). Inochi Futatabi Otaru Unga: Machinami Tatsuno Zemi Kara, 5. *Kankyō Bunka*, 67. Kankyō Bunka Kenkyūjo. Reprinted in Otaru Unga o Kangaeru Kai, ed. (1986b: 111–119).

Miyamoto, K. (1989). *Kankyō Keizaigaku*. Iwanami Shoten.

Mizuno, S. (1986). Seikatsushi Kenkyū to Sono Tayōna Tenkai. In T. Miyajima (Ed.), *Shakaigaku no Rekishiteki Tenkai* (Raiburari Shakaigaku 10) (pp. 147–208). Saiensu-sha.

Morihisa, S. (2016). *"Tomo no Ura" no Rekishi Hozon to Machizukuri: Kankyō to Kioku no Rōkaru Poritikusu*. Shinyōsha.

Morishita, M., Yanagida, R., & Ishizuka, M. (1983a). Otaru no Rekishiteki Kankyō no Saisei ni Kan Suru Kenkyū: Sono 10, Rekishiteki Kankyō no Henyō Jittai (2) (Sengo ni Okeru Ironai Chiku no Suitai Katei). *Nihon Kenchiku Gakkai Hokkaido Shibu Kenkyū Hōkokushū* (Keikakukei), *56*, 173–176.

Morishita, M., Yanagida, R., & Ishizuka, M. (1983b). Otaru no Rekishiteki Kankyō no Saisei ni Kan Suru Kenkyū: Sono 12, Rekishiteki Kankyō no Henyō Jittai (3). *Nihon Kenchiku Gakkai Taikai Gakujutsu Kōza Kōgaishū* (Toshi Keikaku), 2295–2296.

Naitō, T. (2015). *Kita no Shōto "Otaru" no Kindai: Aru Toshi no Denki*. Shumpūsha.

Naitō, T., & Sakuma, M. (2017). *Sengo Otaru no Kiseki: Chihōtoshi no Suitai to Saisei*. Shumpūsha.

Nakanishi, S., & Ueno, C. (2003). *Tōjisha Shuken* (Iwanami Shinsho Akaban 860). Iwanami Shoten.

Nakano, T. (Ed.). (1977). *Kōjutsu no Seikatsushi: Aru Onna no Ai to Noroi no Nihon Kindai* (Sōsho Raifu Hisutori 1). Ochanomizu Shobō.

Nakano, T. (2003). *Seikatsushi* (Nakano Takashi Chosakushū Seikatsushi Shirīzu 1). Tōshindō.

Nakano, T., & Sakurai, A. (Eds.). (1995). *Raifu Hisutori no Shakaigaku*. Kōbundō.

Nakano, Y., Okamoto, T., & Watanabe, T. (2011). Keikan Hyōka to Chiiki Idō: Azuminoshichōsa no Bunseki (Posutā Sesshon Hōkoku Yōshi). *Nippon Shakaigakkai Dai 84 kai Taikai Yōshishū*, 325.

Natsubori, M. (1980). Otaru Unga o Tsubusu na. *Chūō Kōron, 95–3* (Tsūkan 1117), 345–360.

Natsubori, M. (1992). Yameru Kokyō: Unga Sensō no Machi. *Bessatsu Bungei Shunjū, 201*, 428–468.

Natsubori, M. (1997). *Hadaka no Kōshi*. Shinchōsha.

Nishimura, Y. (1997). *Kankyō Hozen to Keikan Sōzō: Korekara no Toshi Fūkei e Mukete*. Kajima Shuppankai.

Nishimura, Y. (2004). *Toshi Hozen Keikaku: Rekishi, Bunka, Shizen o Ikashita Machizukuri*. University of Tokyo Press.

Nishimura, Y., Ōnishi, T., Ōgaki, S., Kishii, T., & Koide, K. (Eds.). (2003). *(Toshi Kōgaku Kōza) Toshi o Hozen Suru*. Kajima Shuppankai.

Nishio, A. (1992). Otaru Unga. *Kōwan, 774*, 24–29.

Nishiyama, U. (1990). *Rekishiteki Keikan to Machizukuri*. Toshibunkasha.

Nishiyama, Y. (1993). Jūtaku, Machizukuri Seisaku to Shimin Sanka. *Toshi Mondai, 84–1*, 49–60.

Nolan, J. L., Jr., & Buckman, T. F. (1998) Preserving the postmodern, restoring the past: The case of Monticello and Montpelier. *The Sociological Quarterly, 39(2)*, 253–269.

Odagiri, M. (1982). (Shūdan no Hakken 22) Otaru Unga o Mamoru Kai. *Gendai no Me, 23–8*, 204–209. Gendai Hyōronsha.

Ogasawara, M. (1986). *Otaru Unga Sensō Shimatsu*. Asahi Shinbunsha.

Ogino, M. (Ed.). (2002). *Bunka Isan no Shakaigaku: Rūburu Bijutsukan Kara Genbaku Dōmu Made*. Shinyōsha.

Okamoto, S., & the Nippon no Minatomachi Kenkyūkai. (2008). *Minatomachi no Kindai: Moji, Otaru, Yokohama, Hakodate o Yomu*. Kyoto: Gakugei Shuppansha.

Okuda, M. (1983). *Toshi Komyuniti no Riron* (Gendai Shakaigaku Sōsho 11). University of Tokyo Press.

Okuda, M. (1993). *Toshi to Chiiki no Bunmyaku o Motomete: 21 Seiki Shisutemu Toshite no Toshi Shakaigaku*. Yūshindō.

Otaru Unga Kenkyū Kōza Jikkō Iinkai. (1979). *Unga no Rekishi Kara Machizukuri Made* (Otaru Unga Kenkyū Kōza Zen 10-kai Kiroku). Otaru: Otaru Unga Kenkyū Kōza Jikkō Iinkai (self-published; 125 pp.).

Otaru Unga Kenkyū Kōza Jikkō Iinkai. (1981). *"Rekishiteki Machinami no Saisei" to "Kōtsu" Kara Otaru no Machizukuri o Kangaeru* (Dai 2-ki Otaru Unga Kenkyū Kōza Zen 10-kai Kiroku). Otaru: Otaru Unga Kenkyū Kōza Jikkō Iinkai (self-published; 153 pp.).

Otaru Unga Kenkyū Kōza Jikkō Iinkai. (1983). *Dono Yō ni Otaru Unga no Saisei o Susumeru ka* (Dai 3-ki Otaru Unga Kenkyū Kōza Zen 7-kai Kiroku). Otaru: Otaru Unga Kenkyū Kōza Jikkō Iinkai (self-published; 126 pp.).

Otaru Unga Mondai o Kangaeru Kai. (Ed.). (1986a). *Otaru Unga Hozon no Undō, Rekishihen*. Sapporo: Otaru Unga Hozon Undō Kankōkai.

Otaru Unga Mondai o Kangaeru Kai. (Ed.). (1986b). *Otaru Unga Hozon no Undō, Shiryōhen*. Sapporo: Otaru Unga Hozon Undō Kankōkai.

Otaru Unga o Mamoru Kai. (Ed.). (1977). *Otaru Unga Sōgō Chōsa Hōkokusho: Chūkan Hōkoku*. Sapporo: Toshi Isan Kenkyūjo (185 pp.).

Otaru Unga o Mamoru Kai. (Ed.). (1981). *Unga o Umeru na: Hozon Yōbō Ikenshū*. Otaru: Otaru Unga o Mamoru Kai (114 pp.).

Otarushi-shi Hensan Iinkai. (Ed.). (c.1994a). *Otarushi-shi Gyōseihen Shiryō* Dai 9-kan Dai 5-hen Dai 8-shō Dōdō Rinkōsen to Otaru Unga (1): Kōhō, Hōdō. Otaru: Otarushi-shi Hensan Iinkai.

Otarushi-shi Hensan Iinkai. (Ed.). (c.1994b). Otarushi-shi Kōhon Dai 9-kan Gyōseihen (ge) Dōdō Rinkōsen to Otaru Unga (1). Otaru: Otarushi-shi Hensan Iinkai.

Otarushi-shi Hensan Iinkai. (Ed.). (1995). *Gyōseihen (ge)* (Otarushi-shi Dai 9-kan). Otaru: Otarushi.

Otarushi Shakaika Fukudokuhon Henshū Iinkai. (Ed.). (1982). *Watashitachi no Otaru*. Otaru: Otarushi Kyōiku Iinkai.

Otarushi Shakaika Fukudokuhon Henshū Iinkai (Ed.). (1992). *Watashitachi no Otaru* (Kaiteiban). Otaru: Otarushi Kyōiku Iinkai.

Ōwada, T. (1986). Dokyumento Otaru Unga (1): Unga Ronsō no Zenshi. *Gijutsu to Ningen, 15–12,* 30–39.

Ōwada, T. (1987a). Dokyumento Otaru Unga (2): Kuiuchi Zenya no Hozon Ronsō. *Gijutsu to Ningen, 16–2,* 94–107.

Ōwada, T. (1987b). Dokyumento Otaru Unga (3): Kui ga Utareru. *Gijutsu to Ningen, 16–3,* 96–105.

Ōwada, T. (1987c). Dokyumento Otaru Unga (4): Otaruhaku to Kakushin Chiji no Ugoki. *Gijutsu to Ningen, 16–6,* 101–111.

Ōwada, T. (1987d). Dokyumento Otaru Unga (5): Chiji no Otaru-iri to Sono Hamon. *Gijutsu to Ningen, 16–8,* 112–119.

Ōwada, T. (1987e). Dokyumento Otaru Unga (6): Rikōru Sengen. *Gijutsu to Ningen, 16–12,* 99–109.

Ōwada, T. (1988a). Dokyumento Otaru Unga (7): Rikōru Sengen no Hamon. *Gijutsu to Ningen, 17–1,* 95–101.

Ōwada, T. (1988b). Dokyumento Otaru Unga (8): Kōji Saikai e. *Gijutsu to Ningen, 17–2,* 96–103.

Ōwada, T. (1988c). Dokyumento Otaru Unga (9): Shūmaku e. *Gijutsu to Ningen, 17–3,* 104–119.

Ōwada, T. (1988d). Dokyumento Otaru Unga (10): S-kun e no Tegami. *Gijutsu to Ningen, 17–6,* 98–113.

Ōyama, N. (2001). *Komyuniti Shakaigaku no Tenkan: Jizoku Kanōna Chiiki Kaihatsu ni Mukete*. Taga Shuppan.

Page, M. (2016). *Why preservation matters* (Why X Matters series). New Haven, CT: Yale University Press.

Pedler, M., Banfield, P., Boraston, I., Gill, J., & Shipton, J. (1990). *The community development initiative: A story of the Manor employment project in Sheffield*. Aldershot, UK: Avebury.

Ren, X. (2008). Forward to the past: Historical preservation in globalizing Shanghai. *City & Community, 7–1,* 23–43.

Rosenfeld, G. D., & Jaskot, P. B. (Eds.). (2008). *Beyond Berlin: Twelve German cities confront the Nazi past* (Social History, Popular Culture, and Politics in Germany). Ann Arbor, MI: University of Michigan Press.
Rossi, P. H., & Dentler, R. A. (1961). *The politics of urban renewal: The Chicago findings*. New York, NY: Free Press.
Sasaki, K. (Ed.). (1982). Machizukuri ni Okeru Kankyō no Kyōikuryoku to Kankyō Ibentogata Shimin Undō no Tenkai ni Kan Suru Kenkyū: Otaru Unga Mondai o Tōshite (Dai 2-kai Toyota Zaidan Kenkyū Konkūru Junbi Dankai Kenkyū Hōkokusho). Toyota Zaidan.
Satō, I. (1998). *Shōgai Gakushū to Shakai Undō: Otona ga Manabu Koto no Imi*. University of Tokyo Press.
Satō, K. (1990). Otaru Unga no Hozon to Kaishū. *Doboku Gakkaishi Bessatsu Zōkan, Kindai Doboku no Hozon to Saisei, 75–14*, 41–47.
Satō, K. (1993). Komyuniti no Naka no 'Komyuniti'. In O. Hasumi & M. Okuda (Eds.), *21 Seiki Nihon no Neo Komyuniti* (pp. 153–176). University of Tokyo Press.
Satō, I. (1992). *Firudowāku: Sho o Motte Machi e Deyō* (Wādomappu). Shinyōsha.
Satō, I. (2002). *Firudowāku no Gihō: Toi o Sodateru, Kasetsu o Kitaeru*. Shinyōsha.
Satō, I. (2006). *Firudowāku: Sho o Motte Machi e Deyō* (Wādomappu). Shinyōsha (revised ed.).
Satō, K. (1994). *Fūkei no Seisan, Fūkei no Kaihō: Media no Arukerojī* (Kōdansha Sensho Mechie 5). Kōdansha.
Seisaku Kenkyūin Seisaku Jōhō Purojekuto. (Ed.). (1998). *Seisaku to Ōraru Hisutorī*. Chūō Kōronsha.
Shakai Idō Kenkyūkai. (2005). *Kindai Toshi no Sōshutsu to Saiseisan: Otarushi ni Okeru Kaisō Kōsei o Chūshin ni* (Monbukagakushō Kagaku Kenkyūhi [Kisō Kenkyū B2] Seika Hōkokusho). Nihon Joshi Daigaku Ningen Shakaigakubu, Shakai Idō Kenkyūkai (419 pp.).
Shibuya, M. (1990). Otaru Unga Sono Go. *Chiiki Jōhōshi Shinaji, 12*, 10–13. Kokudo Chiri Kyōkai.
Shimizu, A. (1990). Meiji Chūki Chihō Seiji no Ichirei: Otarukō Umetate o Megutte. *Hōgaku Ronshū, 40*, 1929–1957. Sapporo: Hokkaido Daigaku.
Shin'chi, M. (1983). Unga Hozon Undō to Shitōkyoku no Taiō. In K. Kihara (Ed.), *Rekishiteki Kankyō* (Jirei, Chihō Jichi 7) (pp. 159–184). Porupu Shuppan.
Shin'chi, M. (1983–1985). Unga Mondai 10-nen no Ayumi (Rensai Kiji). *Asahi Shinbun* (Otaru-ban chōkan), November 25, 1983–March 30, 1985. Asahi Shinbunsha.
Shinozaki, T. (Ed.). (1989). *Chihō Toshi Saikaihatsu Seisaku no Sōgō Kagakuteki Kenkyū: Otarushi Unga Chiku Saikaihatsu o Megutte* (Shōwa 63-nendo Kagaku Kenkyūhi Hojokin [Ippan Kenkyū A60410012] Kenkyū Seika Hōkokusho). Otaru: Otaru Shōka Daigaku Shinozaki Tsuneo Kenkyūshitsu.
Shisei no Ayumi Henshū Iinkai. (1988). *12-nen: Sanki Shimura Shisei no Ashiato*. Otarushi.
Siegenthaler, P. D. (2004). *Looking to the past, looking to the future: The localization of Japanese historic preservation, 1950–1975*. Unpublished Ph.D. dissertation, The University of Texas at Austin, Austin, TX.
Suttles, G. D. (1972). *The social construction of communities* (Studies of Urban Society Series). Chicago, IL: University of Chicago Press.
Sztompka, P. (1986). *Robert K. Merton: An intellectual profile*. London: Macmillan.
Tamano, K. (2005). *Tokyo no Rōkaru Komyuniti: Aru Machi no Monogatari 1900–80*. University of Tokyo Press.
Tamura, A. (1999). *Machizukuri no Jissen* (Iwanami Shinsho Shin Akaban 615). Iwanami Shoten.
Tamura, Y. (2009). *Otaru Unga Monogatari*. Kajima Shuppankai.
Tanada, S. (1996). Otarukō, Rekishi to Bunka no Sōzō. *Kōwan, 823*, 34–36. Nihon Kōwan Kyōkai.
Taniguchi, Y. (1998). Amerika Kankyō Shakaigaku to Paradaimu Ronsō: 'Paradaimu Tenkan Toshite no Kankyō Shakaigaku' Saikō. *Kankyō Shakaigaku Kenkyū, 4*, 174–187.
Tosaki, S. (1979). *Otaru Unga Mondai Zenkai* (Kita no Jichi Tokushū 7). Sapporo: Kita no Jichisha.
Wakabayashi, M. (2012). *Shakai (gaku) o Yomu* (Gendai Shakaigaku Raiburarī). Kōbundō.
Wakita, T. (1993). Otarushi no Sangyō to Tochi Keizai Kara Mita Chiiki Shindan. *Daitō Bunka Daigaku Kiyō* (Shakaikagaku), *31*, 227–254.

Weinberg, N. G. (1979). *Preservation in American towns and cities*. Boulder, CO: Westview Press.

Whyte, W. F. (1955). *Street corner society: The social structure of an Italian slum* (Enlarged Ed.). Chicago, IL: University of Chicago Press.

Yanagida, R., Ishizuka, M., & Morishita, M. (1983). Otaru no Rekishiteki Kankyō no Saisei ni Kan Suru Kenkyū: Sono 13/Shimin Ishiki to Hozon Undō (4) (Ronsō no Kōzō to Machizukuri Undō e no Kanōsei). *Nihon Kenchiku Gakkai Taikai Gakujutsu Kōen Kōgaishū* (Toshi Keikaku), 2297–2298.

Yanagida, R., Ishizuka, M., & Morishita, M. (1984). Machizukuri Ronsō no Tenkai to Kankyō no Kyōikuryoku ni Kan Suru Kenkyū. *Nihon Kenchiku Gakkai Hokkaido Shibu Kenkyū Hōkokushū, 57,* 265–268.

Yasuda, S., & Hara, J. (1982). *Shakai Chōsa Handobukku Dai 3-han* (Yūhikaku Sosho 619). Yūhikaku.

Zenkoku Machinami Hozon Renmei. (Ed.). (1999). *Shin/Machinami Jidai: Machizukuri e no Teian*. Kyoto: Gakugei Shuppansha.

Zerubavel, E. (2003). *Time maps: Collective memory and the social shape of the past*. Chicago, IL: University of Chicago Press.

Zukin, S. (2010). *Naked city: The death and life of authentic urban places*. Oxford: Oxford University Press.

Chapter 2
The Historical Environment as Target: The Rise and Significance of Townscape Preservation Movements

Abstract This chapter first provides a historical survey of Japan's townscape preservation movement and the legal system surrounding preservation, and then situates the Otaru canal preservation movement within this general picture. I argue that Japan's systems of urban planning and cultural asset protection are ill-equipped to address the challenge of preservation. I also review Japan's traditional townscape "etiquette" as a technique for combining private rights with publicness, and contrast this approach with modern urban planning in Japan, which makes no attempt to mediate the clash between public and private rights. Japan's townscape preservation movement recognized that the "legality" of individual buildings obstructed attempts to preserve the broader townscape. This chapter demonstrates how preservation movements that challenge problematic yet "legal" cityscapes and protest the demolition of townscapes that have fallen through the cracks of the system of legal protections, constitute a sharp criticism of contemporary Japan's urban environment.

Keywords Townscape · City planning · Cultural property · Preservation movement · Risk

2.1 The Landscape as a Target of Inquiry

The Japanese landscape has experienced unprecedented change over the past fifty–sixty years—a period during which the country transitioned from rapid economic growth to the post-bubble economy. While a photograph of Venice taken one hundred years ago reveals very little difference from the present day (Franzoi and Smith 1993, 1994), the Tokyo landscape of just sixty years ago is almost unrecognizable today (Tanuma 1990; Katō 1999).

Change has not been limited to most urban landscapes. The land sharking practices endemic to the bubble economy of the late 1980s drove out long-term residents and transformed the familiar appearance of even smaller towns by clearing the way for new office buildings and high-rise apartment buildings. The subsequent collapse of Japan's economic bubble deprived these new buildings of both workers and tenants, leaving behind nothing but vacant rooms. As development plans stalled, properties with no clearly identifiable future use sit abandoned. In the sense that people and

S. Horikawa, *Why Place Matters*,
https://doi.org/10.1007/978-3-030-71600-4_2

occupations are excluded from urban space, this poses an existential problem for Japanese cities. The bubble economy transformed the land from the scene of everyday life, where people worked and lived, into the target of speculation. Landscape change is only the most obvious expression of this phenomenon.

In the face of such unprecedented changes, a number of movements arose to protect landscapes from such unwanted transformation. The 1960s witnessed a rapid rise in "townscape preservation movements" (*machinami hozon undō*), which were sometimes referred to as national trust movements[1] (Zenkoku Machinami Hozon Renmei, ed. 1999). Before turning to the case study of Otaru, let us apply a sociological lens to two questions: Why is landscape change a problem? And why have preservation movements emerged in response?

Before we begin, we must explain the need for a sociological perspective on such issues. The topic of landscape preservation has typically been understood as an area to be dealt with through "scientific" approaches, within fields such as architecture, urban planning, or geography. Why should sociology address the issue? What can really be achieved or discovered through the application of sociological methods and perspectives?

This somewhat daunting question has a concise answer. The landscape issue belongs not only to geography or the natural sciences, but also to sociology. This is because assessments of the aesthetic value of a certain landscape will vary depending upon the historical–social context of the person or groups evaluating it. In other words, the landscape emerges as a focus for sociological analysis when we attempt to understand how it is recognized in connection to specific historical-social contexts or social relationships.

There are other questions to be answered. First, what is the precise "problem" when people discuss the "townscape problem?" Typically, the specific buildings under discussion are legitimate, legally constructed structures—and many are of excellent design. There should be no "problem" with a well-designed and legitimate building. Moreover, if we examine the individual buildings that make up the urban landscape, each and every one is a legal structure. No matter how much we may lament changes to familiar landscapes, this change in itself is not a "problem." What this suggests is that the problem does not pertain to individual buildings, but rather to the "overall urban landscape:" in other words, the townscape that emerges as the composite of these various buildings.

Another issue involves subjective interpretations of the landscape. As discussed in Chap. 1, subjective interpretation makes it difficult to arrive at an unambiguous assessment of a particular landscape. This makes "the landscape" a complicated topic to address. Given this, people often hesitate to address landscape-related issues, concluding that "after all, this is a matter related to each individual's subjective view" and "one can hardly complain about matters of individual taste" (Nihon Bengoshi

[1]These movements are variously referred to as "protection associations," "defense associations" and "preservation movements." Moreover, up until 1945 the term "preservation" was used within the Japanese legal and governmental system. Since then, however, the term "protection" has been used consistently. Given these various uses, the terms "preservation" and "protection" are not strictly distinguished here, and should be considered more or less synonymous.

Rengōkai Kōgaki Taisaku Kankyō Hozen Iinkai, ed. 1995). If we wish to adopt a sociological approach to the issue of landscape change, we must first overcome the two barriers of legality and subjectivity.

Landscape preservation movements have also confronted these barriers. Many landscape preservation campaigns struggled to overcome them by stressing the value of traditional townscapes to "tourism development" (Ōkawa, ed. 1997; Matsumura, ed. 1997) or, in the case of trust movements, attempting to take direct ownership of contested landscapes. Although various movements employed different tactics, all sought to preserve a particular landscape and struggled to overcome the various barriers impeding this goal. In this sense, they can all be considered preservation campaigns.

With the two "barriers" of legality and subjectivity as our guide, let us consider yet another set of questions. How have certain landscapes, such as Japanese townscapes, become an issue of contention? And how do we explain the sudden emergence of townscape preservation movements in Japan? In order to discuss the issue in sociological terms, we must first review the Japanese legal system, trace the historical development of preservation campaigns in Japan, and decode the actual urban design principles at work in Japanese cities.

2.2 Assessing the Depth of the Preservation System: Urban Planning and the Preservation of Cultural Properties

Any landscape preservation movement will inevitably confront the barrier of "legality," as evidenced by the typical rejoinder to preservation efforts: "when all is said and done, this is a lawful, legitimate building, and there can be no basis for challenging it." This begs the question: "Why are redevelopment projects and buildings considered to be problematic by so many residents still lawful?" And: "Does Japan have laws in place to protect the landscape?" The answer to the latter question is, of course, affirmative. There are two sets of pertinent legislation: the City Planning Law and the Law for the Protection of Cultural Properties.

Let us begin with the Japanese system of city planning, which is designed to systematically control the land and space traded and distributed under market principles. Specifically, urban space is zoned into districts, where land use is controlled according to associated "colors, lines, and numbers." There are various problems with this approach. For example, regulations have relaxed previously uniform floor area ratios (FAR), but make no effort to develop a "preferred image of the district," instead of emphasizing only negative provisions. Moreover, residents are not fully recognized or positioned within the city planning process (Igarashi 1991). Of particular significance are the unreasonable floor area ratio designations that create commercializing pressures, resulting in the displacement of residential buildings. Moreover, given the assumption of "private land ownership" and "freedom, in principle, of construction," city planning laws only intervene in a few exceptional situations, as in cases involving "the public welfare," as defined by the Japanese Constitution. Land ownership in Japan is thus almost "absolute," and an urban townscape cannot be preserved without the consent of each relevant landowner.

What is actually protected under Japan's system of city planning laws? Read together, the following observations suggest an answer:

> It is enormously significant that land problems still arise despite the fact that as many as…two hundred and seventy [city and land-related] laws are in effect (or perhaps precisely *because* they are in effect). In other words, the emergence of land-related problems is an entirely natural phenomenon within the legal system… Moreover, even if all of these legal defects were to be corrected, the steep rise in land prices remains a major underlying problem.
>
> Igarashi (1991: 292). Information in parentheses is from the original. Information in brackets has been added by the author.
>
> {…} There are about two hundred laws pertaining to city planning. Despite this preponderance of laws, land use in Japan is in fact far less regulated than in Europe and the United States; indeed, there can be no comparison. {…} Land prices in Japanese cities are so high that they make a mockery of peoples' lives. And yet, somehow the current problem in Japan is the *drop in land prices*. Nothing could provide a clearer indication of how land is treated [i.e., as an asset or speculative venture, not as a place to live].
>
> Fujita (1994: 15–17). Emphasis in the original.

Both Igarashi and Fujita identify the fact that *land-related issues persist even though existing land-related legislation is functioning exactly as intended*. If land problems arise due to a lack of requisite legislation, the remedy would be straightforward enough. In Japan, however, the problem is not the *absence* of land-related legislation, but rather an *excess* of legislation. It is not so much that "even two hundred and seventy laws cannot suffice to protect the landscape from change," but that "Japan's modern city planning legislation has consistently repudiated historic townscapes," and "this is exactly what city planners think of as 'modernization'" (Nishimura 1997: 163). One is forced to conclude that Japan's city planning legislation has legitimized the use of land as an asset or product, and does almost nothing to protect the daily living spaces of ordinary citizens (Igarashi and Ogawa 1993). The common rejoinder to preservationist claims ("after all, this is a lawful, legitimate building, and there can be no basis for questioning it") is thus entirely accurate.

Of course, this does not mean that the risk of a sudden change to the urban environment could not have been avoided. After all, cities are spatially and physically different from rural areas. A wide variety of actors use and share space at high densities and in extremely close proximity; residential buildings, factories, shops, and offices often exist side-by-side. In a single neighborhood, a man returning from his night shift goes to sleep at dawn while an elderly man rises early to make breakfast, and, next door, the machinery of the local factory begins to whirr into action. In the early afternoon, housewives set out to collect their children from kindergarten just as university students yawn, stretch, and prepare to begin their day. This dense and diverse use of space by various actors is fraught with potential conflicts caused by the mutually contradictory use of space. It is thus essential to anticipate and adjust for such "urban risks."

Conflicts arising from contradictory uses of urban space can emerge as disputes over construction. Indeed, architectural conflicts highlight the inherent difficulty of urban land use. Residents of surrounding buildings who assert their basic right to sunlight will inevitably come into conflict with a development company seeking to recover its investment by constructing and selling off high-rise condominiums. This is because, in addition to the continuing context of "development-driven urban planning" (Igarashi and Ogawa 1993), the assessment of the impact of a particular construction project tends to be perceived as a matter of "private opinion," "taste" and in terms of "what is tolerable." If residents lodge a protest with the city, they confront a legal barrier: the city government must approve an application for construction certification if the construction plan itself is lawful. Residents in the vicinity of the new building being contested have no further recourse and are silenced by the law (Horikawa 2001).

In this sort of construction dispute, the residents' assertion of their "right to sunlight" has been recognized, leading to one type of dispute resolution scheme (albeit an imperfect one). Sunlight, however, is just one small component of the broader concept of "environmental rights" gradually developed and advocated by residents in the wake of numerous construction disputes and pollution problems. Apart from the most basic human rights guaranteed by the Constitution, the right to sunlight is the only right of injunction recognized by current judicial precedent. In other words, invoking the right to sunlight is the only socially acceptable form of opposition to specific construction projects.

Although the right to sunlight can be weaponized within a specific dispute, it is powerless to address broader issues pertaining to the regulation of land use and spatial management in the city at large. Instead, "community development" emerged as a framework for restraining unfettered changes to the urban environment.

In the high-growth period of the 1960s, popular movements opposing urban sprawl and associated building disputes prompted local governments to develop certain "guiding principles" (*shidō yōkō gyōsei*) for new development. The 1967 "Guidelines for Residential Land Development Projects" in the city of Kawanishi (Hyogo Prefecture) and 1968 "Guidelines for the Development of Residential Land" in Yokohama (Kanagawa Prefecture) were pioneering examples of this approach. To be sure, the "guiding principles" approach had a number of shortcomings: the guidelines were discretionary and without legal force, the demand that developers shoulder a certain portion of the development burden was potentially unconstitutional, and the guidelines consisted of internal administrative rules that had not been ratified by the local assembly. Nevertheless, these "guiding principles" represented a community development approach to governance that employed the approval rights of local governments (with regard to water supply and road use permissions, for example) as a resource. Up until the early 1980s, at least, these guidelines seem to have fulfilled a certain function and role.

In the mid-1980s, however, the legality of such guidelines came under challenge in the courts. Since then, judicial precedent has established that local government regulations pertaining to land use are without legal basis, and thus unconstitutional. This has forced a retreat among the local governments across Japan that attempted the "guiding principles" approach. Given the growing number of successful lawsuits filed by construction companies and building contractors against local governments for withholding construction permits upon the request of local residents, this was an entirely natural turn of events. A different approach was needed, and local governments and anti-development residents' campaigns subsequently attempted to confer "due process" (due process of law) to the "guideline principles" by enacting them as local ordinances. A series of municipal ordinances from the early 1990s, including Yufuin's "Ordinance for Building a Charming Town" (1990), Setagaya Ward's "Housing Ordinance" (1990), and Kakegawa City's "Land Ordinance for Building a Town of Lifelong Learning" (1991), illustrate the evolution from the "guideline administration" approach to the "community development ordinance" approach (Igarashi and Ogawa 1993).

Yet these ordinances also proved insufficient. Even if enacted, ordinances remained subject to legal and technical restraints: new restrictions must fall "within the scope of the law" and remain "consistent with national law." Moreover, there is no judicial precedent for imposing limitations upon property rights. Considering the constraints inherent to Japan's development-oriented city planning legislation and local attempts to establish "guiding principles" and "community development ordinances," disputes over landscape preservation are not merely inevitable. *The legal system itself* is the source of landscape-related disputes. We can thus conclude that Japan's system of city planning *fails to protect the landscape.* As Nishimura (1997) observes, one could actually argue that Japan's system of urban planning *promotes change and the continual overhaul and renewal of the urban landscape.*

Let us turn now to the Law for the Protection of Cultural Properties, which represents the second legal lineage for Japan's landscape-related regulations. The 1868 "Kamis and Buddhas Separation Order" (the official policy of separating Buddhism and Shintoism) unleashed violent anti-Buddhist forces that led to the "destruction of old objects," as Buddhist temples nationwide were destroyed and ancient works of religious art were sold overseas (Kihara 1982: 2–10). In order to address this situation, the Great Council of State issued a decree in May 1871 entitled "The Proclamation for the Protection of Antiques and Old Properties." This was the first in a series of administrative safeguards for cultural properties in modern Japan. Strictly speaking, the preservation legislation of this era was designed to maintain temple organizations, which had been suspended following the movement to expel Buddhism from the country (Nishimura 1984). The structure of the temple organization was the target of preservation, with the idea that this would protect the physical structure of the temple and prevent the "destruction of the old" and the sale of artifacts overseas. Thereafter, as the concept of "historical evidence" (the internal rules for the preservation of ancient shrines and temples in July 1880) became the standard for preservation, valuable items and structures themselves became the focus of preservation measures. In July 1895, however, the "Regulations on the Application of Funds for the Preservation of Ancient Shrines and Temples" (Ministry of Internal Affairs Ordinance No. 7) changed the wording to "items that have been dedicated to successive generations of the imperial family" (Act No. 7 of 1898, pp. 67–68). In other words, preservation legislation became linked to Japan's emperor system. Put simply, this marked a change in emphasis, from the preservation of "historic sites" to the protection of "holy sites." Following the 1897 Ancient Shrines and Temples Preservation Act, this system would be linked to the 1929 Law for the Preservation of National Treasures (Shimizu 2013).

Let us stop to identify the characteristics of preservation efforts during this period. First, one can identify a certain elitism within this early approach to preservation. Prior to the war, only rare and valuable "excellent specimens," or "holy sites" related to the Imperial Household were selected, while private houses were ineligible as preservation targets.[2] Second, preservation efforts targeted a specific "piece." In other words, the preservation of an "excellent specimen" was unrelated to the environment in which it was located or any details pertaining to the surrounding landscape. This underscores the extremely shallow focus of early preservation efforts.

These characteristics comprised the photographic "negative" of Japan's postwar preservation administration. The effort to reverse this "negative" began with the 1950 Law for the Protection of Cultural Properties (Inagaki 1984; Horikawa 1991; Matsumoto 1997). This expanded the range of preservation targets beyond those connected to the imperial family. Yet one thing remained constant: the target of preservation efforts remained isolated "pieces," divorced from their broader context. This shallow focus was finally addressed by a 1975 amendment to the Law for the Protection of Cultural Properties and the establishment of the "Important Preservation Districts for Groups of Traditional Buildings" system (the "Important Preservation Districts" system), which made it possible to designate a broader area for preservation.[3] In 1996, the Registered Cultural Property System was also introduced (Katagiri, ed. 2000: 1–23). However, the national government will not designate a certain area "of extremely high value" for preservation unless the relevant local government files an application for the designation, based on a local consensus. The conflicting interests of preservation and development will often preclude such an application. Moreover, the hurdle remains high even in the absence of such a conflict, due to the concerted effort required to even file an application. To date, only 118 districts have been designated as "important preservation districts" (for a total area of approximately 3,924.9 ha) in a total of 98 cities, towns, and villages across 43 prefectures.[4] In addition, high land prices in urban areas complicate the selection and designation of preservation areas within cities—designated preservation areas are primarily located in less populated regions where historic buildings remain standing in a cluster. Therefore, most of Japan's land remains outside the scope of the

[2]To be sure, there are some exceptions, such as the survey of the activities in private *minka* houses conducted by the "modernologist" Wajirō Kon. Nevertheless, the preservation system was basically aimed at the preservation of "excellent specimens." See Nishimura (1984) and Horikawa (1991).

[3]Local municipalities may designate "preservation districts for groups of traditional buildings." There is a two-tier system in which the national government may also approve applications from local governments and select "areas deemed to be of great value to Japan" for the higher designation of "important preservation districts for groups of historic buildings."

[4]This figure is generated with data from the website of the Agency for Cultural Affairs, current as of August 17, 2018 (last accessed on July 15, 2019): http://www.bunka.go.jp/seisaku/bunkazai/shokai/hozonchiku/judenken_ichiran.html.

existing legislation for the protection of cultural properties. Furthermore, designated cultural properties are removed from the administrative scope of city planning, and preserved within the prescribed framework for the protection of cultural properties. The flip side of isolating only the most "precious" [structures] for preservation as cultural properties in a highly specific and singular "spot" is that *city planning is left to promote development by default.*

When one stops to think about it, broader landscapes are precisely what city planners ought to control. Yet city planning in Japan remains "development-driven," and does not focus on the goal of "stock rehabilitation" (Igarashi and Ogawa 1993). The most common landscapes have thus been doubly excluded from preservation efforts. They have been omitted from the laws guiding city planning, and they have also been excluded from the ranked system of "cultural properties" and the "important preservation district for groups of traditional buildings" system.

Let us return to the original question: "Are there laws in place to protect the landscape?" We must conclude that while Japan ostensibly has two systems of legal protection, certain landscapes, such as townscapes, have been excluded from both city planning laws and cultural property protection guidelines. In short, these landscapes remain outside the "field of vision" of such laws. This explains the emergence of landscape preservation movements across Japan, as exclusion from the established social system led to a new form of cooperation. What could be a more appropriate subject for sociological investigation?

2.3 The Townscape as Culture

As we have seen, the legal provisions in modern-day Japan are not equipped to resolve the problem of landscape preservation. This state of affairs cries out for sociological analysis. Let us begin with a backward glance, however, and consider responses to urban problems at earlier points in Japanese history. This is not a nostalgic argument for "returning to the past." Rather, it is a strategy for relativizing what is so often viewed as self-evident. Examining a different period in history to relativize the issues of the present has the potential to expand our understanding of current problems and identify potential starting points towards their resolution.

Japan's urban history offers us an important guide. Recent studies in urban history have illuminated a variety of spatial control techniques in pre-modern Japan (Nishikawa 1994; Takahashi and Yoshida, eds. 1989). This is confirmed by studies in both architectural history and city planning (see, for example, Ueda and Tsuchiya, eds. 1975; Kankō Shigen Hogō Zaidan, ed. 1981). Essentially, these techniques were feudal and class-based land distribution systems based on a ruler's grand design and, as such, could never be revived today. Yet both studies vividly depict how, within the limitations imposed by a class-based land distribution system, residents of Japan's pre-modern cities accumulated the norms and know-how to live together within a narrowly delineated site, leading to the establishment of an organic spatial order.

Let us examine the specific nature of the organic spatial order developed by city residents.

This order refers to the "traditional merchant's house (*machiya*) system" as a means of pre-empting or resolving urban problems, the "townscape" (*machinami*) created through this process, and the practice of preservation campaigns to protect this townscape. If we define "culture" as a shared approach to handling oneself and resolving problems with others, which has been internalized by people and institutions to the extent that they unconsciously enact it, then the *machiya* system and resulting townscape can be interpreted as a form of culture. Rather than glorifying the historic townscape, the term "townscape as culture" suggests that traditional responses to the inherent difficulties posed by the urban landscape and use of space is in itself a form of "culture." Decoding this culture is also an opportunity to reconsider the contemporary city.

We will now examine pre-modern systems of spatial control that have been dismissed as "old," "outdated," or "local" and decode their hidden, organic ordering of space. In other words, we will approach preservation as one "method of controlling change" (Gratz 1989, 1993: 224).

2.3.1 Buildings and Sites: The Architecture of Kawagoe

To begin, what exactly is the "traditional *machiya* system?" In this section, let us examine the relationship between buildings and sites, using as our first example the warehouse-style merchant homes (*machiya*) that remain standing today in the city of Kawagoe.

Located in modern-day Saitama prefecture, Kawagoe once played an important role as a northern defense for the capital city of Edo. In 1457, Kawagoe Castle was constructed by Ōta Sukekiyo (Ōta Dōshin) and his son, Ōta Sukenaga (Ōta Dōkan) as a branch castle of Edo Castle. The Kawagoe domain was established in 1590, and prospered economically, as the development of the Shin'gashi River enabled water transport to and from Edo. It was during this time that the area around the castle, lined with warehouses that brimmed with energy and economic activity, became known as "Little Edo."

Two major fires—the first in 1638 and the second in 1893—are important events for our contemplation of present-day Kawagoe. Ironically, the first fire was important because it provided the opportunity for the redevelopment of the castle town by Matsudaira Nobutsuna, which would prove of decisive importance to the long-term prosperity of the domain. The second fire, meanwhile, promoted the subsequent construction of fire-resistant warehouse-style buildings—the origins of the massive clay-walled warehouses in Kawagoe's Ichibangai [street], now a famous tourist attraction.

The warehouses of Kawagoe are much more decorative than those found elsewhere, boasting a sumptuous array of decorative ridge-end tiles (*onigawara*) and disproportionately large roofs. This was because Kawagoe merchants rebuilt their warehouses after the great fire with reference to the large shops of Edo. Their aspirations led to excess, and the Kawagoe merchants ended up creating storefronts even larger and more grandiose than the ones they had seen in Edo. Some readers may find it odd that the "traditional" style of the warehouses that comprise the Kawagoe townscape date not from the Edo period but rather from the mid-Meiji period, but one could argue that Kawagoe already had the historic environment capable of conveying the Kawagoe style. In April 1999, the area was designated as a "traditional architecture preservation district" (and in December of the same year, this designation was upgraded to that of an "important traditional architecture preservation district") and it remains one of the most valuable historic environments remaining in the Kanto region of Japan.

However, the "Kawagoe-style" warehouse should not be evaluated solely on the basis of its unique style, history, and relative scarcity. In order to understand townscape as culture, we must focus on the relationship between buildings and sites in this "Little Edo." In short, we examine the established "cultural practice" of construction.

Since the 1970s, the city of Kawagoe has surveyed the local landscape numerous times (Aramaki 2003; Fukukawa 2003), motivated in part by opposition to the proposed construction of high-rise apartment buildings within the city limits. One of these initiatives was a design code survey, which revealed that, while most of the warehouse-style *machiya* constructed in Ichibanchō seemed, at first glance, to be built in separate and disconnected styles, they were actually built in accordance with a shared set of rules (Kawagoe Ichibangai Shōtengai Kasseika Moderu Jigyō Suishin Iinkai, ed. 1986; Kawagoe Ichibangai Machinami Iinkai, ed. 1988).

This can be explained with reference to Fig. 2.1. Each *machiya* is built in the *hirairi* style, in which the main entrance to the building is located on the side that runs parallel to the roof's ridge. The residential portion of the building is perpendicular to the front shop space, and is followed by a garden space. This is the classic layout of Kawagoe's "warehouse-style" *machiya* architecture (Fukukawa 2003). The Kawagoe *machiya* possessing the eight characteristics enumerated in Fig. 2.1 manage to secure a large commercial space, and, at the southern end of the residential portion, also incorporate a quiet and sunny space surrounded by neighboring buildings, affording its residents privacy, sunlight, and ventilation. Thus, while a street view of the Kawagoe *machiya* townscape presents an uninterrupted line of long, two-story *hirairi* style warehouses, these exteriors mask the existence of the sunny, private spaces behind. Under the difficult conditions imposed by the long, narrow plots of land, this style achieves a successful fusion between business and private life.

The most important aspect of the Kawagoe *machiya*, however, lies in the relationship between individual buildings and their sites. As shown in Fig. 2.2, each site is divided into three zones (spaces) extending inward from the road to the interior: the front shop, the private living quarters, and the garden. The depth of each zone is uniform at four *ken* (a Japanese unit of measure roughly equivalent to 1.9 m) (Kawagoe Ichibangai Shōtengai Kasseika Moderu Jigyō Suishin Iinkai, ed. 1986: 151–169; Fukukawa 2003: 140–143). What is the goal and significance of this architectural rule, which the residents of Kawagoe refer to as the "four *ken*/four *ken*/four *ken*" rule?

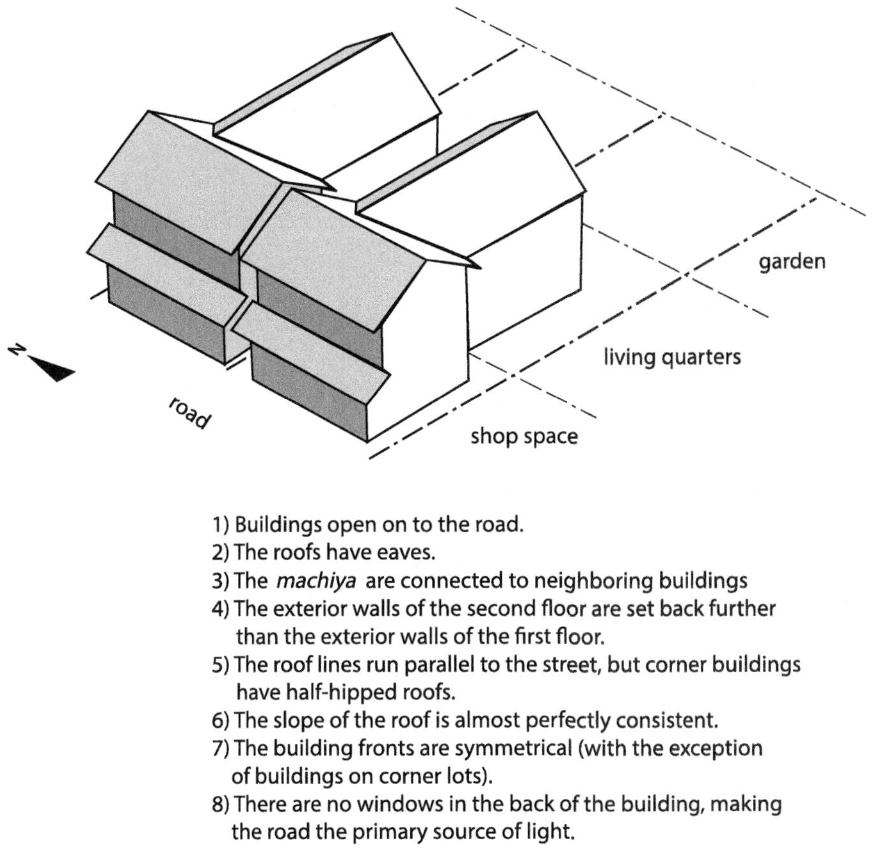

Fig. 2.1 Architectural features of the Kawagoe *Machiya*. *Note* The author's illustration and description of *machiya* features is based upon the work of the Kawagoe Ichibangai Shōtengai Kasseika Moderu Jigyō Suishin Iinkai (1986) and Fukukawa (2003)

If we refer back to Fig. 2.2, we can see that the main focus is on the alignment of the buildings on each site, rather than the length of each "zone." On a long, narrow site, with a narrow frontage and a generous depth, even if a building owner positions his garden as carefully as possible he will lose his access to sunlight if his neighbor to the south builds a two-story warehouse. His careful plan would be immediately rendered meaningless, and his garden consigned to the shadows. Thus, the type of favorable living environment that can be realized on any individual site is very limited, and remains fragile enough to be influenced by a neighbor's decision. This is a real risk that attends urban life.

This gives us the key to understanding the significance of Kawagoe's "four *ken*/four *ken*/four *ken*" rule. Namely, the Kawagoe rule is intended to avert the risk of rapid changes to the daily living environment by aligning shop, residential, and garden zones, thereby achieving a generally favorable environment that could not

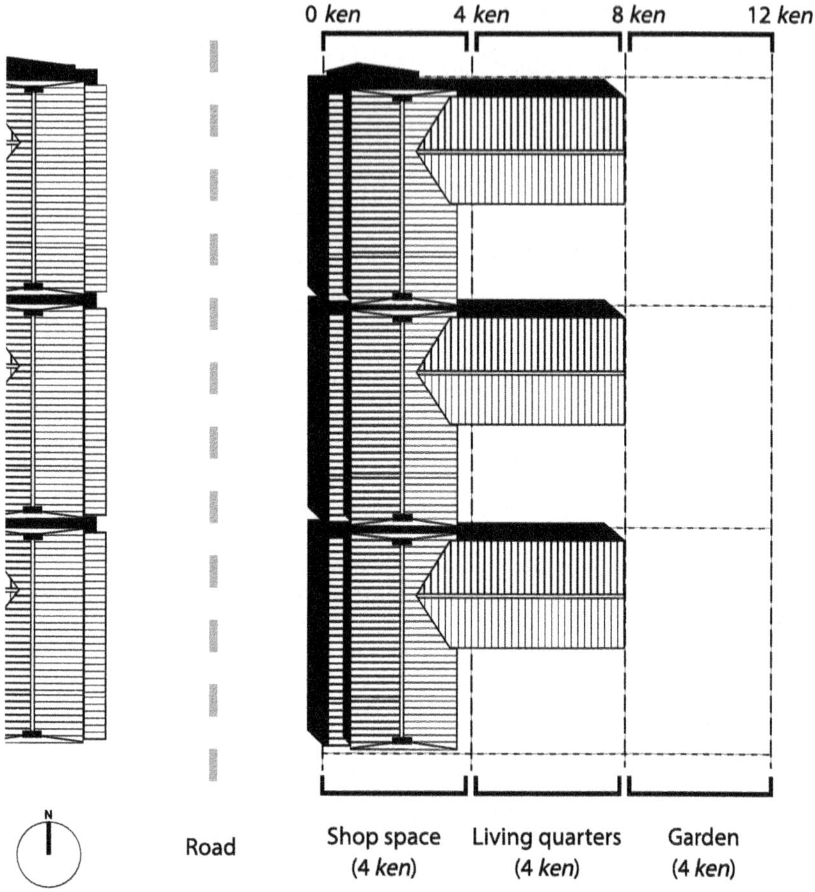

Fig. 2.2 The Rules Governing *Machiya* Layout in Kawagoe. *Note* The author's illustration is based upon the work of the Kawagoe Ichibangai Shōtengai Kasseika Moderu Jigyō Suishin Iinkai (1986) and Fukukawa (2003)

have been realized by any one site alone. In other words, this principle of site use is a method of controlling the structure of individual buildings, and functions as a mechanism for ensuring that the quality of each individual environment will not be affected by the decisions of others. One cannot comprehend this by examining a single site, but a continuous block-by-block examination allows us to conclude that this is indeed "the cultural practice of building" in Kawagoe. Individual buildings and gardens may be privately owned, but they are also a "communal installation" that maintain good environmental conditions for the area as a whole (Horikawa 1998a). The small details of the *machiya* townhouses reveal the elaborate structure by which methods of dwelling and construction deftly connect individual buildings, sites, and blocks—methods developed over years of residence, and passed down from generation to generation (Horikawa 1998b).

Of course, this also helps explain why *machiya* buildings are being destroyed in Kawagoe and many other cities. This process typically unfolds as follows (Fukukawa and Nishimura 1981; Horikawa 1998b):

1. One household destroys its garden in order to expand the residential quarters.
2. The environmental conditions of adjacent houses deteriorate due to this interruption in their access to sunlight and ventilation.
3. The residents of neighboring houses file complaints with the city government, but their claims are rejected by local authorities on the grounds that the neighbor's renovation is "legitimate" in the legal sense.
4. Neighbors thus begin to engage in competitive architectural practices—for example, building in their former garden areas with the aim of restoring former environmental conditions (such as sunlight and natural ventilation).
5. As a result, the environmental conditions of neighboring houses deteriorate, and traditional *machiya* townhouses disappear as this destructive process continues.

The first household to break the rules initiates the process described above, and the neighborhood enters a rapid spiral of loss.[5] Kawagoe's rule is to prevent the first step in this cycle of destruction. In other words, embedded within Kawagoe's "four *ken*/four *ken*/four *ken*" rule is a measure intended to avert the risks that attend high-density urban living.

What is the concrete effect of the Kawagoe rule? Let us consider, for example, a case in which someone new moves into a Kawagoe warehouse-style *machiya*.

New buyers typically decide to purchase a home after examining not only the specific building, but also the types of businesses conducted in neighboring buildings, and the surrounding townscape. Meanwhile, neighbors worry that the newcomers will negatively impact their current living environment. One typical concern might be whether the new occupant plans to destroy the existing warehouse-style building and construct a four-story building of reinforced concrete in its place, obstructing neighbors' access to the sunshine they had always enjoyed. If the Kawagoe rule is actually shared by all parties, however, such concerns are groundless. This is because even if the occupant of the building changes or the building itself is renovated, new entrants understand that a prerequisite of their purchase is their maintenance of the existing environmental conditions, while other residents of the neighborhood are reassured that the status quo will be more or less maintained. We can appreciate how the "four *ken*/four *ken*/four *ken*" rule operates as a guideline for preventing mutual unhappiness and destruction.

Significantly, this rule does not impede the development of individuality or diversity. Broadly speaking, the Kawagoe rule seeks to "balance order and diversity"

[5]These spirals can be analyzed sociologically as "social dilemmas" (Funabashi 1989; Umino 1993). It is this author's view, however, that although "social dilemmas" may be a useful concept for describing the process of *machiya* collapse, different conceptual devices may be necessary to effectively address other aspects of the townscape preservation issue.

(Fukukawa 2003: 130), and the *size and style of individual buildings remain diverse and individual*. Indeed, the actual streets of Kawagoe are by no means uniform, nor is it mandatory for owners to construct buildings according to the same design or plot ratio. On the contrary, as long as the owner observes the warehouse-style *machiya* code of architecture and the "four *ken*/four *ken*/four *ken*" rule, he is free to adopt any style he pleases. Thus, while the townscape created through this shared rule does have a certain set of styles and designs, the result is a kind of "uniform diversity" (Iguchi 2002: 26; Yamazaki 2002: 136–137), in which a variety of buildings are built to reflect the needs and preferences of residents.[6] In essence, this involves "achieving even more freedom by accepting just a little inconvenience." This defines the mutual cooperation that guides and shapes land use in Kawagoe.

2.3.2 *Sites and Blocks: Townhouses (*Machiya*) as Mutual Environmental Protection Systems*

Let us now turn to the traditional *machiya* found in the Kansai region of western Japan, using a survey of Nara conducted by Fukukawa Yūichi and Nishimura Yukio (Fukukawa and Nishimura 1981). Like Kawagoe, the townhouses of Nara are densely situated within the city, and occupy long and narrow sites (typically described as "eel beds") set adjacent to the road. This precludes the possibility of any openings along the sides of the buildings; instead, adequate ventilation and lighting[7] is secured through the skillful combination of two open spaces: the space of the road at the front of the site, and the garden in the back (Fukukawa and Nishimura 1981: 72–75). The uniform depth of the main houses and the alignment of their back gardens allow these hidden gardens to become a continuous space expanding horizontally behind the *machiya* (Fig. 2.3). While each site boasts only a small back garden, the positioning of gardens adjacent to each other creates a generous conduit for ventilation and sunlight. In other words, "environmental conditions that are difficult to maintain in a single building" are nevertheless guaranteed to each household through the "cooperation" between the various *machiya* residents (Fukukawa and Nishimura 1981: 72).

---

[6]This follows the same thought process as what the architect Yoshizaka Takamasa (Ryūsei) once expressed in his urban housing blueprints as "discontinuous unity," in which "individual buildings, while demonstrating their individuality, remain in harmony with the larger group." See Yoshizaka (1975).

[7]Because *machiya* tend to be deep and dark, inner gardens were commonly installed for additional lighting. These inner gardens also secure privacy between the main house and detached rooms, and separate the toilet from the living quarters. They also contain hidden measures to make the hot summer more comfortable. On a hot summer afternoon, residents would sprinkle water (*uchimizu*) on the surface of the garden to cool it, creating a delicate temperature difference between the garden and the surface of the road in front of the sun-scorched entrance. Just a slight difference in temperature creates a wind that ripples gently across the tatami mats. Even when the whole city is still and humid, a faint breeze blows through the *machiya*. The creation of this breeze by sprinkling water on the inner garden was a practical technique developed by *machiya* residents to endure the hot city summers.

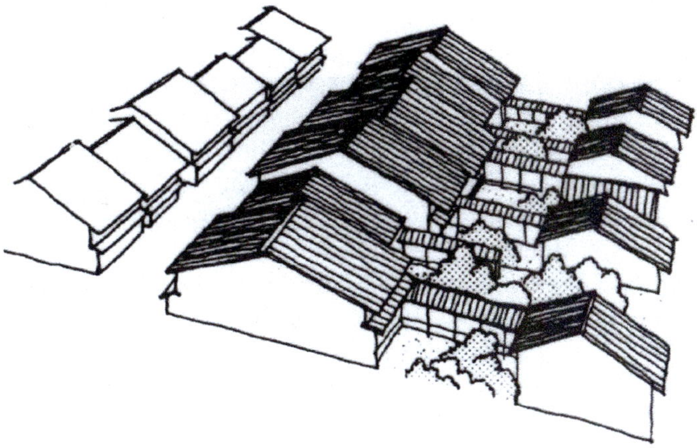

Fig. 2.3 The organic spatial order of the *Machiya* and Inner Garden. *Source* Fukukawa and Nishimura (1981: 79)

This could be described as a shared architectural code or norm, which ensures that everyone gains through cooperation (Fukugawa and Nishimura 1981). Small inner gardens are not only the private property of each individual *machiya* owner, but also a "communal installation" that maintains good environmental conditions (Horikawa 1998a).

Therefore, as in Kawagoe, if one *machiya* owner were to add a building to his or her back garden, the environmental conditions maintained by the group of *machiya* as a whole would collapse, and each household would have to secure lighting and ventilation by means that would ultimately prove counterproductive. This would accelerate the rate at which *machiya* are destroyed (Fig. 2.4). Thus, the design of *machiya* represented the best possible adaptation to the conditions imposed by the long, narrow sites, and "because each individual building observed common norms,

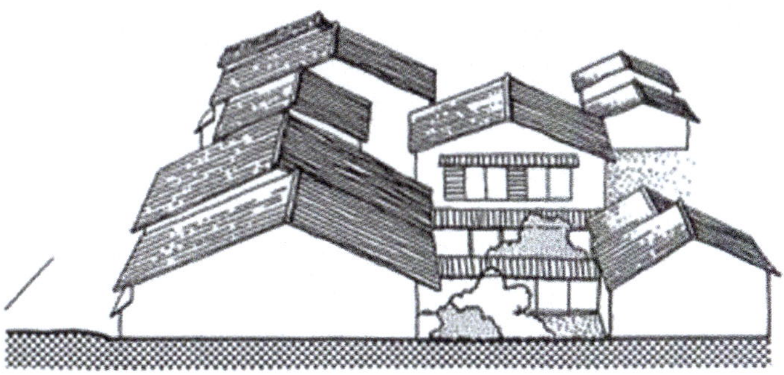

Fig. 2.4 The breakdown of the *Machiya* spatial order. *Source* Fukukawa and Nishimura (1981: 79)

the type of inconsistencies between buildings that tend to arise in densely populated urban spaces are avoided, thereby mutually guaranteeing good environmental conditions" (Fukukawa and Nishimura 1981: 74). One could also describe this as a technique developed by city residents in order to secure the best environmental conditions possible and enjoy a certain degree of freedom within the constraints imposed by feudal land divisions (Horikawa 1998b).[8]

How do local rules, which have controlled land use from individual sites to specific blocks of buildings, attempt to deal with cities as a whole? Is there a rule that can apply to the city at large? To explore these questions, let us turn to the main subject of this volume: the city of Otaru, on the northern island of Hokkaido.

2.3.3 Cities and Blocks: City Design in Otaru

As readers will find a more detailed account of Otaru's history in subsequent chapters, what follows here is a very simple explanation of city design.

Once so prosperous that it was known as the "Wall Street of the North," the city of Otaru straddles a narrow band of flat land between the sea and the mountains; as it grew, the city sprawled up the slopes of the surrounding hills and mountains. Otaru is thus not only a "canal town," but also a "town of hills." The port and ships, the canal, stone warehouses, and cargo barges are visible from atop any of the steep slopes of the city, making the canal and harbor the true symbols of Otaru—and deeply familiar landmarks for all local residents.

During the Meiji period, a growing number of ships passed through the port of Otaru. A canal—the same canal that today is a centerpiece of Otaru tourism—was constructed to facilitate cargo transport through the increasingly busy port. The landscape of the canal and port district, which brought such prosperity to Otaru, is entirely distinct: it is composed of wood-framed stone warehouses that line the canal. The "design principles" informing this landscape are as follows:

1. The exterior walls of the warehouses are built with soft stone imported to Otaru from the area outside Sapporo.
2. The roof height ranges between eight and twelve meters.
3. Nevertheless, the height of the eaves is roughly five to six meters—creating a constant horizontal line.
4. The roof gradient is constant at roughly 16 cm.
5. The roofs are gabled.
6. The front of the warehouse faces the canal, while the rear opens on to a service road (*denuki kōji*).

---

[8]This system is only slightly different from the issue of the suppression of individuals by the collective. As is well known, the liberation of the individual from the restrictions imposed by community has been a principal question for postwar Japanese sociology. Here, we are simply pointing out that the *machiya* system maintained an ambivalent character, even as it performed a certain social function.

Fig. 2.5 Otaru's stone warehouses: design principles. *Source* Kankō Shigen Hogo Zaidan, ed. (1979: 140–141; Fig. 9.13)

Design principles summarized by the author based on Kankō Shigen Hogo Zaidan (1979: 140–141).

The general adherence to these principles creates a rhythmic array of gabled roofs (much like saw teeth) along the gentle curve of the one-kilometer canal. The canal offers a unique view of Otaru, as the town is reflected upside down in the water. Much like Kawagoe, an entirely unique style is evident (Fig. 2.5). Significantly, the individual warehouses did not exist as a single unit unto themselves, but belonged to a larger system that also incorporated the port, the canal, the cargo barges, and so forth. In order to understand what this means, let us turn to a brief explanation of barge-based cargo transport procedures at the Otaru port.

First, the cargo from the ships at the port would be loaded on to a barge that traveled down the canal. From the barge, longshoremen would carry the cargo to the warehouses lining the canal. Once warehouse cargo had been traded or purchased, it was shipped onward to other parts of Hokkaido on large wagon cars or trucks along service roads running behind the warehouses, thereby avoiding any confusion surrounding the flow of cargo or the routes of the various people involved. Together, the canal and warehouses functioned as the interface between the urban area and the port system. In other words, the port, canal, warehouses, and service roads that connected to the city beyond were organically linked into a single entity (Horikawa 1998b, 2001).

What this organically linked system reveals is the wisdom and ingenuity underlying the connection between individual sites and blocks to the whole. There was no confusion in the flow of movement between various points, and the gentle unification of the system did not preclude the freedom of design for each building. Here we can identify "a common architectural norm—a method by which everyone gained from cooperation" (Horikawa 1998a: 138). Significantly, this system did not emerge as a matter of personal "taste," but rather as a method of living and building that was disseminated and accepted across the entire city.

Of course, we cannot turn back the clock. Given the speed of modernization from the Meiji period onward, Japan can no longer rely solely upon these traditional methods of townscape development. There is no choice but to seek new rules, given the fear of being confined in style or form (Nishimura 1997). Yet neither the system of protections developed under the City Planning Law or the Law for the Preservation of Cultural Properties takes townscapes into account. Townscapes remain doubly excluded from consideration as a potential target for preservation.

Exclusion from the system of preservation suggests that townscape change is not in conflict with the legal system, and, in the absence of any specific violation, lawful. This is structurally similar to city planning laws, which only enumerate prohibitions without presenting any positive vision of the ideal city. Yet the case of Kawagoe suggests the shared norms developed by local residents deserve attention, and complicates the assumption that "anything is legal if it is not explicitly mentioned in the law." The city designs of Kawagoe, Nara, and Otaru were not the product of subjective "preferences." Rather, they demonstrate how earlier generations of local residents successfully devised a shared foundation, upon which individuals *could*, within reasonable parameters, express personal preferences. It was because these old methods of city design were ignored, and because such townscapes were excluded from Japan's system of preservation, that grassroots movements to preserve traditional townscapes eventually emerged.

2.4 The Risks of Urban Living and Changes to the Living Environment

Let us return to the present, and to a set of laws that excludes the urban landscape, or townscape, from its field of vision. First, let us embark on another thought experiment. Consider the following scenario: After thinking long and hard, you have finally decided to purchase a home. This purchase will require an investment of more than half of your total life savings, and it will require the repayment of a loan over the course of the next thirty years. After you purchase this home, construction begins on a new high-rise apartment building next door. Suddenly you are deprived of your view and almost all sunlight (Igarashi and Ogawa 1993; Nihon Bengoshi Rengōkai Kōgaki Taisaku Kankyō Hozen Iinkai, ed. 1995).

Is this an entirely natural phenomenon? Or a common risk that attends urban life? If this is not *seen* as one of the risks that accompanies urban life, this is only because it is so widely regarded as *an entirely natural occurrence in the city*. While it would be quite easy to obtain a general consensus among urban residents that "certain measures" are needed to address the external "dangers" that threaten cities—such as epidemics, disasters, and food shortages—the intrinsic risks of urban life, such as the negative impact of high-rise condominiums, is typically explained away as a natural urban phenomenon that must be accepted: "Nothing can be done about something that improves the convenience of city living;" "There is nothing illegal about constructing a high-rise apartment building;" "It falls within the limits of what is tolerable;" "After all, selecting and purchasing real estate is done at one's own risk." Alternatively, far from eliciting sympathy, the concerns of homeowners may be scorned as the "luxurious" worries of those who can afford to purchase a home in the first place. Cities are assumed to be in a constant process of change (Holleran 1998), and risks are understood as an inevitable accompaniment to urban change.

Before we invoke the "convenience of city living" to dismiss the very idea that the negative conditions imposed by new high-rise building constitute a "risk," as so often happens in Japan, we should stop to consider these risks of urban life (which could equally be described as the risks produced by rapid changes to the daily living environment of city residents). But why is a closer consideration of urban risk required?

The construction of high-rise condominiums and other construction practices are regulated by a number of laws, including the Building Standards Law and the City Planning Law. Needless to say, these laws are also the fundamental rules governing the real estate market. The potential for the abrupt "deregulation" of these legal systems, which could render today's "illegal" forms of construction "legal" tomorrow, makes it extremely difficult to anticipate and undertake necessary countermeasures. The consequences of sudden deregulation, moreover, are serious and irreversible. If the purchase of a home was a mere stock investment, one could simply invest in multiple stocks in order to hedge the risk. A personal residence is also a very expensive "commodity," however, and for most it would be impossible to repurchase or mitigate the risk of loss through multiple "stock holdings." This is true anywhere, but all the more so in Japan, where the price of land is particularly high. Thus, in the absence of any regulatory or preventative framework, abrupt and radical changes to the townscape due to the decisions of outsiders (non-residents) pose a fatal risk to residents of the neighborhood. And this poses a risk for anyone living in a Japanese city today, regardless of whether or not she or he actually owns the property.

Fujita (2005) describes this as the "poor living environment of an affluent society." Fujita identifies the scale of "private rights" over land as one cause of the clutter and disorder in Japanese cities (Fujita 2005: 21–24). Fujita argues that in Japan, where private rights and "publicness" or "commonality" are viewed as antithetical concepts, people have clung to private rights as their only means of self-protection. This precluded the full development of the concept of commonality in Japan. From Fujita's perspective, the sudden change in the daily living environment caused by the construction of a high-rise condominium is an inevitable risk.

If this is indeed the case, *why have we Japanese never understood radical urban change as a risk*? This is a question that touches upon the central concerns of this chapter, and the book as a whole. Despite the pervasiveness of the risks attending urban life and their dramatic effect upon the quality of urban life and urban culture, these risks are viewed as self-evident, and thus unproblematic. This book seeks to *problematize the fact that the serious impact of urban change is not understood to be a problem.* As the term "absolute property rights" suggests, without outside intervention the overwhelming power of "private rights" leads to rapid and uncontrolled urban change. Cities are thus in a state of constant change, and residents typically assume that the negative effects of change are necessary evils that can only be endured. The aim of this book is to focus on the practices of those who have refused to view the drastic overhaul of the urban environment as axiomatic and inevitable, and have actively resisted the forces of change.

2.5 The Rise of the Townscape Preservation Movement and Its Implications

2.5.1 The Development of Townscape Preservation Movements Across Japan

As we have seen, Japan's legal system has done nothing to protect the "everyday landscape" of towns and cities. On the contrary, the law frames such landscapes as "broken" and in need of reconstruction. In the 1960s, several landscape preservation movements emerged in response to the steady destruction of the landscape across the length and breadth of the Japanese archipelago. The series of voluntary movements initiated by local residents to preserve the traditional landscape of their local region are described as *machinami hozon undō*, or "townscape preservation movements" (Kankyō Bunka Kenkyūjo, ed. 1978; Kankō Shigen Hogo Zaidan, ed. 1981; Zenkoku Machinami Hozon Renmei, ed. 1999; Katagiri, ed. 2000; Nishimura and Rachi, eds. 2007).

It is difficult to identify the precise origin of Japan's townscape preservation movements, but most scholars would point to a campaign in the city of Kamakura (Kihara 1982). In 1964, Kamakura residents formed an "alliance for the preservation of elegance in Kamakura" to oppose a plan to develop the land behind Tsurugaoka Hachimangu Shrine. The goal of protecting the historic landscape around the shrine gained momentum. This episode is often recalled as one of the first efforts at historic landscape preservation.

In the following year, a symposium on the traditional cultural properties of Kyoto and Nara was held under the auspices of UNESCO. The Law for the Preservation of Ancient Capitals (the Law Concerning Special Measures on the Preservation of the Historic Nature of Ancient Capitals, No. 1, January 13, 1966) was promulgated in 1966. Subsequent ordinances enacted in Kyoto, Nara, Kamakura, and Ikaruga identified historic landscapes for preservation. Following the success of the local movement to protect the area around Tsurugaoka Hachimangu from development, Kamakura was particularly quick to act. All of these cities had groups of traditional buildings, which as a result of rapid industrialization, population growth, and lifestyle changes, were demolished in a sporadic and piecemeal fashion, or updated with modern architectural features, such as colored and corrugated roofs of galvanized steel. These ordinances were enacted as emergency countermeasures to such developments (Ishikawa 1981: 23–25). However, the national fame of these ancient capital cities made it far easier to promote a consensus on the subject of preservation, and explains the early enactment of related ordinances.

It is worth noting that grassroots townscape preservation movements emerged separately from these government-led initiatives. In 1966, residents of the former Kiso valley post town of Tsumago formed a "document preservation society," and in 1968, an "Association of People Who Love Tsumago." In the same year, residents of the village of Shirakawa in Gifu prefecture formed an association for the preservation of Shirakawa's distinctive houses, which feature triangular thatched-rooves. Residents of the town of Imaichō in Nara prefecture also joined together to protect the local landscape, establishing the "Imaichō Preservation Society" in 1971. Unlike Kyoto, Kamakura, and the other areas mentioned above, these preservation initiatives were launched in areas that Japan's postwar development had left behind: villages on the brink of becoming ghost towns as a result of rural depopulation.

Once a prosperous post town along the old Nakasendō highway linking Edo and Kyoto, Tsumago had since become an isolated village, far removed from both railway lines and motorways. Indeed, the steady merger of rural towns and villages threatened Tsumago's very survival. After a lengthy debate, the residents of Tsumago adopted the now-famous "Tsumago charter," and pledged not to "sell, rent out, or destroy" historic buildings. This charter was a simple expression of the spirit of "preservation-based redevelopment"—the idea of "developing something by saving it." At a time when "preservation" meant "cryopreservation" (leaving buildings intact and "frozen in time") and "redevelopment" meant "scrap-and-build," it is striking that Tsumago residents did not seek the construction of a bypass to revitalize the local economy, but instead focused upon tourism development as a means of revival. In another notable development, a few key leaders among the village's residents and the village office took the initiative, with the support of local authorities. The pioneering nature of the Tsumago initiative, at a time when the term "tourism development" was not in common use, is striking (Kihara 1982: 104–119; George 2013).

As its fame spread nationwide, thanks in part to the efforts of Ōta Hirotarō (then an engineering professor at the University of Tokyo), tourists began to flow into Tsumago. The success of Tsumago and the Tsumago motto ("real development is preservation") began to influence other regions. The philosophy and know-how developed in Tsumago was adopted in Naraijuku (also located in the Kiso Valley of Nagano prefecture), the Hokkaido city of Otaru, the city of Yanagawa in Fukuoka prefecture, and the aforementioned Imaichō, in Nara. Information and know-how was shared through exchanges between these various local preservation movements, but of particular significance was the establishment of the National Federation for the Preservation of Historic Landscapes (Zenkoku Rekishiteki Fūdo Hozon Renmei) in 1970. Key members of the 1964 Kamakura campaign to protect the environs of Tsurugaoka Hachimangu led the effort to establish this national federation, which provided a venue for activists to exchange information and opinions on the actual state of local movements and townscapes. Along with the National Association for Townscape Preservation, a "little brother" organization established in 1974, the federation has assumed the role of a "national center" for preservation in the private sector. Both organizations organize nationwide symposiums and seminars (*zenkoku machinami zemi*, or "national townscape seminars"), and often collaborate with the Architectural Institute of Japan and local governments. This cooperative approach to townscape preservation by the public and private sectors is quite different from the example of the British National Trust, a purely nongovernmental organization.[9]

The central government of Japan introduced new measures in response to these developments. In 1968, the Agency for Cultural Affairs was established, and immediately commenced a survey of Western-style architecture from the Meiji period. The Kankō Shigen Hogō Zaidan (Foundation for the Protection of Tourism Resources—more commonly known as the Japan National Trust) was established as an external organization of the Ministry of Transport. The Environment Agency was established in 1971. A number of new, resident-led preservation campaigns also emerged during this period. For example, the Friends of Ikaruga Association and the Association to Protect the Otaru Canal were formed in 1972. In 1973, the Arimatsuchō Community Development Association (Aichi), the Association to Protect Nagasaki and the Nakashima River, and the Association to Save Ōdairajuku (Nagano) were established by local residents.

In 1975, the Law for the Protection of Cultural Properties, which had been prepared by the newly-established Agency for Cultural Affairs, was amended to expand the definition of cultural properties. The text of this amendment reads:

> Article 2
>
> The term "cultural property" as used throughout this Act shall be defined as any of the following:
>
> {…}
>
> 5. Traditional buildings of high value, which comprise a historical and scenic landscape in unison with the surrounding environment (hereafter referred to as "groups of traditional buildings")
>
> Law for the Protection of Cultural Properties, Article 2, Paragraph 5.

This paragraph, which was newly introduced in the 1975 law, expanded the concept of cultural properties from discrete "points" or "individual" sites to broader areas, and "groups" of buildings. This was a revolutionary change from the preservation administration originally developed in the Meiji period. In conjunction with the Order for the Enforcement of the Cultural Properties Protection Act, the City Planning Act, the Building Standards Act, and the Outdoor Advertising Act, this new preservation system for "groups of traditional buildings" has enabled the selection and protection of "traditional building preservation districts." Although "designation" is a common method employed in cultural-property-related legislation, the 1975 act represented a major shift away from the designation method in favor of a "selection" method, whereby local governments apply for recognition from the national government.

The amendment to the Law for the Protection of Cultural Properties prompted many municipalities to enact ordinances in 1976, including Kakunodate (Akita prefecture), Kyoto City, Shirakawa (Gifu prefecture), Hagi (Yamaguchi prefecture),

[9]This is symbolized by the fact that the Kankō Shigen Hogo Zaidan (Foundation for the Protection of Tourism Resources, or the Japan National Trust) was established as an external organization of the Ministry of Land, Infrastructure, Transport and Tourism (formerly the Ministry of Transport).

and Nichinan (Miyazaki prefecture). In 1977, the Agency for Cultural Affairs and the Ministry of Construction launched a joint survey of historic environmental conservation in urban areas, which was later handed over to the Ministry of Construction's "National Townscape List" in 1979. In the same year, even more cities, towns, and villages enacted ordinances to designate "architectural preservation districts." Famous examples include the cities of Kanazawa, Takayama, and Hirosaki.

The most significant event during this period, however, was the very first "national townscape seminar" (sponsored by the National Association for Townscape Preservation), held in the towns of Arimatsu and Asuke. In 1975, Asuke residents had launched an organization to preserve their historic townscape, and began to explore possible means of tourism development. Since joining the National Association for Townscape Preservation, Asuke had gained national recognition, but the local movement had reached an impasse over the question of balancing between "cryopreservation" and the development of the town as a whole. As Ishikawa Tadaomi, an *Asahi Shinbun* reporter who covered townscape preservation movements nationwide, reported:

> From [people] within the movement came a succession of questions: "what's the point of enduring a difficult way of life just for sake of the tourists, and maintaining only the external façade of our buildings?" And: "it's not enough to improve the district designated for historic preservation, if it doesn't make the whole town happy."
>
> Ishikawa (1981: 27).

As local residents grappled with these questions, the first "national townscape seminar" in Asuke fostered an active discussion of various approaches to townscape preservation. At the heart of this debate was the question of who ought to be the "main actor" within the townscape preservation movement. Let us return to Ishikawa's report:

> ...At the first seminar, it was decided that residents, local governments, and cooperating specialists were the main actors in the townscape preservation movement (based on the Arimatsu-Asuke Declaration), and the roles of the three parties were clarified.
>
> Ishikawa (1981: 29). Supplementary parenthetical text is from the original.

At a time when grave pollution issues placed residents' movements and the government in an oppositional framework, the aforementioned Arimatsu-Asuke Declaration affirmed a new path, in which residents and the government would join hands to promote preservation, in cooperation with experts. The framework of resident-led "demand" and "obstruction" campaigns (Nishio 1975) fails to capture this new trend.

At the second national townscape seminar, held in Ōmihachiman in 1979, participants "confirmed the regional diversity of circumstances and character" that allowed for alternatives to the Tsumago method of promoting tourism development through preservation (Ishikawa 1981: 29). Indeed, different types of preservation movements had begun to emerge: in the case of Ōdairajuku, a group of non-residents moved together to the deserted village in the hopes of reviving it.

However, the most significant development during this period was the confluence of efforts to preserve the modern architecture and traditional townscapes. As a result, the scope of the townscape preservation movement expanded beyond the preservation of pre-modern architecture in villages like Tsumago to include buildings constructed in the modern era. Representative examples include the Association to Protect the Historic Hakodate Landscape, which aimed to preserve the Meiji-era Hakodate harbor district, the Nakanoshima preservation campaign in Osaka, and the Association to Protect the Otaru Canal. In fact, the third national townscape seminar was held jointly in Otaru and Hakodate (Ishizuka 1980; Kankō Shigen Hogo Zaidan, ed. 1981).

Through such developments, townscape preservation movements came to advocate for the preservation of historic townscapes as landscapes unique to the local community. After all, activists argued, townscapes are a collective expression of the unique conditions and history of each regional society; they are the shape through which regional society lives and survives. Changes to the townscape alter the contours of lives within the local community. This was a compelling argument for preservation.[10]

Let us return now to the two "barriers" to preservation mentioned at the beginning of this chapter—legality and subjectivity—and examine how townscape preservation movements approached these barriers. First, in order to overcome the first barrier of "legality," townscape movements advocated for the preservation of the *overall landscape*. The legality of individual buildings could not be denied—but nor did their legal status address the problem of townscape change. Therefore, instead of breaking down the landscape into individual elements, preservation activists focused on how to overcome the contradictions between individual buildings and the whole in order to achieve an excellent "scenery as a whole." This concept is clearly conveyed in their choice of the word *machinami*, or "townscape."

The preservation movement addressed the second barrier of subjectivity by rejecting any discussion of the issue in terms of "individual taste." Instead, activists discussed preservation as a means of passing down the history of a community and ensuring its continued survival. In other words, the issue of townscape preservation was framed as a philosophical question (Castells 1983: 215–217; Horikawa 2000a, b): How do we evaluate town scenery as a landscape? What value do we assign to the present townscape as an accumulation of many historical layers? What kind of city and region do we want to build in the future? At the heart of this approach lies the concept of the public nature of urban space—and the question of "who is a city for?" The specific problem that emerges is the destruction of the expertise accumulated through the practice of daily life by local residents, and the memories of local communities, by scrap-and-build redevelopment,[11] which leaves no trace of the past behind. The rise of townscape preservation movements presents a continuing challenge to Japan's growth-driven style of urban management. These movements refuse to view changes to specific buildings as "individual" or "private" concerns,

[10]In the words of Kuwako Toshio, "remaking scenery means simultaneously remaking the self who lives within that scenery" (Kuwako 1999: 99). In this volume, the word "scenery" should be considered to be synonymous with the words "landscape" and "townscape."

[11]For example, Kuwako has argued that "the modern era was created through the repudiation and demolition of the previous significance of space" (Kuwako 1999: 77), and criticized the process of landscape transformation as the "erasure of the personal history of space."

and instead identify them as an issue concerning the overall urban landscape, and related to an overarching "urban ideology."

Townscape preservation movements received a certain amount of legal and institutional support from the Japanese government between the 1980s and the 2000s. Yet from a tactical standpoint, preservation movements still had to demonstrate the effectiveness of preservation in order to gain support. What strategies did members of the preservation movements employ? Broadly speaking, the movements utilized two strategies: "preservation as a resource for tourism development" and "preservation through ownership of the place or building itself." Both strategies are discussed in the following section.

2.5.2 *Landscape as a Resource for Tourism Development*

The first strategy involved transforming cities or towns into tourist destinations by preserving the type of historic buildings that had typically been destroyed during Japan's high economic growth period. Following Japan's rapid population growth (and, in certain rural areas, depopulation) and industrialization, the surviving historic townscapes represented "hometowns of the heart" to which many aspired to return.

It is thus no coincidence that Tsumago became a pioneering example of this strategy. Unintentionally excluded from Japan's modernization and development, the residents of Tsumago sought to turn this misfortune into an asset through "preservation-based redevelopment." Both residents and the local authorities in Tsumago recognized the urgent need to revive their village. Moreover, Tsumago's history as a post town meant that tourism was already the traditional local industry. Although the absence of other resources made "tourism development" Tsumago's only choice, it was nevertheless a reasonable one.

Tsumago's success came as a revelation to depopulated villages across Japan, but this success was also problematic. Tsumago adopted a passive form of development that was constantly influenced by trends in tourism. The tourists visiting Tsumago came to see a traditional way of life in pre-Meiji era Japan. Given this, Tsumago residents could hardly display modern items such as an air-conditioning unit or a bright aluminum sash. Such dilemmas have prompted many to ask: "Who is preservation really for?" And: "what is the purpose of preservation?" The answer to the latter question in many areas, including Tsumago, was: "Tourism development, so that we may continue to live in our local community." This is what "townscape preservation" first meant to depopulated villages, where traditional buildings represented a rare resource for tourism development.

2.5.3 Protection Through Ownership: The Creation and Development of National Trust Movements

By accepting market mechanisms, the tourism development strategy was essentially "passive" in nature. In other words, only preservation "investments" that seemed likely to "pay off" in the tourism market were targeted for preservation. It is no exaggeration to say that in Japan, preservation is only possible for famous and historic buildings that can be converted into tourist facilities (such as restaurants and souvenir stores). Demolition remains the fate of buildings that cannot be so easily converted into "brand name" structures, or which stand outside a district with clusters of historic structures.

Given this, many prefer the idea of buying and preserving such buildings on their own terms. It was this sentiment that triggered the emergence of national trust movements in Japan. Most preservation campaigns target either public spaces, buildings, and waters, or the property of others. It is important to note that the activists' lack of ownership rights over the targets of preservation have delegitimized their claims and involvement. Given that preservation movements involve the exercise of influence over the property of others, private ownership can be described as its most formidable enemy. After all, it is extremely difficult for non-owners to push for the preservation of a place that owners are determined to develop. This is why so many preservation campaigns end in defeat.[12]

National trust movements attempt to use the strength of ownership rights—in which landowners may freely contravene the public welfare in their building choices as long as they are lawful, in an example of "absolute" or "strong" property rights—to the advantage of the preservationist cause. This sets them apart from environmental education lobbies and awareness-building campaigns, or pressure group-based lobbying campaigns for landscape preservation. Trust movements do not simply acquire land. They are entrusted by concerned citizens with preserving the landscape by owning, maintaining, and managing land and buildings targeted for preservation through the mechanism of a trust.

To date, many trust movements have emerged across Europe and America, but the national trust movement originated in England.[13] As those familiar with English history know, the country witnessed two rounds of land "enclosure;" the second round, which began in the latter half of the eighteenth century, resulted in the enclosure and effective disappearance of most open, arable, and theretofore common land. At that time, the demands of population growth necessitated an improvement

[12]Lamenting the destruction of the Kamakura landscape, the famous novelist and Kamakura resident Osaragi Jirō noted: "They (outside land developers) possess property rights to the land. We have no means of defending it. They are free to dispose of the land as they please" (Murakami, ed. 1996: 234; information in parentheses added by the author). Osaragi's words underscore the strength of ownership rights in Japan.

[13]For more information on the British national trust movement, see Fedden (1974, 1984), Kihara (1982, 1984), Murphy (1987), Gaze (1988), Nishimura (1988, 1997), Hara (1989), Fujita (1994), Mukai (1995), Murakami, ed. (1996), Katō and Nomura, eds. (1997), Mizuno (1998), Shioji (1999), and Horikawa (2000a, 2001).

in productivity. New agricultural technologies were incompatible with the open field system, necessitating the large-scale cultivation of arable land. In response to petitions from landowners, acts of Parliament promoted and legalized the extensive enclosure of open land. The General Enclosure Act of 1801 resulted in the enclosure of most open, public land in England and Wales. In response, the Commons Preservation Society (today the Open Spaces Society) was established in 1865 to protect peoples' rights to use and pass through common land. This was the start of the common land preservation movement, but it could not effectively resist the trend toward enclosure, which had the legal backing of the 1801 General Enclosure Act. Moreover, the movement could not acquire land itself as it lacked corporate or legal status.

In 1895, the search for an organizational structure that would allow land and property-holding rights resulted in the creation of the National Trust for Places of Historic Interest or Natural Beauty (hereafter the National Trust). As the organization's official name makes evident, its preservation targets include not only natural landscapes but also historical buildings and historic sites. The National Trust originally acquired nonprofit corporation status under the Companies Act, which enabled it to purchase land. The 1907 National Trust Act made the National Trust a "trust with its own laws" and conferred upon the National Trust the right to declare its holdings nontransferable. In this way, the National Trust has grown into a huge asset-holding organization, and has given national prominence to regional environmental preservation campaigns (Nishimura 1997: 32–80).

Has Japan witnessed similar developments with regard to trust movements? As we have seen, participants in townscape preservation and environmental conservation campaigns across Japan were inevitably overpowered by a formidable enemy: property rights. The British model of a national trust movement was introduced to Japan by the novelist and Kamakura resident Osaragi Jirō, and gradually gained momentum as an effective strategy for combating this "enemy" (Murakami, ed. 1996: 232–250; Kihara 1984, 1992, 1998: 24–34). Notable examples include the Kamakura Trust (*Kamakura Fūchi Hozon Kai*) (Kihara 1982; Hara 1989), which, as we have already seen, remained dedicated to the protection of the Oyatsu mountains behind Kamakura's Tsurugaoka Hachimangu shrine, the Ten'jinzaki Nature Conservation Society (Toyama 1995), which promoted the purchase of Cape Ten'jinzaki near the city of Tanabe in Wakayama prefecture to prevent its development as a resort, and the "One Hundred Square Meter Movement" for the preservation of Shiretoko national park (Kihara 1984, 1992, 1998) by the town of Shari in Hokkaido prefecture. While these movements cooperated with local governments to differing degrees, they all involved the voluntary purchase of land and assets by local citizens in order to preserve the landscape.

Nevertheless, the national trust movement in Japan has not proved as successful as its British predecessor: unlike the growth-control city planning of Europe and the United States, Japan's land policies are intrinsically pro-development, and Japan's political structure remains dependent upon public works projects and high land prices. Despite the establishment of an NPO/NGO system in Japan, the Japanese tax system continues to present enormous difficulties to new entrants. Successful examples of trust initiatives to date have been limited to areas where land prices are relatively low—such as depopulated areas, natural landscapes, and virgin forests—rather than crowded urban townscapes. High land prices continue to make it extremely difficult to preserve land in large urban areas with a highly concentrated population. The state of preservation in "central areas" such as Tokyo is far from encouraging. The overwhelming pressure for development in central Tokyo can be described as the "momentum" propelling the city as a whole, and the overriding concern remains achieving a floor space index that can counterbalance land prices.[14]

From this overview, we can conclude that "tourism development" and "national trust" strategies do not necessarily produce positive results. The Otaru case[15] is an example of the former strategy for preservation—an attempt to overcome the "barriers" through "tourism development" and a movement to preserve the legacy of modernization—not in a rural area, but in a city. We will explore the movement and its underlying logic in subsequent chapters.

2.6 Sociology and the Landscape

With the townscapes of Kawagoe and Otaru as our guides, we have traveled from specific building sites to city design as a whole. In the process, we have concluded that city design need not be discussed in terms of "state-of-the-art architecture," but rather in terms of the question of how the know-how and "placeness" accumulated by local residents through the practice of daily life can be linked to the city as a whole. From the control of individual sites in Kawagoe and the overall design of Otaru, we have confirmed the existence of a local, resident-based system of spatial control that stands in contrast to Japan's centralized city planning administration and market-driven forms of urban development. In other words, we have seen how townscapes reveal skillful techniques for combining private rights with a public nature—unlike modern Japanese city planning, which makes no attempt to mediate conflicts between different sets of private rights. This has brought into relief the problems inherent to Japan's city planning laws, which attribute responsibility for disputes regarding private rights solely to the individual—tacitly approving development as a result. Japan's townscape preservation movements refused to address

[14]For example, with regard to the Ōkawabata River City redevelopment project in the Akashichō neighborhood of Tsukiji, a former foreign settlement whose existing community was moved across the river to Tsukishima, Okuda Michihiro writes: "the project was explained to me by people involved (in the redevelopment) project, but, after all, this was a building scheme to adjust the "floor space index" to match and compensate for the "high land price." It is impossible to imagine that any provisions were made with regard to urban landscape or environmental design." (Okuda 1989: 74; quotation marks are from the original text, information in parentheses has been added by this author).

[15]The canal preservation movement in Otaru also developed a trust movement aimed at the purchase of historic warehouses, but high prices forced the movement to abandon the effort. A detailed discussion of this process can be found in Horikawa (2001).

changes to specific buildings at the level of individual preference, and instead sought an "urban philosophy" that grappled with the question of how to create a city with a truly public nature.

The warehouse-style *machiya* that comprise the townscape of Kawagoe imposed certain limitations without suppressing individual expression. Otaru's various small communities, each of a different character, remained connected to the sea through the intermediary of the canal. Both cases demonstrate that the real challenge of urban design is to explore how we can mitigate and avoid the risks of unfettered construction without limiting individual freedom. In this sense, the prescience of Japan's townscape preservation movements, which as early as the 1960s began advocating for a new urban renaissance drawing upon the traditional methods of establishing an urban order—from individual sites to the city at large—deserves recognition and reappraisal. Townscape preservation movements identified the problem with approaching preservation in terms of the legality of specific buildings. By looking at the built environment of the entire townscape, preservation movements identified how an urban landscape *could be both entirely legal and problematic*—and how legal loopholes allowed for the full-scale destruction of existing townscapes. They also called into question the subjective approach to the landscape as a matter of personal preference—a matter of "liking or disliking." As the examples of Kawagoe and Otaru demonstrate, in the sense that sharing a minimal set of norms gave greater freedom of expression to the owners of individual buildings, cultural townscapes may actually provide the very infrastructure required for the expression of personal taste. This offers an opportunity to contemplate the relationship between freedom and the public nature of the city, and explains why a study of townscape preservation must focus upon preservation movements. It is also what transforms the historical townscape into a subject of sociological inquiry.

References

Aramaki, S. (2003). Shimin Soshiki ni yoru Rekishiteki Machinami Keikan Hozen no Katsudō: 'Kawagoe Kura no Kai' o Jirei ni. *Toshi Mondai, 94(7)*, 77–92. Tokyo Shisei Chōsakai.

Castells, M. (1983). *The city and the grassroots: A cross-cultural theory of urban social movements*. Berkeley, CA: University of California Press.

Fedden, R. (1974). *The National Trust: Past and present*. London: Jonathan Cape. [Yotsumoto, T. (trans.). (1984). *Nashonaru Torasuto: Sono Rekishi to Genjyō*. Jichōsha.]

Franzoi, U., & Smith, M. (1993). *Canal Grande*. Venezia: Arsenale Editrice srl. [Jinnai, H., & Nakayama, E. (trans.). (1994). *Venezia Dai Unga*. Yōsensha.]

Fujita, H. (1994). Nihon Toshi no 'Keikaku' to 'Mukeikaku,' *Sōbun, 356*, 14–17. Sōbunsha.

Fujita, H. (2005). Nihon no Toshi to Machinami: 'Keikaku' to 'Mukeikaku' no Hazama de. In K. Kasahara, K. Nishihara, & T. Miyauchi (Eds.), *Shakai Kōzō no Tankyū: Genjitsu to Riron no Intāfeisu* (Shimoda Naoharu Kyōju Tsuitō Ronbunshū) (pp. 205–227). Shinsensha.

Fukukawa, Y. (2003). Toshi Komyuniti no Hozen. In Y. Nishimura, T. Ōnishi, S. Ōgaki, T. Kishii, & K. Koide (Eds.), *(Toshi Kōgaku Kōza) Toshi o Hozen Suru* (Chapter 5). Kajima Shuppankai.

Fukukawa, Y., & Nishimura, Y. (1981). Machinami to Jūkūkan no Saisei. In Kankō Shigen Hogo Zaidan (Ed.), *Rekishiteki Machinami Jiten: Furusato no Machi, Sono Hozon to Saisei no Tame ni* (pp. 72–81). Kashiwa Shobō.
Funabashi, H. (1989). 'Shakaiteki Jirenma' Toshite no Kankyō Mondai. *Shakai Rōdō Kenkyū, 35(3–4)*, 23–50. Hosei Daigaku Shakaigakubugakkai.
Gaze, J. (1988). *Figures in a landscape: A history of the National Trust*. London: Barrie and Jenkins.
George, T. S. (2013). Furusato-zukuri: Saving home towns by reinventing them. In C. Gerteis & T. S. George (Eds.), *Japan since 1945: From postwar to post-bubble* (pp. 27–46). London: Bloomsbury.
Gratz, R. B. (1989). *The living city*. New York, NY: Simon and Schuster. [Hayashi, Y., Tomita, Y., & Miyaji, M. (trans.). (1993). *Toshi Saisei*. Shōbunsha.]
Hara, M. (1989). *Rekishiteki Fūdo no Hozon: 'Kamakura Shimin' no Hibi* (Akansasu Kenchiku Sōsho 2). Odawara: Akansasu Kenchiku Kōbō (Shinshoban, 256 pp.).
Holleran, M. (1998). *Boston's "Changeful times:" Origins of preservation and planning in America* (Creating the North American Landscape). Baltimore, MD: The Johns Hopkins University Press.
Horikawa, S. (1991). Taishōki Bunkazai Hozon o Meguru Gyōsei to Minka Chōsa: 'Ten' Toshite no Bunkaizai Hozon. In T. Kawai (Ed.), *Kindai Nihon Shakai Chōsashi (II)* (pp. 243–278). Keio Tsūshin (Keio University Press).
Horikawa, S. (1998a). Toshi Kūkan to Seikatsusha no Manazashi. In A. Ishikawa, K. Satō, & K. Yamada (Eds.), *Mienai Mono o Miru Chikara: Shakai Chōsa to Iu Ninshiki* (pp. 133–149). Yachiyo Shuppan.
Horikawa, S. (1998b). Rekishiteki Kankyō to Chiiki Saisei: Machinami Hozon ni Okeru 'Bashosei' no Sōtenka. In H. Funabashi & N. Iijima (Eds.), *Kankyō* (Kōza Shakaigaku 12) (pp. 103–132). University of Tokyo Press.
Horikawa, S. (2000a). Torasuto Undō. In Chiiki Shakai Gakkkai (Ed.), *Kīwādo Chiiki Shakaigaku* (pp. 318–319). Hābesutosha.
Horikawa, S. (2000b). Unga Hozon to Kankō Kaihatsu: Otaru ni Okeru Toshi no Shisō. In S. Katagiri (Ed.), *Rekishiteki Kankyō no Shakaigaku* (Shirīzu Kankyō Shakaigaku 3) (pp. 107–129). Shinyōsha.
Horikawa, S. (2001). Keikan to Nashonaru Torasuto: Keikan wa Shoyū Dekiru ka. In H. Torigoe (Ed.), *Shizen Kankyō to Kankyō Bunka* (Kōza Kankyō Shakaigaku 3) (pp. 159–189). Yūhikaku.
Igarashi, T. (1991). *Tochi Kaikaku no Puroguramu: Toshi e no Kenri*. Nihon Hyōronsha.
Igarashi, T., & Ogawa, A. (1993). *Toshi Keikaku: Riken no Kōzu o Koete* (Iwanami Shinsho Shin Akaban 294). Iwanami Shoten.
Iguchi, K. (2002). 'Kiwadatsu' Dezain Kara 'Osamaru' Dezain e. In Toshibi Kenkyūkai (Ed.), *Toshi no Dezain: 'Kiwadatsu' Kara 'Osamaru' e* (pp. 7–28). Kyoto: Gakugei Shuppansha.
Inagaki, E. (1984). *Bunka Isan o Dō Uketsugu ka* (Toshi no Jyānarizumu). Sanseidō.
Ishikawa, T. (1981). Machinami Hozon no Jūmin Undō. In Kankō Shigen Hogo Zaidan (Ed.), *Rekishiteki Machinami Jiten: Furusato no Machi, Sono Hozon to Saisei no Tame ni* (pp. 23–29). Kashiwa Shobō.
Ishizuka, M. (1980). (Otaru Hōkokukai) Otaru, Unga, Soshite Watashitachi no Ayumi. *Kankyō Bunka, 47*, 16–18. Kankyō Bunka Kenkyūjo.
Kankō Shigen Hogo Zaidan (Ed.). (1981). *Rekishiteki Machinami Jiten: Furusato no Machi, Sono Hozon to Saisei no Tame ni*. Kashiwa Shobō.
Kankyō Bunka Kenkyūjo (Ed.). (1978). *Rekishiteki Machinami no Subete*. Wakaki Shobō.
Kankō Shigen Hogo Zaidan, (Ed.). (1979). *Otaru Unga to Sekizō Sōkogun* (Kankō Shigen Chōsa Hōkoku 7). Kankō Shigen Hogo Zaidan.
Katagiri, S. (Ed.). (2000). *Rekishiteki Kankyō no Shakaigaku* (Shirīzu Kankyō Shakaigaku 3). Shinyōsha.
Katō, M. (1999). *Tokyo Kieta Machikado*. Kawade Shobō Shinsha.
Katō, I., & Nomura, Y. (Eds.). (1997). *Rekishiteki Isan no Hogo*. Shinzansha.
Kawagoe Ichibangai Machinami Iinkai (Ed.). (1988). *Kawago Ichibangai Machizukuri Kihan*. Kawagoe: Kawagoe Ichibangai Machinami Iinkai. Reprinted (1999) (A4 ring binding, 152 pp.).

Kawagoe Ichibangai Shōtengai Kasseika Moderu Jigyō Suishin Iinkai (Ed.). (1986). *Kawagoe Ichibangai Shōtengai Kasseika Moderu Jigyō Hōkokusho, Komyuniti Māto Kōsō Moderu Jigyō* (A4, 297 pp.). Kawagoe: Kawagoe Ichibangai Shōtengai Kasseika Moderu Jigyō Suishin Iinkai.

Kihara, K. (1982). *Rekishiteki Kankyō: Hozon to Saisei* (Iwanami Shinsho Kiban 216). Iwanami Shoten.

Kihara, K. (1984). *Nashonaru Torasuto* (Toshi no Jānarizumu). Sanseidō. Reprinted (1992) *Nashonaru Torasuto* (Sanseidō Sensho 168). Sanseidō. New edition (1998).

Kuwako, T. (1999). *Kankyō no Tetsugaku: Nihon Shisō o Gendai ni Ikasu* (Kōdansha Gakujutsu Bunko 1410). Kōdansha.

Matsumoto, Y. (1997). Nihon ni Okeru Bunkazai Hogo Seisaku, Rippō no Tenkai: Meijiki = Hakai Yori Hozon e no Michi. *Kikan Kagoshima Keizai Daigaku Shakaigakubu Ronshū, 15(4)*, 1–40. Kagoshima: Kagoshima Keizai Daigaku Shakaigakubu.

Matsumura, K. (Ed.). (1997). *Sanson no Kaihatsu to Kankyō Hozen: Rejā Supōtsuka Suru Chūsankan Chiiki no Kadai*. Tokyo: Nansōsha.

Mizuno, S. (1998). Seiki Tenkanki Igirisu no Kankyō Hogo Katsudō: Nashonaru Torasuto Sōsetsu o Meguru Aratana Tenkai. *Seiyōshigaku, 191*, 22–41.

Mukai, K. (1995). Sanka to Kōryū ni Yori Chiiki Shigen no Hozen to Sōzō: Igirisu no Nashonaru Torasuto Undō. In N. Imamura & K. Nagara (Eds.), *Chiiki Shigen no Hozen to Sōzō, Keikan o Tsukuru to wa Dō Iu Koto ka* (Zenshū Sekai no Shokuryō, Sekai no Nōson 9) (Chapter 1). Nōsangyoson Bunka Kyōkai.

Murakami, M. (Ed.). (1996). *Ningen to Bunmei o Kangaeru: Mizu no Oto* (Osaragi Jirō Essei Serekushon 2). Shōgakukan.

Murphy, G. (1987). *Founders of the National Trust*. London: Christopher Helm. [Yotsumoto, T. (trans.). (1992). *Nashonaru Torasuto no Tanjō*. Ryokufū Shuppan.]

Nihon Bengoshi Rengōkai Kōgaki Taisaku Kankyō Hozen Iinkai (Ed.). (1995). *Kaete Mimasen ka, Machizukuri* (J-JEC Kankyō Sōsho Shirīzu 5). Jikkyō Shuppan.

Nishikawa, K. (1994). *Toshi no Shisō* (NHK Bukkusu 694–695). Nippon Hōsō Shuppan Kyōkai.

Nishimura, Y. (1997). *Kankyō Hozen to Keikan Sōzō: Korekara no Toshi Fūkei e Mukete*. Kajima Shuppankai.

Nishimura, Y. (1984). Kenzōbutsu no Hozon ni Itaru Meiji Zenki no Bunkazai Hogo Gyōsei no Tenkai: 'Rekishiteki Kankyō' Gainen no Seiseishi Sono 1. *Nippon Kenchiku Gakkai Ronbun Hōkokushū, 340*, 101–110.

Nishimura, Y. (1988). Eikoku Nashonaru Torasuto no Dōtei to Kankyō Hozen Undō no Tenkai 1–4. *Toshi Mondai, 79(9)*, 87–100; *79(10)*, 85–99; *79(11)*, 83–99; *79(12)*, 69–90.

Nishimura, Y., & Rachi, M. (Eds.). (2007). *Shōgen, Machinami Hozon*. Kyoto: Gakugei Shuppansha.

Nishio, M. (1975). Gyōsei Katei ni Okeru Taikō Undō: Jūmin Undō ni Tsuite no Ichikōsatsu. *Seiji Sanka no Riron to Genjitsu* (Nenpō Seijigaku 1974), 69–95. Iwanami Shoten.

Ōkawa, N. (Ed.). (1997). *Rekishiteki Isan no Hozon, Katsuyō to Machizukuri*. Kyoto: Gakugei Shuppansha.

Okuda, M. (1989). Hābādo G.S.D. Seminā o Oete. *SD, 8908*, 74.

Shimizu, S. (2013). *Kenchiku Hozon Gainen no Seiseishi*. Chūō Kōron Bijutsu Shuppan.

Shioji, Y. (1999). Heritēji to Ikiru: 29, Ingurando Kantorīsaido o Shikō Suru Shakai, Igirusu. In H. Satō (Ed.), *Sumai wa Kataru* (Shirīzu Kenchiku Jinruigaku, Sekai no Sumai o Yomu 3) (pp. 63–80). Kyoto: Gakugei Shuppansha.

Takahashi, Y., & Yoshida, N. (Eds.). (1989). *Kūkan* (Nihon Toshishi Nyūmon 1). University of Tokyo Press.

Tanuma, T. (Ed.). (1990). *Kimura Ihei no Shōwa* (Chikuma Raiburarī 39). Chikuma Shobō.

Toyama, Y. (Ed.). (1995). *Tenjinzaki no Shizen o Taisetsu ni Suru Kai Nijushūnen Tsūshi*. Tanabe, Wakayama: Tenjinzaki no Shizen o Taisetsu ni Suru Kai.

Ueda, A., & Tsuchiya, A. (Eds.). (1975). *Machiya: Kyōdō Kenkyū*. Kajima Shuppankai.

Umino, M. (1993). Kankyō Hakai no Shakaiteki Mekanizumu. In N. Ijima (Ed.), *Kankyō Shakaigaku* (pp. 33–53). Yūhikaku.

Yamazaki, M. (2002). Machinami o Mamori, Sodate, Tsukuru. In Toshibi Kenkyūkai (Ed.), *Toshi no Dezain: 'Kiwadatsu' Kara 'Osamaru' e* (pp. 123–148). Kyoto: Gakugei Shuppansha.
Yoshizaka, T. [Ryūsei]. (1975). Furenzoku Tōitsutai. *Toshi Jūtaku, 94*, 10–11.
Zenkoku Machinami Hozon Renmei (Ed.). (1999). *Shin/Machinami Jidai: Machizukuri e no Teian*. Kyoto: Gakugei Shuppansha.

Chapter 3
A City with a Grudge: A History of the Port City of Otaru

Abstract Social issues do not materialize out of thin air. The community development effort that sprung up in the 1970s and centered upon the preservation of Otaru's historic townscape can only be understood within the context of the city's unique history and environment. This chapter thus offers an essential prelude to the analysis that unfolds in subsequent chapters. I give an overview of Otaru's history and explain how "revitalization" emerged as a policy imperative for local officials. After political policies that prioritized Sapporo's development led to the decline of Otaru's port and commerce functions, the once prosperous port city of Otaru was left to become "a city with a grudge." Readers will gain a clear view of the historical process by which the "canal controversy" of the 1970s and subsequent years came about.

Keyword Otaru · Sapporo · Preservation · Community development · Revitalization

3.1 A Resentful Murmur

Social issues do not materialize out of thin air. The community development effort that sprung up in the 1970s and centered upon the preservation of Otaru's historic townscape can only be understood within the context of the city's unique history and environment. This chapter is thus an essential prelude to the analysis found in subsequent chapters. Yet we begin on an unsettling note.

> I still believe that in the heart of this town (Otaru) there's this *deep*, deep-seated resentment of Sapporo. We have this complex (about Sapporo), so to speak.
>
> Interview with Akama Hajime in Otaru, September 4, 1998. Information in parentheses has been added by the author.

Today, Akama Hajime lives in Sapporo. After living in Otaru as a university student and later as a businessman involved in tourism development, however, he feels a deep affection for his former home and continues to consider himself an "Otaru man." As such, Akama's description of the city's deep-seated resentment of Sapporo is not the casual analysis of an uninformed outsider. It is an honest self-assessment.

S. Horikawa, *Why Place Matters*,
https://doi.org/10.1007/978-3-030-71600-4_3

Where does this resentment stem from? When did it begin? What processes instigated and perpetuated this resentment? The phrase "a city with a grudge" is a fitting epigraph to the story of the Otaru canal, for Otaru's changing position within Japan's political and economic history has shaped the outlook and sentiments of the city's residents.

3.2 An Overview and Periodization of Otaru's History

Otaru is a port city on Japan's northernmost island of Hokkaido. The city is located in the north Shiribeshi subregion, 30 km northwest of Sapporo on Ishikari Bay (Fig. 3.1). Protected from large waves by rocky cliffs, Otaru occupies an excellent location for a port. The city is surrounded by mountain ranges on three sides, and the hills of Otaru slope precipitously down to meet a narrow stretch of flat land along the coastline (Fig. 3.2). Indeed, Otaru's physical configuration is like that of a concert hall, with its hilly slopes the seats that fan outward and upward at a steep incline from the "stage" or harbor. The sea and the port are visible from any vantage point in this city of hills.

The port has been the cornerstone of the city's growth. In the Edo period, Otaru was a destination for the cargo ships known as *kitamaebune* (literally, "northern bound ships") traveling from Osaka across the Inland Sea and up the western coast of Japan

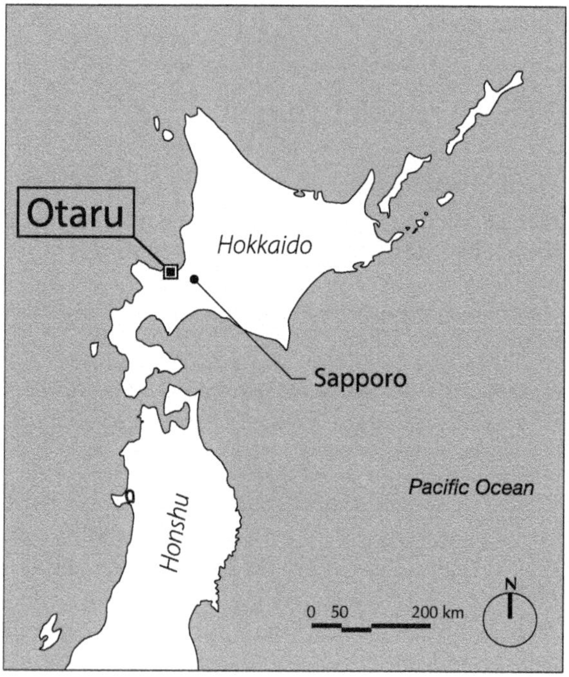

Fig. 3.1 Location of Otaru, Hokkaido (Author's illustration)

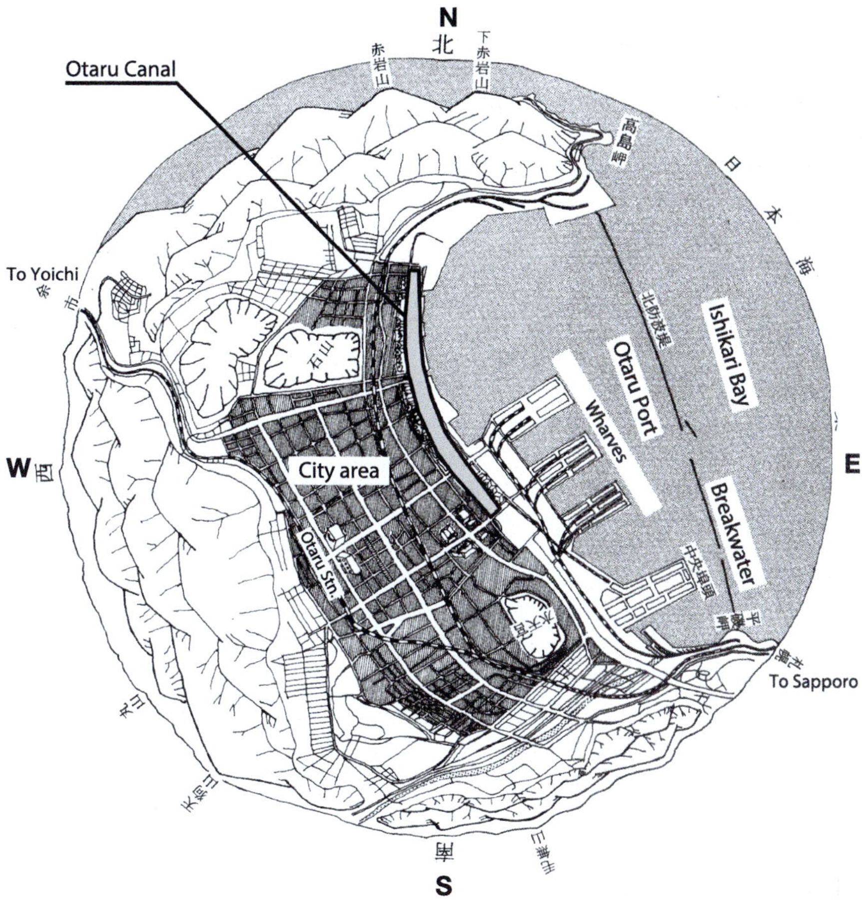

Fig. 3.2 Topography of Otaru: a fisheye lens view (Circa 1978). *Source* Kankō Shigen Hogo Zaidan, ed. (1979: 14), with supplementary details added by the author

to Hokkaido. Later, foreign freighters arrived at the port. While the Hokkaido port of Hakodate became a seafood market, the chief exports from Otaru's port were coal and fertilizer. Wholesalers concentrated in the city of Otaru, which developed into an important market for legumes and other agricultural products (Fukase 2003; Nakanishi 2013).

The name "Otaru" derives from the Ainu word *otarunai*, meaning "rivers running across a sandy beach." This is an apt description of the local terrain, as several rivers flow across the sandy coastal lowlands before reaching the ocean. Old documents and maps also refer to the area as Otarunai. By 1635, Shukutsu (the present-day Shukutsu-cho in Otaru) was "already a well-known hamlet in Ezo (the old name for lands north of the Japanese "mainland" of Honshu)" (Otarushi 1958: 63). During the Kyōhō era (1716–1736), the area began to prosper under the rule of the Matsumae clan, as merchants from Ōmi Province (modern-day Shiga Prefecture) were contracted to conduct trade with the local Ainu population. Thereafter, Otarunai moved from Matsumae control to the direct control of the Tokugawa shogunate, as it developed into an active trading base known as "Otaru *basho*" and "Takashima *basho*."

The *basho* system was unique to the Matsumae domain. The land in Hokkaido was not cultivated for rice farming, making the type of rice tribute practiced in other feudal domains impossible. Instead, Matsumae vassals were allocated certain areas in which to trade with the Ainu and used their profits from this trade as a tribute. In the Matsumae domain, the word *basho* thus referred to a place where trade with the Ainu was permitted. Between 1789 and 1800 (the Kansei Period), the number of *basho* increased to forty-two in western Ezo and forty-one in eastern Ezo, for a total of eighty-three (Otarushi 1958: 55–56). This number included Otaru *basho* and Takashima *basho* (both located within the area of present-day Otaru): busy trading posts in the salmon and herring trade conducted by fishermen on boats owned by Ōmi merchants.

Eventually, the Tokugawa shogunate placed Otaru under its direct control as a safeguard against a potential Russian advance from the north. Otarunai became an official village in 1865. In 1869, Otarunai was renamed Otaru, while Ezo became Hokkaido. In 1870, just five years after becoming a village, Otaru's status was raised to that of a town, which conferred a higher level of administrative power. Thus began the urbanization of the area around the Katsunai River—one of the "rivers running across the sandy beach."

This is but a brief prehistory of the "dawn"[1] of Otaru. Our survey of the city's history will continue based on the following periodization (which I have based on the history presented in Kankō Shigen Hogo Zaidan, ed. 1979: 14–25):

1. 1868–1885: Otaru as Sapporo's outer port
2. 1886–1898: Development and expansion
3. 1899–1921: Otaru's zenith and the first "canal war"
4. 1922–1939: Reorganization of the port
5. 1940–1945: The wartime controlled economy
6. 1946–1966: A city in decline
7. 1967–1984: The second "canal war" period
8. 1985–2001: Otaru as a tourist destination
9. 2002–Present: The beginning of the end

One should note that what follows is not a comprehensive overview of Otaru's history. The periodization and focus are specific to the main subject of this book: the battle over the fate of the Otaru canal, and subsequent efforts at community development. This is particularly true of the periodization from the 1960s onward: 1966 is an important historical juncture because it was the year of the Otaru city

[1] Of course, from the perspective of the Ainu, Hokkaido's indigenous people, the terms "prehistory" or "dawn of a new era" are entirely inappropriate and refer to nothing more than the history of Hokkaido's settlement and development by the *Wajin* (ethnic Japanese). However, the Ainu perspective on Hokkaido's history is outside the scope of the present study, and this limitation is reflected in some of the terminology.

government's planning decision regarding the Rinkōsen road project. A new era began in 1985, the year after the canal preservation movement splintered and died (Kankō Shigen Hogo Zaidan, ed. 1979; Niikawa 1986; Ogasawara 1986). A final period began in 2002, as the tide of tourism began to ebb.

3.3 From Port City to Tourist Destination: A History of Otaru City

3.3.1 1868–1885: Otaru as Sapporo's Outer Port

In 1869, the new headquarters of the Hokkaido Development Commission (the Kaitakushi) was established in Sapporo, and Otaru was designated as its outer port. This designation carried enormous significance for the humble trading post of Otaru. The 1873 discovery of coal in Horonai and the city's proximity to trade with Karafuto (known to the Russians as Sakhalin) were also important factors, but Otaru's growth can be attributed largely to the Meiji government's selection of Sapporo as the administrative and political base of its effort to colonize Hokkaido. While the very construction of the Hokkaido Development Commission headquarters in Sapporo boosted the Otaru economy, it was the relationship between the two cities that proved decisive. This relationship reflected the typical approach to urban development in early modern Japan, which split political and economic functions between two closely linked cities. While the Meiji government cultivated certain cities as centers for administration and development, it did not open them up to the outside world. Instead, an adjacent port city was often used as an intermediary. The strategy was to divide economic, administrative, and political functions between these "sets" of cities: Kyoto and the Port of Kobe, for example, or Tokyo and the Port of Yokohama.[2] In this case, central administrative agencies were assigned to Sapporo, while complementary port functions were concentrated in Otaru. The selection of Otaru over Hakodate as Sapporo's port established the Sapporo-Otaru line as the heart of the Kaitakushi's operations and transformed Otaru from a simple trading *basho* into a commercial port city of central importance.

As the single point of entry for cargo destined for Sapporo, Otaru flourished as Hokkaido's distribution base. Horonai coal also left through Otaru's port, and in 1877, the Temiya pier was built at the port to facilitate the process of loading coal onto waiting ships. Rail service began in 1880. The Temiya Line was not only the first

[2]One could also cite a more pragmatic reason for the division of labor between Tokyo and Yokohama: Tokyo Bay is too shallow for large ships to enter. But the point may be clearer if we think of these sets a bit differently: Kyoto and the Port of Kobe—Kobe Higher Commercial High School (the forerunner of Kobe University); Tokyo and the Port of Yokohama—Tokyo Higher Commercial School (the forerunner of Hitotsubashi University); or Sapporo—Otaru—Otaru Higher Commercial School (the forerunner of the Otaru University of Commerce). For more on the origin and evolution of commerce and business education in Japan, see Tachibanaki (2012).

railroad in Hokkaido, but also just the third in all of Japan. This fact reflects Otaru's importance to the Meiji government and the Hokkaido Development Commission.

Notwithstanding the remarkable development of Otaru and Sapporo during these years, Hakodate, the old seat of Matsumae power, remained "the first city" of Hokkaido. In 1882, the Kaitakushi was dissolved and Hokkaido was divided into "three prefectures and one bureau": Hakodate-ken (the area around the city of Hakodate), Sapporo-ken (the area around Sapporo, including Otaru), and Nemuro-ken, which encompassed the rest of eastern Ezo. Government enterprises such as coal-mining, railways, and factories were controlled by the Hokkaido Project Management Bureau (under the Ministry of Agriculture and Commerce). The cornerstone of Sapporo's success had been laid with its designation as Hokkaido's administrative center, but the city itself was still developing. And while the prospects for Sapporo's "outer port" city also looked bright, administrative classifications did not yet reflect Otaru's future importance.

3.3.2 1886–1898: Development and Expansion

In Otaru, the period following the 1886 establishment of the Hokkaido Agency in Sapporo was characterized by remarkable development and a number of land reclamation projects (Figs. 3.3 and 3.4). This period also witnessed a population explosion. As shown in Table 3.1, the population of Otaru more than tripled during this twelve-year period, making Otaru the second most populous area of Hokkaido, after Hakodate.

In the same period, Otaru pulled past Hakodate in terms of export profits (Kankō Shigen Hogo Zaidan, ed. 1979: 17L). In 1889, the Meiji government named Otaru a "special export port," and in 1894, a "special trading port" (Inayoshi 2014: 96–100)—designations that were accompanied by greater levels of government investment and protection. This remarkable growth was driven by the steadily increasing Horonai coal trade and the development of the Ishikari Moor.

A second defining feature of this period was the land reclamation work conducted along Otaru's coastline, which addressed the growing demand for expanded port facilities. In 1889, an area of roughly 125,780 m^2 was reclaimed from the coastline between the Okobachi River and Temiyamachi. In 1891, another area of 29,790 m^2

Fig. 3.3 A view of Otaru Port from Temiya (Around 1892). *Source* Otaru Kuyakusho (1911: 3)

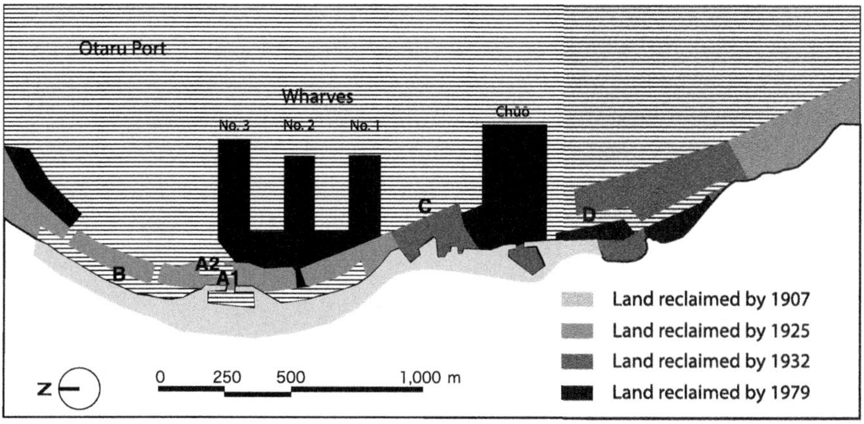

Fig. 3.4 Land reclamation and changes to the Otaru Coastline (1907–1979). A1: Earliest moorings. A2: Moorings immediately following the canal's construction. B: Canal. C: Sakaimachi quay. D: Second stage of the canal. *Note* The illustration is by the author, based upon the work of the Kankō Shigen Hogo Zaidan (1979: 17–21)

was reclaimed between Sunazakichō and Arihorochō. In total, an area of approximately 155,570 m^2 was reclaimed from the sea along the Otaru coastline during this period (Kankō Shigen Hogo Zaidan, ed. 1979: 16–17). Public boat moorings were also established at Minamihamachō and Arihorochō in 1891. In fact, almost the entire area east (on the seaward side) of Ironai Ōdori in modern-day Otaru is built on land reclaimed from the sea during this period. The steady reclamation of land, frequent large fires, and the ever-increasing prosperity of the port gradually nudged Otaru's center northward, away from the Katsunai River. Instead, Sakaimachi, Minatomachi, Kitahamachō, and Minamihamachō became the central commercial districts of Otaru (Takabatake 1899: 209–210; Kankō Shigen Hogo Zaidan, ed. 1979: 16–17; Okamoto and Nihon no Minatomachi Kenkyūkai 2008).

3.3.3 1899–1921: Otaru's Zenith and the First "Canal War"

In 1899, the rapidly expanding Otaru joined Sapporo and Hakodate to become a self-governing ward of Hokkaido. It was during this period that Otaru reached its zenith as the most important base for trade along the Sea of Japan—a region invigorated by Japan's victory in the Russo-Japanese war and subsequent acquisition of Karafuto (South Sakhalin), the brisk cereals trade during the first world war, and the Japanese government's plan to systematically develop the Hokkaido region as a source of exports for the European market. Japan's possession of Karafuto was a boon for Otaru, as all goods to and from the territory passed through its port. As war threw European markets into chaos, Hokkaido-grown legumes and grains came to dominate the global market: for a while it was even said that Otaru exports could influence

Table 3.1 Population changes in Otaru (1868–1898)

Year	Population
1868	2,230
1869	2,611
1870	3,169
1871	4,020
1872	4,242
1873	4,552
1874	4,849
1875	5,176
1876	5,533
1877	6,170
1878	7,341
1879	8,301
1880	9,551
1881	10,634
1882	12,783
1883	13,264
1884	12,078
1885	12,822
1886	15,882
1887	15,461
1888	18,788
1889	20,162
1890	21,383
1891	26,597
1892	31,472
1893	34,259
1894	39,644
1895	46,982
1896	50,717
1897	54,966
1898	56,961

Source Takabatake (1899: xii)

market prices in Europe. Nineteen different banks, including the Bank of Japan, maintained branches in Otaru, and in 1916, the ratio of Otaru bank holdings to that of Hokkaido as a whole was the highest of any ward in the prefecture (Kankō

Shigen Hogo Zaidan, ed. 1979: 18L).[3] Otaru had surpassed Hakodate to become the undisputed center of the Hokkaido economy.

This period also witnessed the completion of the northern breakwater and a drastic increase in the number of commercial warehouses. In 1900, a total of 2,817 ships (with a gross tonnage of 856,954 t) entered Otaru's harbor. By 1921, that number had risen to 4,857 (with a gross tonnage of 4,768,515 t). In twenty years, the number of ships had almost doubled, while tonnage had increased more than five-fold (Watanabe 1974: 312–314). In 1900, Otaru had 161 warehouses, covering a total of 10, 650 *tsubo* (a traditional unit of measure equivalent to 3.31 m^2). In 1921, that number had risen to 218 (for a total of 22,419 *tsubo*) (Kankō Shigen Hogo Zaidan, ed. 1979: 18L). The ports of Otaru and Hakodate had divided the economic realm of Hokkaido between them and competed fiercely, but by this point the ascendancy of Otaru's port was complete. An 1899 history of the Otaru port adopted an optimistic tone, noting the port's "promising future" (Takabatake 1899: 170).

Certain changes would be necessary to justify this optimism. Otaru's "naturally good harbor" required further fortification and modernization to allow larger ships to anchor safely in all weather. For one, the breakwater against high waves was insufficient. There was not enough open land adjacent to the port for cargo sorting. The moorages were too narrow. There was not enough space for barges or other vessels to pull up to shore and transfer cargo. These were all important issues, but the construction of a new breakwater and a canal was deemed the highest priority. The breakwater constructed during this period is still in use a century later, but a more complicated fate awaited the canal. Given the canal's significance to Otaru, let us take a closer look at the history of its construction.

The need for a canal to deal with the rapidly increasing number of ships arriving at the port was first identified in the second section of the 1903 Explanation of Redevelopment Plans for the Otaru Port (Otarushi 1963: 568). At that time, ships would unload their cargo onto barges for transfer to the waterfront or nearby moorings. There, longshoremen would carry the cargo on their backs to the warehouses that lined the shore. After negotiations for their sale had been completed, the cargo would leave the warehouses by wagon (or later by truck) along *denuki kōji* (narrow alleys that ran behind the warehouses) for distribution across Hokkaido. In short, the port, barges, warehouses, and *denuki kōji* functioned as a single system of interlocking parts (Horikawa 2005). This form of cargo transport was both time-consuming and (as cargo ship owners paid for moorage and labor) expensive. The port's infrastructure had to be updated in order to promote a faster, more efficient form of cargo transport (Figs. 3.5 and 3.6).

The construction of a canal through off-shore land reclamation was intended to increase the number of landing stages for boats, and, by "linking to a large number of warehouses," enable the "direct transfer of freight from barges to the warehouses situated along both sides of the canal." The wording of the 1903 plan makes it clear that the canal would not be dug out of existing land, but rather created through the

[3] There were ten head and branch bank offices in Sapporo, and sixteen in Hakodate during the same period. Kankō Shigen Hogo Zaidan, ed. (1979: 18R).

Fig. 3.5 Otaru Warehouses (Constructed between 1890–1894) and Barges at the Wharf (1911). *Source* Otaru Kuyakusho (1911: 11)

Fig. 3.6 A street scene in Otaru (1911). *Source* Otaru Kuyakusho (1911: 4)

reclamation of land just off-shore. There were two clear advantages to this method of construction: it would allow warehouses to be located along *both* sides of the canal, and it would avoid chipping away at the already narrow strip of land between the coastline and the hills of Otaru. Construction of the canal began in August 1914 and was completed in September 1923. The new canal and surrounding warehouses became a vital new component of the port's interlocking system of cargo transport by barge.

The first proposal for a canal had been made in 1903, yet two full decades passed before the canal's completion. Why did it take so long for the canal to be completed? In short, Otaru's remarkable economic growth created new and often conflicting interests within the local business community. Three successive political confrontations subjected the original plan to endless changes and significantly delayed the canal's construction:

> ...Otaru has fought three tough fights over the canal. First came the dispute between the district and Hokutan [Hokkaido Colliery and Railway Company]. The second revolved around the question of whether to construct a canal or a wharf. Finally, there was the Reform Club's revised proposal. All three were reportedly fierce conflicts, which consumed the political world, business circles, and all the ward residents.
>
> Endō and Moriyama (1979: 27L). Information in brackets has been added by the translator.

The first of these "tough fights" centered upon a dispute regarding the coastal boundary line between land reclamation work to be performed by the influential semi-governmental coal company, Hokkaido Colliery and Railway Company (the forerunner of the Hokkaido Colliery and Steamship Company) and land reclamation work undertaken as a public works project of the Otaru District. In 1904, the two sides reached an agreement. Unlike this simple "border dispute," a second disagreement involved the actual goal of renovation. While all parties (primarily local business interests and their political representatives) agreed on the need to quickly improve port facilities, they split over the question of whether a canal or a new wharf would most improve cargo handling at the port. There were advantages and disadvantages to both options, and the interests of the various factions were divided.

Constructing a new wharf perpendicular to the coastline would allow cargo ships to pull up directly to the wharf to unload and take on new freight. The primary advantage of this approach was that it would eliminate the need for barge transport altogether: as long as the wharf extended into the deeper waters of the harbor, cargo ships could dock there directly. By contrast, the canal option was a way of optimizing the existing system of barge transport. It also offered better protection from choppy waters and would allow more warehouses in the immediate vicinity of the port. Hiroi Isamu, a professor from Sapporo Agricultural College and a consulting engineer to Otaru, settled this second dispute in favor of the canal option, as records from this period attest:

> Doctor Hiroi Isamu is one of our country's experts on port development and the consulting engineer to Otaru District... {...} [Doctor Hiroi] gave his opinion that, given the types of cargo and cargo crating at the port of Otaru, for the present it would be more convenient to use barges for their transport, and adopt a canal system. In accordance with his opinion, the

district revised its plan by replacing the wharf with a canal, and submitted a third application for a permit in November 1909.

Otaru Shōkō Kaigisho, ed. (1928: 5).

This passage underscores the extent of Hiroi's influence, as well as the rationale for adopting the canal plan: "given the types of cargo and cargo crating at the port of Otaru, for the present it would be more convenient to use barges for their transport, and adopt a canal system."

Permission for the canal's construction was granted on March 27, 1914, and construction began in the August of the same year. However, a subsequent local election ushered in a new majority party (the Reform Club, or *Kakushin Kurabu*), which aligned with business interests opposed to the canal plan. The Otaru district assembly was again thrown into confusion. This third conflict over the canal died down following a change in leadership (the transfer of the ward's chief administrator); ongoing construction also made it increasingly difficult to reverse course. The canal was finally completed in September 1923. This turbulent early history of the canal explains why Otaru has often been referred to as "a city of political conflict." Indeed, the Reform Club's 1914 attempt to overturn the canal construction plan is sometimes referred to as the first of Otaru's "canal wars."

Designed to facilitate the smooth flow of people and goods through the port, the canal and surrounding warehouses functioned as an interface between the urban area of Otaru and the port and cargo transport zone. With the completion of the canal and the new warehouses that lined its banks, all the components that would shape postwar Otaru were in place: the busy shoreline; the warehouses where the cargo was loaded and unloaded; behind the warehouses, the *denuki kōji* service roads, along which cargo was dispatched from the warehouses for onward transport; the adjacent Ironai district, where the wholesalers were concentrated; and finally, the "Wall Street of the North:" a street lined with the various banks that supported economic activity at the port. This "skeleton" of postwar Otaru was already fully formed in the early 1900s. As we shall see in Chap. 4, however, this "skeleton" would later constrain the city's growth.

3.3.4 1922–1939: Reorganization of the Port

The canal was completed in 1923, a year after Otaru finally attained the status of municipality (Figs. 3.7 and 3.8). The new canal measured 1.3 km in length and 40 m in width and had a depth of 2.4 m. These dimensions allowed fully loaded barges to pull up to warehouses on both sides of the canal simultaneously, without obstructing the passage of other barges down the canal. According to Watanabe (1974: 312–214), the Otaru port quickly set new records for the number of ships (6,428 ships in 1924) and gross tonnage (11,769, 983 t in 1929). These records remain unbroken to this day. What this suggests is that, contrary to the claims of local travel guides, the canal was never the engine of Otaru's prosperity. Rather, the canal seems to have supported,

Fig. 3.7 The busy port of Otaru (Early Showa period, 1920s–1930s). *Source* From a set of sixteen postcards, *Yama ni Megumi Haretaru Shokō no Otarushi*, printed in the early Shōwa period (Author's personal collection)

Fig. 3.8 The Bustling Yanagawa Commercial District (Early Showa period). *Source* From a set of sixteen postcards, *Yama ni Megumi Haretaru Shokō no Otarushi*, printed in the early Shōwa period (Author's personal collection)

Fig. 3.9 The Otaru Canal in 1976. Geospatial Information Authority of Japan (GSI) aerial photo of "Otaru," August 1976 (original scale 1: 8,000)/CHO-76-3 C4B-21

and perhaps extended, the port's "golden age" between 1923–1930 (Figs. 3.9, 3.10 and 3.11).

Indeed, the canal is better described as the *product* of Otaru's prosperity. Otaru's good fortune can be attributed to the relative advantages conferred by national policies that designated Otaru as Sapporo's port and Otaru's importance to trade along the Sea of Japan (particularly with Karafuto). Otaru's prosperity, in turn, enabled renovations to the port—including the construction of the canal and a breakwater that remains in use a full century later.[4] Just as these renovations were completed, however, Otaru's fortunes began to change.

The opening of the Seikan Ferry service in 1924, which linked railway lines in Aomori and Hakodate and enabled freight car transport, was a harbinger of Otaru's changing fortunes (Toshima 1986: 65–67), as it signaled a major change to the

[4]Together, Otaru's designation as the outer port serving Sapporo and the "north breakwater" (begun in May 1897 and completed in 1908 during the first period of port construction) designed by Dr. Hiroi Isamu, a professor at Hokkaido Agricultural College and an engineer for the Hokkaido government, contributed to the growth of Otaru's port. The breakwater referred to here is the "south breakwater and island breakwater" designed by Itō Chōemon during the second phase of construction at the port (begun in 1908 and completed in 1921).

Fig. 3.10 An old advertisement still visible on a warehouse door. Photograph taken by the author (March 1984)

traditional method of cargo transport. Truck and train-based land transport were becoming more common, and sea transport increasingly involved the use of cranes to transfer cargo to and from large ships docked directly at a wharf or quay. By comparison, moving cargo with barges and human labor was more time-consuming and expensive. In an August 1923 address entitled "My Impressions of Otaru," the influential port engineer Niwa Sukihiko (1868–1955) noted:

> Indeed, at the port of Otaru all of the cargo is loaded and unloaded by barge. This method of cargo handling makes it an anomaly among today's commercial ports. {…} Transporting cargo by barge is risky, as the cargo can get wet or damaged. That is why the general trend today is for ships to unload cargo directly at a pier, or moor alongside the wharf for unloading.
>
> Otarushi (1926: 24–25).[5]

[5]"At the third general meeting of the Port Association in Hokkaido in August, we had the opportunity to listen to the excellent opinions and theories of a number of leading experts in the field regarding their inspections of Hokkaido ports {…} This report has been compiled…to facilitate their inspection of the Otaru port" (Otarushi 1926: Preface). This report was compiled by the city

Fig. 3.11 A Cutter and the Hokkai Seikan Warehouse. Photograph taken by the author (September 2006)

Niwa's point was clear: Otaru's approach to cargo handling was already an "anomaly" in a world that had converted to wharf-side cargo transfer. Indeed, the superior efficiency of unloading a ship's cargo by crane for immediate transfer to a railcar or truck waiting at the wharf was all too obvious.

Otaru hastened to catch up with the new trend in cargo handling. In 1928, work began on a second phase of land reclamation work at the port, which was completed

government to celebrate the prosperity of Otaru's port, but Niwa's comments regarding the port's "anomalous" character were an ominous portent of the future.

in 1932. Extending from Sakaimachi to the mouth of the Katsunai River, this land reclamation project was undertaken to enable ships to pull up directly along the quay to unload cargo; today it remains as the Sakaimachi Quay. In 1937, a new wharf was constructed at its northern end. This was an effort to optimize the cargo loading and unloading process and make Otaru more competitive with other commercial ports. Inevitably, these new initiatives diminished the canal's role within the Otaru port system. Moreover, the port's transition to wharf-based cargo handling transformed the canal into the relic of a former age: the canal's "antiquity" was further confirmed by the construction of the new wharf in 1937. Watanabe Teinosuke states as much in his history of the Otaru canal: "the canal lost its function with the end of barge-based cargo transport" (Watanabe 1979: 118). Originally the product of the port's prosperity, the canal was now left behind as authorities sought to rationalize the port's cargo handling system.

To be sure, Otaru's growth had been supported in large part by local capital, rather than investments from major companies in Tokyo or Sapporo, and coal exports from the Otaru port remained brisk. Thus Otaru's economy did not collapse overnight. Still, broader changes to port logistics and cargo transport cast a silent but ever-lengthening shadow over the city's future.

3.3.5 1940–1945: The Wartime Controlled Economy

With the onset of the second world war, all economic activity was placed under the direct control of the national government. Every city was affected by the implementation of this wartime planned economy, but the impact on Otaru was particularly grave. Sapporo's postwar ascendance was cemented when the national government assigned oversight of controlled commodities to Hokkaido's political center, Sapporo, rather than to the port city of Otaru. Additionally, in a heavy blow to Otaru's wholesalers, government-regulated unions were entrusted with trade in all controlled commodities. Deprived of their traditional stock, Otaru's wholesalers began dealing in non-controlled commodities; however, many firms were forced to suspend operations or close down altogether (Kankō Shigen Hozon Zaidan, ed. 1979: 21L). In the course of just a few years, the Otaru economy underwent an enormous and wrenching transformation as a result of national policies that forcibly relocated many of the city's primary functions to Sapporo. Otaru's decline was thus precipitated by a wartime economy that altered the structures by which Otaru had achieved its economic supremacy.

Otaru's loss was Sapporo's gain. Sapporo was already Hokkaido's administrative center, but the war transformed it into the prefecture's economic center as well. Akama's observation that "in the heart of this town (Otaru) there's this *deep*, deep-seated resentment of Sapporo" is thus a pithy account of Otaru's decline. The war deprived Otaru of its central role in Hokkaido's development. With no other outlet, the collective anger and nostalgic yearning for the past shared by Otaru's citizens developed into a silent but persistent "grudge" against their more fortunate neighbor.

3.3.6 1946–1966: A City in Decline

The war had destroyed local economies across Japan. Nationally, however, the dominant story became one of the transitions from "postwar reconstruction to rapid economic growth." Let us move beyond the national narrative to examine the processes unfolding in the city of Otaru. While Otaru's population continued to increase (although just barely), the rate of increase slowed year by year, and by 1965 the city's population began to decline (Table 3.2).

Meanwhile, 1960 was a major turning point in terms of the volume of cargo freight passing through the port of Otaru—a key indicator of the local economy (Table 3.3).

Up until 1960, the economic downturn had been offset by an increase in coal exports, leading to an export growth rate of 17.9 percent. In other words, "the momentum of the Otaru port continued to grow, up until the start of the high economic growth period" (Kanekura 1986: 93). Together, these two indicators suggest that the years between 1960 and 1965 marked the real transition point in the postwar economy of Otaru.

In its heyday, Otaru had produced many riches. At the height of the herring trade, "herring palaces" were built by families who had made their fortune in the industry, while Otaru's dominance in the bean trade created "bean barons." There were also "the Wall Street of the North" and bustling Ironai Ōdori, where the wholesalers were concentrated. The city boasted a thriving literary scene, which included the poet Ishikawa Takuboku and the novelist, poet, and translator Itō Sei. Indeed, Otaru's urban glamor was such that the architect and "modernologist" Kon Wajirō, who typically observed changes to cityscapes and people in modern life from the streets of Ginza, once made a special observation trip to Otaru (Horikawa 1998, 2001: 170). However, the wealth accumulated during Otaru's "golden age" was rapidly diminishing. From this period onward, the same descriptive phrase would invariably

Table 3.2 Population Growth in Otaru (1925–1975)

Year	Population	Rate of increase (%)
1925	134,469	–
1930	144,887	8
1935	153,587	6
1940	164,282	7
1950	178,330	8
1955	188,448	6
1960	198,511	5
1965	196,771	−0.9
1970	191,856	−2.5
1975	184,406	−3.9

Notes Compiled from annual census reports
Source Hoshino (1986: 6, Table 4)

Table 3.3 Changes to the types of cargo handled at the Otaru Port (1939–1984)

	Growth rate					Component ratio (%)				
	Export	Import	Export (within Japan)	Import (within Japan)	Total	Export	Import	Export (within Japan)	Import (within Japan)	Total
1939	100.0	100.0	100.0	100.0	100.0	11.6	1.6	67.4	19.6	100.0
1960	45.4	352.2	130.3	99.3	117.9	4.4	4.9	74.2	16.5	100.0
1970–74 (annual average)	21.6	1,195.5	48.6	163.8	86.7	2.9	22.4	37.6	37.1	100.0
1975–79 (annual average)	11.3	1,144.8	24.1	176.1	70.6	1.8	26.4	22.9	48.9	100.0
1980–84 (annual average)	11.1	849.3	16.5	191.0	63.6	2.0	21.7	17.5	58.8	100.0

Note Data for all but ferry cargo calculated from annual statistics reported in the 1985 *Otarukō Tōkei Nenpō*
Source Kanekura (1986: 92; Table 1)

precede any mention of Otaru: *shayō no machi* (a town in decline—or more literally, a "sunset town").

What were the reasons for Otaru's decline? We have already observed how Otaru's wholesale industry was largely destroyed by the wartime economy. We have also noted how outdated infrastructure had cost the port its competitive edge. In this period, several additional factors came into play. First, as Japan entered an era of rapid economic growth, its primary source of fuel changed from coal to oil. This shift dealt a crippling blow to the Otaru economy, which relied heavily on coal exports. From 1960 onward, the volume of freight passing through the port of Otaru declined steadily (see Table 3.3).

Second, the relative importance of the Otaru port diminished as it lost trading partners in Europe and Karafuto. This was related not only to changes in the type of freight entering and leaving the port, but also to changing distribution routes. In the time of the *kitamaebune*, the Sea of Japan route had been the main route for trade. Now the Pacific route dominated, prompting a rapid increase in ports along Hokkaido's Pacific coast. The opening of the Tomakomai port in 1962 sounded a death knell for the port of Otaru: "the wholesalers in the cereal and seafood trades who had long supported the economy of the commercial city of Otaru" were "ruined or forced into decline by this major change to distribution routes" (Kankō Shigen Hogo Zaidan, ed. 1979: 21R). Bank branches began to disappear from Otaru. Mitsui, long a central presence in Otaru, was dissolved during the postwar dissolution of the financial conglomerates known as *zaibatsu*. This led to an exodus of capital, as funds were merged, renamed, and funneled to Sapporo. In 1921, there were nineteen banks with branches in Otaru. In 1960, there were just thirteen.

A slump in the herring trade was the third factor in Otaru's decline. Encircled by mountains and the sea, Otaru lacked the space to develop a substantial industrial area, and thus relied heavily upon commerce and fishing. But changes to the herrings' seasonal migration patterns led to a series of poor catches, which eventually became chronic. Before long, herring had almost disappeared from the seas around Hokkaido.[6]

Otaru was thus left behind as postwar Japan hurtled toward rapid economic growth. Also left behind were the canal and stone warehouses that symbolized the city's lost prosperity. The survival of Otaru's historic landscape was hardly by design. The city's historic area had emerged from the war unscathed only to be excluded from municipal strategies for reviving Otaru's fortunes as a commercial port and through urban redevelopment. The city of Otaru could not afford to demolish these old buildings, so it simply abandoned them. In short, the canal and the stone warehouses were not intentionally preserved—they simply *remained*.

For the city government, meanwhile, the most pressing task was to revive the local economy through the redevelopment of the port, along the cornerstone of Otaru's prosperity. In order to compete with ports along Hokkaido's Pacific coast, the port

---

[6]Some forty years later, herring have returned to the seas near Otaru and are visible during spawning season (most herring lay their eggs near the beach). While herring are now caught every year, their volume does not compare to that during the peak years of Otaru's herring trade.

needed to meet the demands of truck-based cargo transport. In other words, the port needed more wharves to facilitate cargo loading and unloading. Additionally, the city sought to address the increase in automobile traffic passing through Otaru and a better transportation system to link Otaru with Sapporo and central Hokkaido. With the city's fortunes waning, local leaders set about to transform Otaru to meet the demands of a very different age.

The first goal involved outfitting the port to enable freight transport by truck and wharf-side cargo handling, through the construction of Wharf No. Two, Wharf No. Three, Ironai Wharf, and Katsunai Wharf, and the addition of transporters, pneumatic unloaders, and a container yard. City officials hoped that modernizing the port infrastructure would allow Otaru to compete with the Pacific ports and the planned Ishikari Bay New Port.[7] Ironically, the barge and canal cargo transport system developed after such fierce debate was now an impediment to Otaru's economic recovery. The Otaru canal was seen as a relic of a bygone era that should be erased by redevelopment. A second wharf was duly completed in 1950, and a third in 1967.

The second goal was positioned as means of improving the city's transportation system, but it was not entirely unrelated to the port: the new highway was part of the plan to create a unified transport system linking the port to the rest of the city and beyond. Soon, two specific road plans appeared to further these two goals: one for a Sapporo-Otaru bypass, and the other for the construction of a new harbor road, known as the Rinkōsen (Shinozaki 1989: 4–6).[8] Both plans originated in the 1961 "comprehensive plan for the [greater] Sapporo region," and were carried over within the 1962 "basic plan for industrial revival in Otaru city." The Otaru bypass was a means of linking Otaru with neighboring regions. The Rinkōsen, meanwhile, was designed to alleviate the pressures of increasing traffic volume in Otaru—and carry trucks and cargo to and from the port.

In July 1966, a committee led by Inoue Takashi (1917–2001), then a professor at the University of Tokyo and a former Ministry of Construction bureaucrat, drafted a blueprint for the Rinkōsen. The Minister of Construction approved this blueprint in August (Shinozaki, ed. 1989: 5; Ogasawara 1986). This signaled the launch of a massive new project to stimulate the local economy. While the Otaru Chamber of Commerce and Otaru's citizens may well have been interested in the potential economic effects of such stimulus measures, they remained entirely indifferent to the question of how such projects would alter the physical landscape of Otaru. Indeed, the fact that the Rinkōsen would be constructed on land reclaimed from the Otaru canal went unheeded by the city's residents.

---

[7]For more on the Ishihikari Bay New Port, see Otaru Kaihatsu Kensetsubu Otaru Kōwan Kensetsu Jimusho, ed. (1987), Ishihikarikō Shinkōshi Henshū Iinkai, ed. (1991), and Kikuchi (2012).

[8]The Ministry of Construction approved the project on August 25, 1966 (Bulletin number 2912). The official name of the new road was the Toshi Keikaku Highway No. 3.2.4 Rinkōsen (identifying the highway as one created under the Toshi Keikaku (City Planning) Law, but this book refers to the six-lane road simply as the "Rinkōsen.")

3.3.7 1967–1984: The Second "Canal War"

The years between 1960 and 1965 cemented Otaru's decline. From 1967 onward, a number of "food and textile trading companies and city banks withdrew" from the city, and "between 1960 and 1970, the number of wholesalers in Otaru dropped by 226" (Shinozaki, ed. 1989: 3). Otaru's clout as a major commercial port city was a thing of the past.

Naturally enough, the theme of economic stimulus pervaded the city. Both city officials and Otaru citizens understood the goal as that of *re*stimulation, or "revitalization." This word choice reflected local pride in Otaru's history at the center of the Hokkaido economy—and local resentment of the city's usurper, Sapporo. As we have seen, the specific content of Otaru's "revitalization" effort was the modernization of port facilities and the construction of the six-lane Rinkōsen. The construction of new wharves to facilitate cargo handling could make Otaru more competitive with other Hokkaido ports, but it would also render the canal redundant. The canal would thus be paved over to make way for a new road that could carry trucks to and from the Otaru port. From a city planning perspective, these were highly conventional measures for the time.

However, as the construction of the Rinkōsen proceeded in the direction of the canal's southern end, a few Otaru residents reacted with alarm. In 1973, the destruction of the gabled-roofed Arihoro warehouses prompted concerned local citizens to take action. In the same year, twenty residents launched the Association to Protect the Otaru Canal (APOC).[9] The APOC was founded by residents from all walks of life, united by the goal of "saving the irreplaceable Otaru canal." The APOC would later become the parent body for a more far-reaching organization, the Committee of One Hundred for the Otaru Canal, which incorporated a broad swathe of the local business community and drastically expanded support for the movement among ordinary Otaru residents. Public support for the cause eventually intensified the decade-long confrontation between preservation activists and the city government.

In the first phase of the movement, activists sought to keep the canal, which they saw as the "very spirit of our hometown," untouched—frozen in time. Over the course of their long stand-off with the local administration, however, some activists began to argue that the canal should be preserved as an essential resource for community development. These activists rejected the notion that their campaign was the nostalgia-driven and "frivolous" project of mere "dilettantes." Instead, they identified the canal and the stone warehouses of the canal district as an important resource for local tourism and argued that Otaru's recovery hinged upon remaking the city as a tourist destination. In a nutshell, the goal of the preservation movement changed over time, from "cryopreservation" to community development.

Initially, the preservation movement submitted multiple petitions to the municipal and prefectural governments. Their momentum dwindled, however, as Otaru city officials steadfastly refused to suspend a major public works project funded

---

[9]This section offers a brief overview of the APOC's campaign to save the Otaru canal. A more detailed account can be found in subsequent chapters.

by national subsidies. Moreover, as long as the preservation movement adopted the confrontational tone of an "anti-administration" opposition movement, it could not hope to garner support among more conservative residents. Activists thus developed a different rationale for preservation: tourism development. They argued that Otaru's historic canal and townscape should not be understood as old and inconvenient features of the city, but as important resources that, if properly developed, held the key to Otaru's rebirth. This new rationale for preservation dispelled the "anti-development, anti-authority" image of the preservation movement and suggested that preservation and development could actually go hand-in-hand.

This softer "tourism development" message enabled more conservative-minded members of Otaru's business community to support the preservation effort. Local citizens also warmed to this exceedingly pragmatic, non-ideological argument for preservation. In 1979, the city government responded to the growing force of the preservation campaign by offering a compromise solution: narrowing the southern portion of the canal to allow the Rinkōsen to run alongside—rather than over—the canal. While activists rejected the city's proposal, there was a general surprise that the city had been forced to offer any sort of compromise.

The rapid growth of the preservation movement introduced new and often opposing interests within the campaign. Ultimately, the preservation movement splintered and collapsed in the autumn of 1984. The city government moved ahead with the compromise plan it had offered in 1979: half of the canal was reclaimed, and the six-lane Rinkōsen was constructed alongside the narrowed portion of the canal. The canal that we see today is the result of this process (Fig. 3.12). Ironically, Otaru's canal gained national fame as a result of the long dispute over its fate and went on to become the centerpiece of local tourism: at its peak, Otaru welcomed more than nine million visitors in a single year (Table 3.4).

The second "canal war" period can thus be understood as one dominated by the question of whether Otaru's recovery should be achieved through "public works" (port redevelopment and road construction) or by shifting the city's key industry to tourism. This was the debate at the heart of the 1967–1984 "canal war."

3.3.8 1985–2001: Otaru as a Tourist Destination

The Rinkōsen road was completed in 1986. The subsequent period was defined by the success of local tourism. In the mid-1980s, Otaru, which had hovered perilously close to the bottom of a popular magazine ranking of Japanese cities, went from a declining port city to a booming tourist destination, known for its "retro townscape" and "romantic and nostalgic atmosphere." Meanwhile, national media reports on the long stand-off between preservation activists and the Otaru city government had transformed the canal into a famous landmark. As Table 3.4 demonstrates, Otaru's name recognition and popularity put the city on par with Sapporo as a tourist destination. Talk of Otaru was no longer prefaced with that old phrase, *shayō no machi*,

Fig. 3.12 A view of the Otaru Canal, Footpath, and Rinkōsen Road. Photograph taken by the author (September 2006)

which had dogged the city during its long decline. The sun was no longer setting on Otaru.

Otaru's rebirth as a tourist destination was not the result of a deliberate course of action undertaken by municipal authorities or local citizens. Rather, it was the unintended consequence of the protracted canal war. Following the break-up of the preservation campaign, the activists who had fought to save the canal retreated from public life and did not play an active role in Otaru's subsequent transformation into a tourist city. The owners of the souvenir shops that sprung up like weeds throughout the historic canal district had no connection to the earlier preservation movement. Meanwhile, the city government lacked any substantive tourism policy or local tourism office for some years.

This unregulated tourism was met with resistance and apprehension by many Otaru residents. The city government faced the question of how to achieve a locally led approach to managing tourism, while the destruction of historic buildings and the construction of high-rise apartment buildings adjacent to historic preservation districts also became pressing issues (Figs. 3.13, 3.14, 3.15 and 3.16). A proposal to invite the major shopping mall chain MyCal to Otaru prompted some former preservation activists to reemerge into public life in protest. Notwithstanding the failure of their campaign, their community development efforts had been aimed at creating "the tourist city of Otaru." The arrival of a national chain store was antithetical to the goal of citizen-led community development, rooted in a historic landscape unique to Otaru.

Table 3.4 The ebb and flow of Otaru Tourism (1962–2015) (Unit: 1,000)

Year	Visitors	Year	Visitors
1962	1,551	1989	3,845
1963	1,327	1990	4,362
1964	1,394	1991	4,933
1965	1,870	1992	5,374
1966	2,169	1993	5,086
1967	2,269	1994	5,860
1968	2,253	1995	5,624
1969	2,257	1996	5,511
1970	2,297	1997	6,064
1971	2,339	1998	6,656
1972	2,381	1999	9,729
1973	2,411	2000	8,593
1974	2,442	2001	8,933
1975	2,342	2002	8,476
1976	2,383	2003	8,002
1977	2,377	2004	7,540
1978	2,483	2005	7,560
1979	2,453	2006	7,696
1980	2,438	2007	7,405
1981	2,437	2008	7,144
1982	2,662	2009	6,871
1983	2,562	2010	6,677
1984	3,392	2011	6,036
1985	2,724	2012	6,599
1986	2,733	2013	7,107
1987	2,945	2014	7,447
1988	3,400	2015	7,949

Source Compiled by the author using data from the annual reports of the Otaru City Tourism Division ("Kankō Irikomi Kyakusū no Suii")

Yet the former preservation activists who had previously emphasized "community development rooted in the historic environment" now had to develop a broader-ranging vision that incorporated plans for the city's commercial districts and addressed transportation concerns. While some erstwhile activists launched a new opposition movement and others chose to address their grievances through legal channels, several key members of the canal preservation campaign developed an entirely new approach by building new relationships with local officials and using their influence from positions within the city government, the city council, and third

Fig. 3.13 A historic building and modern apartment building. Photograph taken by the author (September 2006)

sector organizations to make policy proposals. This period can thus be characterized by the renewed involvement of certain preservation activists in local affairs and the transformation of "a protest movement launched in the outfield by twenty-somethings" into "a movement linking the infield and outfield, launched by mature fifty-somethings."

The Otaru city government responded with new measures for tourism and community development and revised its port policies, based on the understanding that development should remain rooted in Otaru's history. Whether explicit or implicit, there

Fig. 3.14 A tourist street in Sakaimachi. Photograph taken by the author (September 2008)

was a growing acceptance of the fact that Otaru's "selling point" was not the capacity of its port, but rather the historic townscape. Otaru's distinctive scenery came to be understood as a collective good and a source of income for local residents. In some cases, local residents voluntarily built new homes in the architectural style of "old Otaru." And while the exterior of historic buildings continued to be used to attract tourism, the interiors were often renovated and put to new uses. From the 1990s onward, this has been the basic design trend, as the survey data presented later in this volume demonstrates. These policies and design trends reflect how local citizens and the city government have come to make sense of the protracted "canal war."

Fig. 3.15 A scene from the aging city of Otaru. Photograph taken by the author (September 2009)

At its peak, Otaru welcomed 9,729,000 tourists in a single year—an extraordinary number for a regional city with a population of less than 140,000. These numbers forced [the Otaru city government and residents alike] to recognize that, despite Otaru's longstanding pride in its status as a "port city," the city's economic wellbeing now depended on tourism, not the port. The mass media and researchers also saw Otaru as a shining example of tourism-based development. The long "canal war" ended up allowing the "declining port city" of Otaru to rise up from the ashes and

Fig. 3.16 The mixed blessing of tourism. Photograph taken by the author (September 2009)

transform itself, phoenix-like, into a "city of tourism." This was the common narrative of Otaru's post-1960s history.[10]

[10]Consequently, the wounds of the "canal war" and the heated and often slanderous exchanges that characterized the final stages of the preservation campaign have been consigned to oblivion. Also forgotten are the questions that dominated that final stage of the preservation effort: How does one live in a city? Why preserve? Why construct a new six-lane road? How can public works projects be reassessed? The canal war is only alluded to in passing (if at all) as a "necessary evil" preceding Otaru's grand rebirth. The real story of the canal war has been erased from the narrative of Otaru's

3.3.9 2002–Present: The Beginning of the End

The significant decrease in tourists—the city's lifeline—signified yet another change of era. Tourism peaked in 1999 when more than nine million tourists visited Otaru. Considering Otaru welcomed a million tourists between 1962 and 1965, and just two million tourists between 1966 and 1983, this was a truly remarkable result for a single year. In 2000, however, the number of tourists dipped to 8,593,000. Tourist numbers briefly increased to 8,930,000 in 2001, but have steadily declined since 2002. The Great East Japan Earthquake of 2011 prompted a precipitous drop in tourism. While tourist numbers recovered somewhat in 2012, the majority of tourists to Otaru are now from overseas. The Japanese have lost interest in Otaru as a tourist destination.

Meanwhile, local myths have been destroyed. The local belief that "sushi restaurants never go out of business" has been upended, as bankruptcy has forced a number of local sushi restaurants to close their doors. Sushi has been replaced by confectionery: Otaru is now a "city of sweets," not the "city of sushi." LeTAO (a local confectionery chain that claims to represent "nostalgic modern" Otaru, although it lacks any historic connection to the city) has become Otaru's most famous tourist brand—supplanting Kitaichi Glass, a local glassmaker with deep roots in Otaru. This is a stark refutation of the history-based community development efforts that focused on the unique and historic aspects of the local region. Taken together, these changes signal "the beginning of the end" of Otaru tourism—and also the city's incipient depopulation (Horikawa 2012).

3.4 The Beginning of the End—And a Second Chapter

The dying city of Otaru sought its rebirth through a road construction plan. Counter to expectations, it ended up becoming a popular tourist destination. The city must now grapple with some important questions. Will Otaru seek to reconfigure its local tourism industry on its own terms or shift to other development strategies—attracting major chains to the city, for example, or constructing more high-rise buildings? And what importance will today's Otaru decide to place on its historic landscape? The struggle to resist the "beginning of the end" and once again remake the city of Otaru has begun a second chapter.[11]

Now it is time to ask how the first chapter of this story was experienced by Otaru's citizens. How did the city's residents actually live through the history described in

renaissance; its lessons have not been put to good use. This volume is an attempt to extract the lessons consigned to oblivion.

[11] The "first chapter" was, of course, the confrontation over the fate of the Otaru canal. In that sense, the current challenge facing Otaru represents a "second chapter." The preservation movement itself has also used the "chapter" analogy in their narrative of Otaru's community development. An August 20, 1996 symposium on Otaru future was titled: "Towards Tomorrow: The Second Chapter in Otaru's Community Development Activism." See Otaru Shinpojiumu Jikkō Iinkai, ed. (2009: 49).

these pages (Tamano 1997)? It was not a mere sentiment that compelled the people of Otaru to support preservation (or development); a range of structural and subjective factors informed their actions. In Chap. 4, we shall examine the logic of the actors (primarily within the Otaru city government) who set out to reverse the fortunes of this "city with a grudge" through a new road project. In Chap. 5, we turn to the philosophy and lived experiences of the men and women who fought to save the canal and Otaru's historic townscape. Having briefly reviewed the city's history, it is time to turn our attention to the *lived* Otaru.

References

Endō, A., & Moriyama, G. (1979). Otaru Unga no Keisei to Sono Haikei. In Otaru Unga Kenkyū Kōza Jikko Iinkai (Ed.), *Unga no Rekishi Kara Machizukuri Made* (Otaru Unga Kenkyū Kōza Zenjūkai Kiroku) (pp. 23–33). Self-published (125 pp.).

Fukase, K. (2003). Otaru. In Y. Maruyama, H. Kokaze, & N. Nakamura (Eds.), *Nihon Kōtsushi Jiten* (pp. 157–158). Yoshikawa Kōbunkan.

Horikawa, S. (1998). Toshi Kūkan to Seikatsusha no Manazashi. In A. Ishikawa, K. Satō, & K. Yamada (Eds.), *Mienai Mono o Miru Chikara: Shakai Chōsa to Iu Ninshiki* (pp. 133–149). Yachiyo Shuppan.

Horikawa, S. (2001). Keikan to Nashonaru Torasuto: Keikan wa Shoyū Dekiru ka. In H. Torigoe (Ed.), *Shizen Kankyō to Kankyō Bunka* (Kōza Kankyō Shakaigaku 3) (pp. 159–189). Yūhikaku.

Horikawa, S. (2005). Toshi Seikatsu to Seikatsu Kankyō Hendō: Rōkaruna Kūkan Seigyo Shisutemu, Saikō. In H. Fujita & M. Urano (Eds.), *Toshi Shakai to Risuku: Yutakana Seikatsu o Motomete* (Shirīzu Shakaigaku no Akuchuariti: Hihan to Sōzō 8) (pp. 173–204). Tōshindō.

Horikawa, S. (2012). Kankyō Shakaigaku ni totte 'Higai' to wa nani ka: Posuto 3.11 no Kankyō Shakaigaku o Kangaeru Tame no Ichi Sozai Toshite. *Kankyō Shakaigaku Kenkyū, 18,* 5–26.

Hoshino, M. (1986). Otaru (Hokkaido) to Tsuruga (Fukuiken): Jinkō kara mita Sobyō. *Toshi Mondai, 77–5,* 3–15.

Inayoshi, A. (2014). *Kaikō no Seijishi: Meiji Kara Sengo e.* Nagoya: University of Nagoya Press.

Ishikariwanshinkō Henshū Iinkai (Ed.). (1991). *Ishikariwanshinkōshi.* Sapporo: Hokkaido Kaihatsu Kyōkai.

Kanekura, T. (1986). Kōwan Keizai no Henka to Kadai. *Toshi Mondai, 77–5,* 88–106.

Kankō Shigen Hogo Zaidan (Ed.). (1979). *Otaru Unga to Sekizō Sōkogun* (Kankō Shigen Chōsa Hōkoku 7). Kankō Shigen Hogo Zaidan.

Kikuchi, T. (2012). Chiiki Kankyō, Chiiki Shigen no Katsuyō o Jushi Suru Chiiki Kaihatsu Seisaku no Tenkai: Hokkaido Ishikariwanshinkō Chiiki o Jirei Toshite. *Keizai Chirigaku Nenpō, 58,* 299–308.

Nakanishi, S. (2013). *Kitamaebune no Kindaishi: Umi no Gōshōtachi ga Nokoshita Mono* (Kōtsu Bukkusu 219). Kōeki Zaidan Hōjin Kōtsu Kenkyū Kyōkai.

Niikawa, T. (1986). Toshi Gyōsei no Dōkō to Kadai: Toshi Keikaku o Megutte. *Toshi Mondai, 77(5),* 16–37.

Ogasawara, M. (1986). *Otaru Unga Sensō Shimatsu.* Asahi Shinbunsha.

Okamoto, S., & The Nippon no Minatomachi Kenkyūkai. (2008). *Minatomachi no Kindai: Moji, Otaru, Yokohama, Hakodate o Yomu.* Kyoto: Gakugei Shuppansha.

Otaru Kuyakusho. (1911). *Tōgū Gyōkei Kinen Otaruku Shashinchō.* Otaru Kuyakusho.

Otarushi. (1926). *Otarushi Toshi Keikaku Sankō Shiryō* (B5, 12 pp.). Otaru Shiyakusho.

Otaru Kaihatsu Kensetsubu Otaru Kōwan Kensetsu Jimusho (Ed.). (1987). *Ishikariwanshinkō Kensetsu no Ayumi: Dai 1-sen Nyūkō Made.* Hokkaido Kaihatsukyoku Otaru Kaihatsu Kensetsubu Otaru Kōwan Kensetsu Jimusho.

Otarushi. (1958). *Otarushishi Dai Ikkan*. Otarushi.
Otarushi. (1963). *Otarushishi Dai Nikan*. Otarushi.
Otaru Shōkō Kaigisho (Ed.). (1928). *Otarukō ni Kan Suru Chōsasho*. Otaru Shōkō Kaigisho.
Otaru Shinpojiumu Jikkō Iinkai (Ed.). (2009). *Otaru Unga to Sekizō Sōkogun no Hozon Undō Kara Nani o Uketsugu ka: Chiiki ni Iki, Chiiki o Mamoru...Machizukuri Undō no Senkusha Mineyama Fumi-shi ga Tsutaeru Koto Shinpojiumu Kaisai Hōkoku* (A4, 76 pp.). Sapporo: Nihon Kenchiku Gakkai Hokkaido Shibu.
Shinozaki, T. (Ed.). (1989). *Chihō Toshi Saikaihatsu Seisaku no Sōgō Kagakuteki Kenkyū: Otarushi Unga Chiku Saikaihatsu o Megutte* (Shōwa 63-nendo Kagaku Kenkyūhi Hojokin [Ippan Kenkyū A60410012] Kenkyū Seika Hōkokusho). Otaru: Otaru Shōka Daigaku Shinozaki Tsuneo Kenkyūshitsu.
Tachibanaki, T. (2012). *Sanshōdai Tokyo, Osaka, Kobe: Nihon no Bijinesu Kyōiku no Genryū*. Iwanami Shoten.
Takabatake, G. (1899). *Otarukōshi, Otaru Iwauchikan Kyūgunshi*. Ishikarikoku, Sorachigun: Takabatake (self-published).
Tamano, K. (1997). Toshi Shakai Undō to Ikirareta Kūkan: Seikatsushi Chōsa no Chiken Kara, *Ryūtsū Keizai Daigaku Shakaigakubu Ronsō*, 7-2, 71–107.
Watanabe, T. (1974). *Otaru Bunkashi: Shiseishikō 50 Shūnen Kinen*. Otarushi (republished in 1975).
Watanabe, T. (1979). *Otaku Unga Shi*. Otarushi.

Chapter 4
The Case for Change: The "Canal Issue" from the Perspective of the Otaru City Government

Abstract This chapter focuses upon the Otaru city government, which played a central role in the canal preservation controversy, and analyzes the development rationale of administrative authorities—the "agents of change" in Otaru. Specifically, I use interviews with the former mayor and other officials, the minutes of city council proceedings, and other public records to identify how local authorities understood the significance of the canal and the canal controversy. In short, local authorities saw the canal as a functional "space." As such, it made sense for the canal to be reclaimed as part of the broader shift away from the port's anachronistic canal freight system. Moreover, city officials' decision-making was held captive by the shared assumption that any change to their plans to pave over the canal—a public works project funded by the national and prefectural governments—would jeopardize their city's eligibility for future government subsidies.

Keywords City government · City planning · Change · Continual discontinuity

4.1 The Forces of Change

Laid low by postwar revolutions in distribution and energy, by the late 1950s, the once prosperous port city of Otaru was in decline. City officials pinned their hopes for a comeback on a regional development plan that would link Otaru more closely to the prefectural capital, Sapporo. Specifically, they identified the overhaul of port facilities and the construction of a new highway between the two cities as potential motors of economic stimulus. These massive and costly public works projects were premised on the belief that public investments in construction-related industries would ripple through the local economy, benefiting everyone. This assumption was hardly unique to Otaru: the regional stimulus policies that sustained Japan's period of high economic growth from the mid-1950s to the early 1970s stemmed from a common faith in "the ripple effect of public works projects."

Who were the chief actors behind the effort to transform Otaru through a regional stimulus policy, and what was their specific "logic" for change? Who led the effort to reclaim and pave over the Otaru canal? Conversely, who opposed these efforts

S. Horikawa, *Why Place Matters*,
https://doi.org/10.1007/978-3-030-71600-4_4

and fought to save Otaru's historic townscape? What was their "logic" of preservation? The first two questions are the subject of this chapter, while the latter two are addressed in Chap. 5. Viewed from either perspective, however, the origin of the protracted "Otaru canal controversy" remains the same: the fight over the fate of the canal began with the city's plan to construct the Rinkōsen, a new six-lane road that would run through the port district of Otaru.

Before the need to preserve and reuse historic environments was widely recognized in Japan, change was generally viewed positively. New things were believed to be best—the most sophisticated and efficient. In the early 1960s, for example, references to the "release of a new model" carried far greater significance than one might imagine: the very phrase signified change, and inspired excitement and anticipation. The 1960s was a time when the Japanese greeted change, progress, and "newness" itself with a simple—and perhaps naïve—sense of expectation. Allusions to the "1964 Tokyo Olympics" resonated with the Japanese in a way that references to the "2020 Tokyo Olympics" have not, for example. It was within this historical context that the Otaru city government and city council members, with the support of local industrial associations and interest groups,[1] became the principal actors seeking to transform the canal and port district. To be sure, many local citizens also supported these changes, but the central role of the Otaru city government makes it the primary focus of this chapter.

4.2 The Canal as Road Construction Site: The First Phase of the Rinkōsen Plan

There were two chief objectives to the reconstruction plan drafted by the Otaru city government: restoring economic activity at the port through the development and re-outfitting of the entire port district, and developing the local industrial base through the construction of a new highway (Niitani, ed. 1987: 66–67). At the port, city bureaucrats targeted aging infrastructure: the time and labor-intensive transport of cargo by barge needed to be replaced by the far more efficient method of quayside cargo handling and truck-based distribution. They also recognized the need to connect the port to inland consumers: a new highway could link the port of Otaru directly to the lucrative hinterland around Sapporo, Hokkaido's flourishing capital city. The mutually reinforcing nature of these two goals made the construction of the six-lane Rinkōsen, which would run through the port and canal district, a natural proposition. Discontinuing cargo transport by barge would render the Otaru canal useless—and the canal occupied the ideal site for the new road (Table 4.1).

[1] For example, representatives from the warehouse and distribution industries formed the "Association to Promote the Construction of the Rinkōsen." Similar neighborhood-level associations were established by leaders of each neighborhood association.

Table 4.1 An overview of the Rinkōsen Road Plan from formulation to completion

Date	Event	Developments at the city and prefectural level		Notes
		Otaru City	Hokkaido prefecture	
May 19, 1966	Otaru submits a plan for road changes and additions to the Hokkkaido city planning council			Ministry of Construction, Hokkaido branch, Notification #124
July 25, 1966	The Ministry of Construction (Hokkaido branch) Notification #124 is included as agenda item #5 to the council			Signed by [Hokkaido Governor] Machimura Kingo, chairman of the Hokkaido regional city planning council
August 25, 1966	The existing road route is abolished and new measures, including those regarding road breadth, are again adopted.		Notification of change to plan (Ministry of Construction Bulletin 2912)	
July 7, 1972	Road number change		Notification of change to plan	
July 20, 1972		17th meeting of the city planning council (Q&A)		Meeting duration: 15:06–17:00
August 18, 1972	Revisions widen a portion of the new road (through the addition of the Chūō Futōsen)		Notification of change to plan	
October 22, 1979		43rd meeting of the city planning council (first report)		10:00–12:30. This meeting could not be completed in one session, and extended to a second day.
November 13, 1979		43rd meeting of the city planning council (second report)		13:30–18:30

(continued)

Table 4.1 (continued)

Date	Event	Developments at the city and prefectural level		Notes
		Otaru City	Hokkaido prefecture	
December 4, 1979		Original plan submitted to prefecture		
July 30, 1980			Hearing with the Otaru mayor	
August 5, 1980		47th meeting of the city planning council (Q&A 1)		13:30–17:10. Meeting held over two days.
August 7, 1980		48th meeting of the city planning council (Q&A 2)		11:00–13:00
September 20, 1980	Revisions widen a portion of the road (canal section)		Notification of change to plan (Hokkaido Bulletin 2361)	
August 10, 1981	Revisions widen a portion of the road (near Inakita crossing)		Notification of change to plan	
May 15, 1986	Boundaries revised (to create a new park)		Notification of change to plan	
May 8, 1986				Certain portions of the Rinkōsen open
October 1988				Construction of the Rinkōsen is completed

Table created by the author using the following sources:

1. Official documents released by the Otaru city government in response to a request from the author on August 24, 2010.

2. Official documents released by the Hokkaido prefectural government in response to a request from the author on September 3, 2010.

3. Official documents appended to Notification #78 from the city planning section of the Otaru city government ("3.2.4-gō Rinkōsen ni Kakaru Toshi Keikaku Shingikai tō Kaisai Keii ni Tsuite") on September 6, 2010.

The first iteration of this plan was made public in September 1961, "Comprehensive Plan for the Sapporo Region," developed by the Hokkaido prefectural government. Further details emerged in 1962, "Basic Vision for the Revival of Industry in the City of Otaru," which included the construction of industrial and residential complexes between Sapporo and Otaru and a central expressway linking the two cities (Shinozaki 1989: 5) This highway was variously referred to as the "Sapporo-Otaru Bypass" and the "Otaru Bypass" in the 1964 Basic Plan for the Construction of New Industrial Cities in Central Hokkaido. In 1966, the new highway plan was formally adopted by [Otaru] city planners and referred to as the Sapporo-Otaru Highway and the Otaru Rinkōsen (literally, the "Otaru harbor road"). The former referred to the highway that would link Otaru to Sapporo and other cities in Hokkaido's interior, while the latter referred to the portion of the new road that would run through the city of Otaru itself, before connecting to the intercity highway.

In 1965 and 1966, the "Otaru bypass" was discussed at the Otaru City Council and the city's ad hoc budgetary committee. The issue had not yet been politicized: minutes of council proceedings reveal only brief references to the proposed road, with various speakers noting that the canal area would be reclaimed for its construction, that the city would seek funding for the project from the prefectural and national governments, and that the proposal could not be publicized because it was still under review by the city planning consultative council (Otaru Shigikai 1966: 108–109).

The following excerpt from a July 1966 meeting of the budgetary committee illustrates how council members understood the significance of the new highway plan to the city of Otaru:

> Director of Civil Engineering: {…} At present, the majority of traffic generated by activity at the Otaru port flows between Otaru and Sapporo on the national highway. {…} In a single day, approximately 36,000 vehicles travel between the [train] station and the Sangyō Kaikan (Industry Building). Approximately 24,000 vehicles pass through the vicinity of the Inakita police station. Approximately 33,000 vehicles travel from the Sangyō Kaikan to the eastern areas. These roads are operating at full capacity, and our forecasts for traffic volume in 1985 vastly exceed [current capacity]. Traffic in Otaru will be paralyzed unless a new main road is constructed to replace these routes. Industrial activity will also be paralyzed. We have no choice but to situate a "replacement road" along the shoreline. {…} The city will not have to pay a single yen of the construction cost. Two-thirds of the cost will be funded by the central government, and the remaining one-third will be funded by … the Hokkaido regional government."
>
> Excerpted from a July 8, 1966 reply recorded in the minutes of the second ordinary session of the special budgetary committee, 1966 min of the Otaru Special Budget Committee, Registration no. 134. Information in parentheses has been added by the author, information in brackets by the translator.

In short, future traffic volume was projected to outstrip the capacity of National Highway 5, the single main line passing through Otaru. The "replacement road" necessitated by these predictions would be the new Rinkōsen, and "situat[ing] this 'replacement road' along the shoreline" necessitated the reclamation of the Otaru canal, which ran parallel to the Otaru waterfront. It is clear that the city government was chiefly concerned with resolving the severe traffic congestion in Otaru, and addressing the future increase in traffic volume. Former Otaru mayor Shimura Kazuo later gave a concise summary of the city's chief concerns:

> Growth has been spontaneous [and uncontrolled] in Otaru, and looking at traffic conditions, we needed another main road along the coast, in addition to National Highway 5. [The

> idea was to] build one more highway and then have two main roads {...}, thereby creating a ladder-style traffic system. The new highway needed to perform three functions. First, it had to alleviate traffic congestion in the city. {...} It also needed to enhance transport functions between the port and the interior regions. And further...it had to handle a portion of pass-through traffic.
>
> Interview with former mayor Shimura Kazuo in Otaru, September 3, 1998. Information in parentheses has been added by the author, information in brackets by the translator.

In short, if the proposed Rinkōsen fulfilled the three functions noted by Shimura, it would avert the impending "paralysis" of Otaru's traffic and industrial activity without burdening the city with "a single yen" of the construction cost. Resolving a pressing local issue entirely with national and prefectural subsidies was enormously significant—and an opportunity the financially strapped city of Otaru could not afford to pass up. City officials thus entrusted the drafting of the Rinkōsen proposal to the University of Tokyo professor Inoue Takashi, a former Ministry of Land, Infrastructure, and Transport bureaucrat. For Otaru, an economically weak local government entirely dependent on national subsidies, Inoue's connections to the central ministries were of vital importance. City planners readily adopted his blueprint, which was referred to as the "Inoue Plan."

The citizens of Otaru were initially uninterested in the specifics of the Inoue Plan. In fact, "for several years after the plan's adoption, there was no opposition from citizens, and the plan moved forward with little difficulty" (Niitani, ed. 1987: 71) (See Figs. 4.1, 4.2 and 4.3). Yet it is worth asking why city planners felt they had "no choice but to situate the replacement road along the shoreline," and over the canal itself.

The initial blueprint for the Rinkōsen was based on the projections of future traffic volume made by the Otaru City Committee for Formulating Integrated City Planning Policy. This committee estimated that by 1985, 90,000 vehicles would traverse city roads on a daily basis. The construction of a new six-lane road was thus considered essential.[2] According to Inoue, who also chaired this committee: "Taking future considerations into account, my draft is premised on the need to secure enough space" (Inoue 1981: 44). Otaru is a hilly city. With flat land in short supply, Inoue identified the canal as the ideal location for the new road. A city pamphlet acknowledged the canal had once "played a leading role in the loading and unloading of cargo by barge," but went on to explain: "Another main thoroughfare is needed in addition to the [existing] national highway. After considering how to position [this new road], and given the decline in barge cargo transport, [the city established a] plan in which a portion of the canal would be used to create the new Rinkōsen" (Otarushi Dobokubu 1986: 2).

A 1987 report assessing predictions made for transportation planning purposes, compiled by a research group headed by Niitani Yōji, who had also helped draft the city's original plan, concluded that the "route passing through the area of warehouses and canal in the port district" had been deemed "the most feasible and appropriate

[2]The committee also proposed building an elevated road above the six-lane Rinkōsen, but this proposal was not incorporated into the final plan.

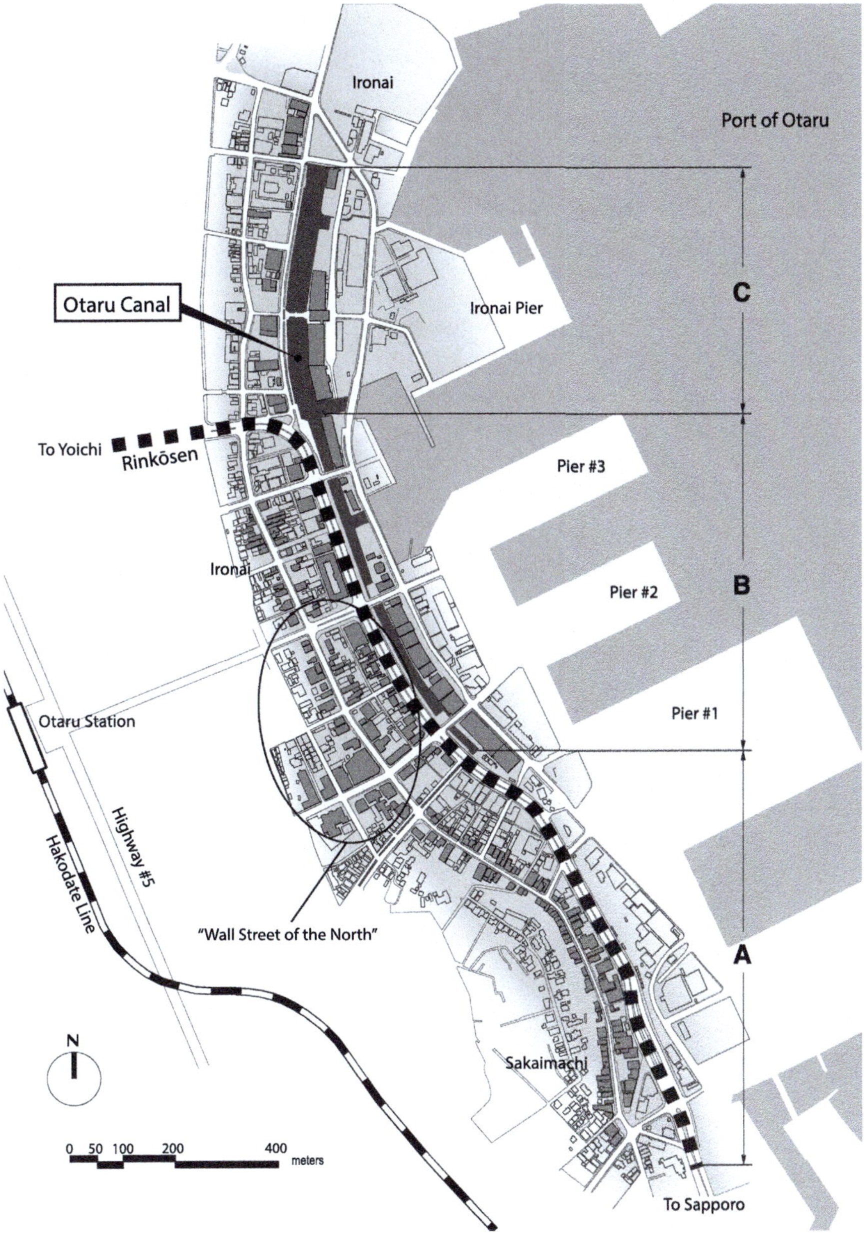

Fig. 4.1 The Canal District and the Rinkōsen. **A**: Area completed by 1973 **B**: Area contested by preservation activists between 1973 and 1984 **C**: Area remaining at the original width (Kitahama) (Author's illustration)

Fig. 4.2 An arial view of the Canal District (May 1975). *Source* Ogasawara (1986: 10)

route," given "the level of activity at the warehouses and canal at the time," and because of the "typical attitude toward historic buildings at the time" (Niitani, ed. 1987: 71). This was an indirect way of explaining that the canal had outlived its purpose (Figs. 4.4, 4.5 and 4.6).

Desperate to revitalize the stagnant local economy, the Otaru city government identified the obsolete canal as an area of considerable size on which to construct the new road. A city document that reviews this period further illuminates the administration's rationale:

Fig. 4.3 The Desolate Otaru canal (Autumn 1975). *Source* Slide provided by Jun Horikawa, then a resident of Sapporo

Fig. 4.4 A Birdseye view of the canal from the roof of the Hokkai Seikan building (Autumn 1975). *Source* Slide provided by Jun Horikawa, then a resident of Sapporo

Fig. 4.5 A rotted boat slowly sinks into the canal (Autumn 1975). *Source* Slide provided by Jun Horikawa, then a resident of Sapporo

Fig. 4.6 Otaru residents fishing from Barges on the canal (Autumn 1975). *Source* Slide provided by Jun Horikawa, then a resident of Sapporo

> The canal area was the only possible location close to both the city center and the port that could support changes to the way freight is moved to and from the port (namely, the increase in truck transport). Up until now, the canal itself had functioned as a method of transportation. *Its new mission was to become a road.* Given the absence of obstructing buildings, *construction could also be completed easily.* This was the city administration's way of thinking.
>
> Shisei no Ayumi Henshū Iinkai (1988: 200–201). Information in parentheses and emphasis have been added by the author.

"Up until now, the canal itself had functioned as a method of transportation. *Its new mission was to become a road.*" The city administration saw the canal as nothing more than a means of transportation. Moreover, the reference to the lack of "obstructing buildings" underscores another compelling justification for the siting of the new road. The canal was a public waterway that already belonged to the city. There were no buildings to be relocated or demolished, and no landowners to be compensated. This was a crucial factor in the city's assessment, as it would be for any public works project. The city's decision was thus an exceedingly rational and unsurprising one, and entirely in keeping with the conventional approach to public works projects at the time.

Meanwhile, the reference to the "easy" construction process underscores the salience of the city's budgetary concerns. A major point in question was whether the massive road construction project would be funded by the city itself or through subsidies from the national and prefectural governments. When they were later confronted with calls for the canal's preservation, city leaders ruled out the possibility of changing or suspending the project, explaining this "would require the city to repay government subsidies received for construction work to date. Re-routing [the new highway] would make it difficult to acquire budgetary allocation for work already completed." Moreover, "reversing a previously adopted plan would discredit the Otaru city administration and impact future operations" (Shisei no Ayumi Henshū Iinkai 1988: 214). In the subsequent conflict over the fate of the canal, city officials maintained that once the highway project had been greenlighted for funding by higher levels of government, it was too late for the city to reverse the decision on its own. Moreover, any reversal or change would negatively impact the city's ability to secure future subsidies.

It is difficult to overstate the importance of the latter concern: city officials were guided by their assumption that any change to the construction plan would negatively impact their ability to secure future subsidies. It is worth remembering that while city employees were also Otaru residents, their response to the canal issue was shaped and constrained by this overriding assumption. The following excerpt from a meeting of the city council's construction committee illustrates the centrality of this shared belief:

> I doubt the central government would be quick to allocate a budget if the route [of the Rinkōsen] were to change. I worry the central and prefectural governments might even withdraw [from the project].

City Councilman Takahashi, speaking before the Standing Committee on Construction, June 5, 1978. Record set: March 1975-April 1979 (2), Registration no. 403. Information in brackets added by the translator.

Takahashi's brief statement conveys the collective fear that subsidies would be denied, and that the withdrawal of the national and Hokkaido prefectural governments could derail the road project entirely.

The administration's position can thus be summarized in four points: (1) The canal district had outlived its original purpose; (2) As a public waterway, the canal would be relatively easy to redevelop into a major road; (3) The city of Otaru would not pay for the new road, as construction would be funded by the central and prefectural governments; and (4) Any change to the construction plan would negatively impact Otaru's eligibility for future subsidies. According to the Otaru city administration, the principal agent of change, the canal district was an "easy" construction site for the new road, and any deviation from the construction plan was unacceptable.

4.3 From Highway to "Canal Park"

Armed with a blueprint drafted by a prominent and well-connected engineering professor, the promise of funding, and the "ideal" construction site, city officials must have expected the construction of the Rinkōsen to proceed with little difficulty. Their expectations were confounded by subsequent developments. Between 1971 and 1973, warehouses in the Arihoro district of Otaru were demolished to clear the way for the future path of the Rinkōsen. The destruction of the Arihoro warehouses inspired concerned local citizens to establish the Association to Protect the Otaru Canal (Otaru Unga o Mamoru Kai, hereafter the APOC). In the face of mounting opposition, the city was gradually forced to change its approach. Meanwhile, successive "oil shocks" slowed work on the Rinkōsen project to a standstill. In 1978, the city commissioned Iida Katsuyuki, an associate professor at Hokkaido University, to revise the Rinkōsen plan. The revised blueprint (known as the Iida Plan) was made public in 1979. Iida's revisions to the original plan were modest: most of the canal would still be reclaimed, but Iida proposed retaining a narrow portion of the canal alongside the new road. Significantly, however, the new blueprint reflected changes to the administration's understanding of the canal district. For the first time, the city's plan incorporated the language of the preservationists, with references to Otaru's history and cultural heritage. Before turning to the specific contents of the Iida Plan, let us trace how this evolution came about.

On February 15, 1974, a petition from the APOC was put before the city council's construction committee. APOC chairman Koshizaki Sōichi first explained the aim of Petition 161. A question and answer session followed, during which officials again explained the outline and rationale for the Rinkōsen construction project. The city's civil engineering director emphasized that without the Rinkōsen, "we will be unable to cope with Otaru's traffic demands in 1985" (Standing Committee on

Construction Record set: 1971–1974, Registration no. 157, February 15, 1974). This was not only because of the need to deal with pass-through traffic, but also because the new highway had to link to existing roads "in a dense and coordinated" fashion. The following excerpt from the official minutes of the meeting illustrates the city's rejoinders to the various alternative proposals put forth by the APOC:

> Civil Engineering Director: "Next I'd like to offer our opinion on the three alternate proposals submitted by the Association to Protect the Otaru Canal.[3] {...} Regarding the first proposal for an underground route, the diagram passed out by [Deputy Mayor] Shimura is at a scale of 10,000:1, and is thus a rough sketch that makes detailed conjecture impossible. But based on this, [it seems to call for] a six-lane tunnel with a width of between 25 and 30 meters, and a length of 1.4 kilometers. Simply considered in terms of relocating [existing buildings], it would be difficult to widen National Highway 5 into an eight-lane [road, as national regulations would demand]. Furthermore, the approach from both Hon-dori or the national highway terminus would (cut off) traffic from two or three municipal roads.[4] {...}
>
> The second proposal, for an elevated port highway, would require the relocation of warehouses and other buildings around the wharves (and is thus not desirable). Moreover, this proposal calls for an elevated road to span a portion of the harbor.[5] The Rinkōsen is intended to operate as a trunk road with dense connections to vertical routes through the city, such as the Asakusa road. Traffic patterns make this second proposal unthinkable.
>
> The third proposal is for a harbor Akaiwa bypass. [But] the planned Rinkōsen road is a general road intended to address the future demand for transport within the city and inter-city travel. It is not a bypass. {...}
>
> Takeda: Is this your position as the director of civil engineering, or is it the result of a more comprehensive review conducted together with other departments?
>
> Director of Civil Engineering: This is both my position as director of civil engineering and the city's position. I haven't discussed the (APOC) proposals with other departments, but the Otaru City Comprehensive Plan for 1968-177 touches upon them, albeit briefly, and my explanation here today is exactly the same.
>
> Makino: If the city has established a single position [on the issue], how will this be shared with citizens in the future?
>
> Director of Civil Engineering: Our first priority has been to explain our position before this committee. We would then like to consider some form of PR [campaign].
>
> Morioka: The movement to save the canal is becoming a far-reaching citizens' campaign. The gap between the city and citizens will only widen if the city just says that nothing can be done. Don't you think there's a need for some form of dialogue with citizens?
>
> Director of Civil Engineering: [The Rinkōsen plan] was first reported in the November 19, 1973 edition of the *Asahi Shinbun*, and we met in early December with Mr. Fujimori (Shigeo) to explain the city's basic position. Next, we'd like to meet with Mr. Koshizaki (Sōichi) at the earliest opportunity to explain and gain his understanding. I'd like to give more thought to the matter of a PR [campaign] aimed at citizens.

[3] A diagram showing the APOC's three alternate proposals for siting the Rinkōsen can be found in Chapter 5.

[4] The director's point is that technical requirements regarding road elevation, width, and the angle of incline would make it impossible for existing roads to link up to the route proposed by the APOC. One primary purpose of the Rinkōsen, however, was to alleviate traffic congestion by providing as many connections to city roads as possible.

[5] The APOC had proposed constructing a double-decker road, utilizing an existing bridge that spanned a portion of the harbor.

Morioka: I doubt that justifying your position by referring to a planning decision adopted eight years ago will be enough to satisfy the citizens' movement. Do you?

Director of Civil Engineering: That is why I didn't touch on that point in my explanation here today, either.

Takeda: I think there needs to be a more comprehensive deliberation by the city [administration]. As he explained here today, Mr. Koshizaki is not thinking only about 1985, but about twenty or thirty years in the future. What is more, he raises the issue not just as a technical problem, but also from a different perspective: he is concerned with protecting our cultural heritage. I believe these points should be incorporated into [the city's] deliberations.

Director of Civil Engineering: I believe that transportation issues should be planned step-by-step.[6]

{…}

Committee Chairman: In consultation with the council, I have decided to carry over all matters submitted to this committee for further deliberation. The session is hereby adjourned.

Minutes of the Standing Committee on Construction, February 15, 1974. Record set: 1971-1974, Registration no. 157. Information in parentheses has been added by the author, occasionally to correct the transcription of the session. Information in brackets has been added by the translator.

In this exchange, the civil engineering director stresses the unsuitability of the APOC's various "counterproposals," and his opinion that the original plan remained the best. In his view, the third counterproposal submitted by the APOC could not even be considered an "alternative," as it deviated from the purpose guiding the original plan. His admission that no other city departments were involved in negotiations demonstrates that, at this point in time, the city administration approached the issue as a transportation issue—a technical issue to be handled by the civil engineering and city planning departments. Indeed, as far as the city was concerned, there was no "canal issue" in 1974. At this point in time, there was only the "Rinkōsen issue." The attitude of city officials is also worth noting: they did not refuse to engage with preservation activists on the basis of "an eight-year old planning decision" ("that is why I didn't touch on that in my explanation here today"), but continued with polite and detailed explanations to win acceptance of their plan. While they did not reject dialogue on the issue, they discussed the Rinkōsen plan as a settled matter.

The growing momentum of the canal preservation movement prompted city officials to harden their stance. The APOC-sponsored canal poster exhibition at a local department store and the Hokkaido Board of Education's recommendation of a survey of the canal and surrounding stone warehouses motivated city officials to quickly resolve the issue. On December 23, 1974, the committee handed down its decision.

Session resumed at 11:51 pm.

Uproar in the conference room, many people speak.

Committee Chairman: I hereby reconvene the committee.

Kobayashi: I move to put the matters referred to this committee to an immediate vote.

---

[6]Here the Director of Civil Engineering is responding to Takeda's comments by implying that such "big picture" concerns as cultural heritage or historic preservation have no place within the bureaucratic, step-by-step planning process.

Committee Chairman: Those in favor of the motion, please stand. [*Counting votes*] There is a tie; I vote to adopt the motion. All those voting to reject Petition 161 and to adopt Petitions 165, 170, 203, and 209, please stand. There is a tie; I vote to reject Petition 161, and to adopt all others. Measure 23 is adopted, Petition 196 is adopted, Petition 196 is adopted, Petition 20 and Appeal 224—partially adopted. All others carried over without objection. (*Uproar in the conference room, many people speak at once*).

Session concluded.

Otaru City Public Management Committee, December 23, 1974. Record set: 1971-1974, Registration no. 157.[7]

This session had been convened at 1:30 p.m., and ended more than ten hours later, at 11:52 p.m. The committee had recessed at 4:45 p.m. to allow committee members to confer, and intermittent negotiations ensued. The session was reconvened just before midnight and, in a matter of sixty seconds, the committee rejected the APOC's Petition 161, and approved a separate petition filed in support of the Rinkōsen. The APOC protested the result of the vote and accused the committee of steamrollering. In the new year, preservation activists grew more aggressive as they put their case before the Hokkaido Prefectural Assembly and the Hokkaido Board of Education. Back in Otaru, they held a general meeting to mark the formal establishment of the APOC.

Proponents of the Rinkōsen project responded by going on the offensive. Two petitions (95 and 96) supporting the road construction project came before the December 1975 meeting of the city council's construction committee. The following is an excerpt from the committee's deliberations:

Shimano: What sort of contact do you have with the APOC?

Director of Development: We've talked with them once since the third session (of the regular city council). We don't agree. I'd like to have a more substantive discussion with the APOC, and hope to meet with them before the end of the year.

Shimano: What is your personal opinion on the subject of the canal's preservation?

Director of Development: Personally, I would like to keep the canal if that were possible. However, the city planning laws have designated that area as the site of the Rinkōsen, and so the canal must be reclaimed to support Otaru's growth."

Minutes of the Standing Committee on Construction, December 19, 1975 (Fourth general session of the 1975 assembly). Record set: March 1975-April 1979, Registration no. 402. Information in parentheses has been added by the author.

"The canal must be reclaimed to support Otaru's growth." The goal of "growth" trumped the director's personal position ("I would like to keep the canal if that were possible").

Since August 1975, the canal issue had been complicated by the question of surveying the canal and surrounding area. Representatives from the city and the APOC met intermittently to discuss the possibility and potential structure of a joint survey, but no ready solution emerged.

---

[7]Minutes of committee proceedings are typically signed by the committee chairman and two additional council members. The minutes for this session, however, were never signed. One can only speculate as to whether this singular deviation from protocol was related to the manner in which the vote was conducted or the result of the vote.

Shimano: What type of discussions are you conducting with the APOC?

Director of Development: Since the start of the year (1976), we've had two conversations. The APOC contends that leaving the canal intact need not be a precondition [of the joint survey], and that conducting the survey should take priority.

Shimano: So the city can't make any progress unless we push back [against the APOC's demands] by collecting all the [documentation to prove] that a technical survey isn't possible?

Director of Development: That's exactly right. What's more, both sides have some misgivings regarding a survey. What I mean is: we're worried that a survey will conclude that the canal should be left [intact], and [the APOC fears that] even if a survey is conducted, it won't conclude that the canal should remain. This matter requires a bit more discussion.

Minutes of the Standing Committee on Construction, March 25, 1976 (January 1976 regular session). Record set: March 1975-April 1979 (1), Registration no. 402. Information in parentheses has been added by the author, information in brackets by the translator.

The administration worried that a survey would "conclude that the canal should be left [intact]." Conversely, APOC members worried that "even if a survey is conducted, it won't conclude that the canal should remain." Both sides were searching for a way of gaining the tactical advantage.

As talks wore on, city officials noted the numerous petitions filed in support of the Rinkōsen (March 22, 1977), and began to worry that a survey of the canal district could delay or even halt construction (July 22, 1977). Officials finally abandoned the idea of conducting a joint survey (June 5, 1978). The minutes of proceedings at each of these three junctures offer greater insight into how this process unfolded.

March 22, 1977

Shimano: How are you dealing [with the issue]? What has been the progress of your talks with the APOC?

Director of Development: {…} On the Rinkōsen issue, the city agreed to conduct a survey with the APOC. But—and I think this is due to organizational issues within the APOC—even if we reach an agreement to proceed in a certain way, that has all changed by the next time [we meet], and so there has been no progress on the [joint] survey. Our basic position is that in order to build the road quickly, we should talk about limiting the survey's scope to the question of whether or not the road could be constructed anywhere besides the current route at [our] next meeting. I don't mind taking some time if we can convince [the APOC] that the existing plan is the only viable option for situating the Rinkōsen. That's the approach I'd like to take.

Shimano: There are now petitions from the Otaru business community supporting [the road construction]. What is your reaction to these?

Director of Development: I realize that many citizens are asking for the road to be constructed at an early date. The twenty-one petitions on the issue [submitted during this session] make the matter all the more pressing. I hope to reach an arrangement with the APOC and conduct the survey as soon as possible.

Minutes of the Standing Committee on Construction, March 22, 1977. Record set: March 1975-April 1979 (1), Registration no. 402. Information in brackets has been added by the translator.

July 22, 1977

Sasaki: What are the prospects for the early completion of the Rinkōsen, and what sort of budgetary request has been submitted to the Hokkaido [prefectural government]?

Director of Civil Engineering: The current stance of the [prefecture] is that if the highway is constructed up until the canal [portion of the route], and then construction cannot be undertaken for one year, or two years, the project could be discontinued or suspended. In that event, even if all the relevant parties later reach a consensus to reclaim the canal and move to resume construction, the prefectural road budget would make this almost impossible. In order to protect our budget, our only option is to slow down the pace of construction.. It is essential that we resolve the canal issue with no further delay, and we will continue to negotiate with the APOC.

Sasaki: Discontinuing or suspending the project would create a huge problem, so please conduct vigorous discussions with the APOC and make every effort to achieve the early completion [of the road construction project]. What is the prefecture's position on this point?

Director of Civil Engineering: During a press conference, the APOC unilaterally presented their way of thinking about the survey and survey participants—despite the fact that these [remain] points of contention at our meetings regarding the [joint] survey. I have grave doubts about whether the APOC has a sincere desire to work with us to reach a consensus. I will consider convening another meeting with them on the [subject of the joint] survey to rebuild trust. If that doesn't seem possible, the city should proceed [with the survey] alone. In that event, the survey would cover three items: the road, the canal, and [surrounding] buildings. We need to quickly work out whether the route of the highway can be changed, and then officially conclude that the canal is our only option. If we can obtain the consent of the APOC and the majority of Otaru citizens, that would enable us to make a budgetary request to the prefectural government.

Minutes of the Standing Committee on Construction, July 22, 1977. Record set: March 1975-April 1979 (1), Registration no. 402. Information in brackets has been added by the translator. Emphasis added.

June 5, 1978

Committee Chairman: I hereby convene this meeting. {...} The mayor and deputy mayors have asked to report on the findings of the investigation into alternative plans for the Rinkōsen. Please go ahead.

Director of Civil Engineering: During the first session...the mayor ultimately issued a statement to the effect that he would "continue to examine the route of the Rinkōsen." I'd like to report on the findings of this examination. We have concluded that from an engineering and a city planning perspective, there is no viable alternative to the proposal that was deliberated during the first session. Please see the accompanying "Review of Alternative Proposals for the Rinkōsen."

{...}

Yokō: Does this mean that you have concluded your review of all alternate proposals? Or is this just the first phase?

Director of Civil Engineering: We have covered every conceivable route. There is no other route that is even remotely viable as an alternative.

{...}

Honma: There is a considerable difference between the line of thought at the time the Rinkōsen plan was adopted and the current understanding of the cultural value [of the canal]. The composition of the APOC also seems to have changed considerably. How about meeting once more to talk?

Director of Civil Engineering: The road is just as necessary today as when the [original] construction plan was adopted in 1966. The Rinkōsen is essential to relieving traffic congestion in Otaru. Upon review, we've concluded that we have no alternative but to build the road on the land reclaimed from the canal. We've tried to explain this from the perspective of Otaru's economic growth and alleviating traffic congestion, but no matter how many times we talk with the APOC they say they haven't received a full explanation [from us]. I think the APOC is trying to evade the issue.

Honma: There will be trouble later if you proceed while the APOC situation remains unresolved. I'd like you to meet with them one more time and proceed once you've achieved a mutual understanding.

Deputy Mayor: As the sequence of events up until now shows, the city has made efforts to continue discussions, but we have no choice but to bring these to a close. We will not conduct a joint survey with the APOC. If the APOC continues to demand further explanation, we'll do that to the full extent.

Takahashi: From the city's position, the review of conceivable routes [for the Rinkōsen] has come to an end, correct?

Director of Civil Engineering: That is correct.

{…}

Takahashi: This report has certainly brought the scale of the road and construction-related issues into sharp relief. We are in a position to make a decision on how to proceed. If the road is to be constructed, we must think about how to complete the project at the earliest possible date in order to respond to the earnest wishes of [Otaru] citizens. I doubt the central government would be quick to allocate a budget if the route [of the Rinkōsen] were to change. I worry the central and prefectural governments might even withdraw [from the project].

Director of Civil Engineering: There is no suitable route that could replace the one in our proposal. If we proceed with the current plan, the canal will be reclaimed. But if we are prevented from [reclaiming land from the canal], the Rinkōsen project will be suspended. In that event, we can't take responsibility for future traffic congestion in Otaru.

Minutes of the Standing Committee on Construction, June 5, 1978. Record set: March 1975-April 1979 (2), Registration no. 403. Information in brackets has been added by the translator.

At the March 22, 1977 committee meeting, misgivings were expressed regarding whether or not the APOC was negotiating in good faith. The director of development stated his desire to satisfy the citizens supporting the road's construction: "many citizens are asking for the road to be constructed at an early date. The twenty-one petitions on the issue make the matter all the more pressing." Budgetary concerns loomed larger at the July 1977 meeting, with one official predicting that if "construction cannot be undertaken for one year, or two years, the project could be discontinued or suspended. In that event, even if all concerned parties later reach a consensus to reclaim the canal and move to resume construction, the prefectural road budget would make this almost impossible." "Due to budgetary considerations," city officials wanted to avoid a suspension of the project at all costs. Indeed, they were willing to slow down the pace of construction to avoid an outright suspension of the project. By the June 1978 meeting, city officials had reviewed "every conceivable route," and concluded that "there is no other suitable route that could replace the route in our current proposal." What is more, the deputy mayor concluded that "the city has made efforts to continue discussions, but we have no choice but to bring these to a close. We will not conduct a joint survey with the APOC." While city officials had persevered in their discussions with preservationists, long after the controversial December 1974 committee vote, these talks were now at an end.

Needless to say, the decision to end discussions with the APOC was influenced by growing misgivings about "secur[ing] funding from the national government if the route were to change," and the collective "worry that the national and prefectural governments might withdraw [funding for the project]." City officials sensed an impending crisis: if they failed to execute the Rinkōsen project as planned, "the project will be suspended. In that event, we can't take responsibility for future traffic congestion in Otaru."

Given the gravity of their concerns, why did city officials continue to engage with representatives of the canal preservation movement? They continued talks with the APOC even after the vote to approve to Rinkōsen project had been pushed through committee and, as we shall see, eventually submitted a revised blueprint for the Rinkōsen that inspired even more discussion. *Even after announcing that their negotiations with the APOC were at an end*, city officials continued to engage with preservation activists. But why?

There are two reasons for the tenacity with which city officials sought to reach a consensus with the preservation movement. The first had to do with the basic policy of Mayor Shimura Kazuo, who took office midway through the negotiation process and valued dialogue with local citizens. Years later, Shimura, who described his twelve years in office as "beginning and ending with the canal" (Interview with Shimura Kazuo in Otaru, September 4, 1997) still stressed the importance of consensus-building:

> I'm personally convinced that we achieved the coexistence (of the road and the canal) to the greatest possible extent in our creation of a new cityscape. It was the best approach. I think it goes without saying that consensus with residents is essential to various operations. However, depending on the operation—and particularly when it comes to major undertakings—it can be hard to reach a consensus with one hundred percent of residents. It's difficult. So what does one do in that case? How much effort did you put into achieving consensus? I think that's what matters when it comes to the question of consensus. In my view, we did our utmost to reach a consensus, and so (with regard to the construction of the Rinkōsen) we [eventually] went ahead with the established policy {…}
>
> Interview with Shimura Kazuo in Otaru, September 4, 1997. Information in parentheses has been added by the author. Information in brackets has been added by the translator.

Even in the absence of a consensus with "one hundred percent" of Otaru residents, what mattered was "how much effort" the administration had made. Shimura remained convinced that "we did our utmost to reach a consensus." His approach offers one explanation for the city government's persistence in continuing discussions with the APOC.

The burgeoning preservation movement provided an even more compelling reason. A timeline of events between 1975 and 1979 demonstrates the growing momentum of the movement to save the canal:

> The Otaru Board of Education held a hearing on the status of the Otaru canal issue at the Hokkaido Prefectural Board of Education (February 1975)
>
> The Hokkaido Prefectural Assembly adopted a petition submitted by the APOC (February 1975)
>
> The APOC marked its formal establishment (June 1975)

An opinion survey conducted by the Otaru University of Commerce showed that 46% of Otaru residents supported the preservation of the canal, while only 27% opposed it. (September 1977)

Shimamoto Torazo, a Lower House member [representing a Hokkaido district], referred to the Otaru canal issue while speaking at the Diet, and expressed his support for the canal's preservation. (May 1978)

After Koshizaki's sudden death, Mineyama Fumi assumed the leadership of the APOC. (May 1978) Under the leadership of Mineyama, a highly respected figure in the local community, the preservation movement gained new momentum and broader public support.

The Kankō Shigen Hogo Zaidan formed a team to survey Otaru, headed by Kyoto professor Nishiyama Uzō, a leading figure in the national townscape preservation movement. (June-August 1978)

The Association to Think About the Otaru Canal Issue (Otaru Unga Mondai o Kangaeru Kai) was launched in Sapporo. In Otaru, one hundred thousand people turned out for the Port Festival, organized by local preservation activists. (July 1978)

Diet member Ogasawara Sadako raised the canal issue during a question and answer session of an Upper House Budgetary Committee meeting. (March 1979)

This timeline demonstrates the issue's growing national reach: the fate of a canal in Otaru was no longer a mere local concern—it was being discussed at the prefectural and national levels of government. Moreover, public opinion increasingly favored preservation. It is thus reasonable to assume that city officials felt compelled to continue their dialogue with preservation activists in the hopes of somehow securing their agreement to the Rinkōsen construction project.

Of particular influence was the outside world's growing recognition of the canal's cultural significance. Nationwide, townscape preservation movements were gradually bolstered by new developments, including an expansion of the Law for the Protection of Cultural Properties (through a July 1975 amendment to Chap. 5, Sect. 2, Article 83-2), which introduced the designation of "preservation districts for groups of traditional buildings." Moreover, the opinions of the Hokkaido Board of Education, the Agency for Cultural Affairs, and the Minister of Education (who oversees and directs Japan's educational and cultural affairs) carried far greater political influence than the protests of local citizens. This type of outside pressure gave further momentum to the canal preservation movement.

At the Hokkaido Prefectural Assembly, the Superintendent of Education Nakagawa Toshiwaka replied to a question from assembly member Funayama Hiroji (of the Japan Socialist Party) as follows:

Superintendent of Education Nakagawa Toshiwaka takes the podium. (*Many people speak*). I will reply to the question posed by assembly member Funayama. The stone warehouses and canal in Otaru are a testament to the role of Otaru's port in Hokkaido's industrial and economic development. I believe that scenery and atmosphere have cultural value. {…} I've had various discussions with the Otaru city government up to this point, and in any case, the preservation of traditional buildings according to the Law for the Protection of Cultural Properties is different from the preservation of other individual cultural assets in that the will of the local municipality, and of the local land and building owners, are decisive factors. For that reason, we must await the judgement of the relevant municipality. For now, I'd like to wait and see what (Otaru) does.

Minutes of the second general session of the the Hokkaido Prefectural Assembly, July 13, 1978: 218-219. Information in parentheses has been added by the author.

Nakagawa's reply may seem innocuous enough, but the very fact that Hokkaido's superintendent of education had declared "the scenery and atmosphere" of "Otaru's stone warehouses and canal" to be of "cultural value" was significant. Nakagawa's reference to the "decisive" importance of the will of the "local municipality and land and building owners" established the prefecture's position on the matter: given the importance of local consensus, the prefecture would take a "wait and see" approach. The message conveyed to Otaru city officials was this: given the canal's cultural value, the prefectural government was closely following developments. The city should work with the preservation movement to achieve a consensus.

The same message was repeated from even higher levels of government, increasing the pressure on the city administration. The Upper House Budgetary Committee was the scene of the following exchange:

Ogawasara Sadako: {...} I'd like to pose a question from the position of wanting to save this canal and the stone warehouses at any cost. {...} I believe this (the "canal issue") is also an important issue for authorities responsible for preserving Japan's cultural assets. And {...} the issue is not limited to a single regional government. I believe at this stage the issue must be addressed from a broader perspective. I've made many inquiries [on this matter], but a real academic investigation [into the matter] has not yet been conducted. Naturally enough, the Agency for Cultural Affairs has already acknowledged the cultural legacy and significance [of the Otaru canal]. I'd like to clarify (this once more). {...}

Ministerial Aide Yoshihisa Katsumi: I am of the very same opinion regarding the cultural value of the Otaru canal and the stone warehouses. {...} Based on our judgement of [the canal's] extremely high cultural value, in 1975 we directed the city of Otaru to undertake a survey {...}

Ogasawara Sadako: You may answer this briefly. If, for example, the city (of Otaru) reaches a consensus with residents and requests [official designation as a] preservation district, you would naturally designate the area as a Group of Traditional Buildings, correct? There is no time remaining, so you may answer with a simple yes or no.

Ministerial Aide Yoshihisa Katsumi: We have every intention of doing so.

National Minister (Minister of Education) Naitō Takasaburō: {...} We are talking about that canal and group of warehouses, so the loss of that cultural asset would be a great loss for Japan. {...} The problem, as you know, is the local area. We must gain the understanding and cooperation of the local people, and if necessary, I'd like to visit the local area (Otaru) to talk with everyone and ask [for their cooperation]. I'd like to work as one with the Agency for Cultural Affairs and also seek the cooperation of the Ministry of Construction.

Minutes of the 4th Subcommittee of the 87th Upper House Budgetary Committee, March 28, 1979 (1): 15-18. Information in parentheses has been added by the author, information in brackets by the translator.

This straightforward question from Ogasawara, a Japan Communist Party Diet member, elicited an equally direct reply from government officials. The fact that the central government recognized the area's merit as a cultural asset and had "every intention" of designating the canal and surrounding buildings as a Group of Traditional Buildings should it receive such a request are both significant. Naitō's acknowledgment that the loss of "that canal and group of warehouses" would be "a great loss for Japan" was also a far more powerful statement than the views expressed at the Hokkaido Prefectural Assembly; moreover, it had been issued by the education minister himself. Such a message could hardly be ignored by the Otaru city administration. The ramifications of changing a district that the national government was prepared to designate as a Group of Traditional Buildings would be far too grave for one lowly local government to contemplate.

In fact, a meeting of the Otaru City Planning Commission on November 13 of the same year underscores the immediacy with which the national-level discussion of the canal permeated the local debate:

> [Name redacted]: On the subject of the survey required by the City Planning Law {…} At the time the survey was conducted (1972), what [was the status of the] canal and the stone warehouses? I assume it was different than today, when the Agency for Cultural Affairs has clearly acknowledged their cultural value. At that time, how did the survey identify the canal and the stone warehouses?
>
> {…}
>
> Director: You are asking how we identified them at the time of the survey.
>
> [Name redacted]: Just the canal and the stone warehouses.
>
> City Planning Director: Categories of buildings include 'commercial,' 'residential,' and 'industrial.' My understanding is that they were categorized as 'warehouses' and 'factories' under municipal management.
>
> [Name redacted]: The survey of the canal and stone warehouses at issue was not conducted in order to evaluate their cultural value—isn't that correct? And so, in this blueprint, the commercial warehouses have just been colored in like so—that's how they were identified? Just as a way of showing that this type of warehouse exists. And so the 1972 survey made no attempt to assess the cultural significance of the canal and stone warehouses, as many people today might expect it would. That's the type of survey the prefecture desired at the time.
>
> Minutes of the 44th City Planning Council Session, November 13, 1979. Certain names were redacted by the Otaru city government before it disclosed the minutes of this meeting. Information in parentheses has been added by the author, information in brackets by the translator.[8]

In this exchange, a city council member points out that a 1972 city survey of the canal had made no attempt to assess the cultural value of the canal and its environs. Now, in 1979, the Agency for Cultural Affairs had officially recognized their cultural significance, yet the Otaru and prefectural governments were attempting to proceed as if nothing had changed—something this council member could not accept. The position of the Agency for Cultural Affairs is what gives this speaker's words their force: words spoken before the Upper House budgetary committee were being

[8]On August 24, 2010 the author filed a request to review the minutes of the Otaru City Planning Council, and received paper copies of the same on September 6, 2010. Despite the passage of thirty-one years, the mayor of Otaru still approved the decision of the City Planning Section to withhold the names of the speakers at this meeting, citing privacy concerns (Communication #78 from the Otaru City Planning Council to Horikawa Saburō on September 6, 2010). The city's redaction of remarks made more than three decades ago, during the execution of official business in the public forum of the city council, is not only hugely detrimental to academic research, but also reflects a grave lack of accountability on the part of the Otaru city government.

weaponized back home in Otaru to discredit the conclusions of the city government's survey.

A revised blueprint for the Rinkōsen, known as the Iida Plan, became the city's response to the mounting demands for the canal's preservation. Let us turn to the substance of the Iida Plan and the process surrounding its implementation by the city government.

The Iida Plan identified the canal district as an "entirely dilapidated area that many people hope could be restored to life by a new dynamism." For this to occur, "The road plan supported by citizens and the historical legacy of this district need to be joined together in a new way." At the same time, there was "little point" in turning the district into a "lifeless history museum." The canal need not be preserved in its current form (Otaru City and Hokudai Iida Kenkyūshitsu 1979: 10). Iida's reference to a "lifeless museum" can be understood as a rebuttal to the preservation movement's goal of saving "the entire canal." The reference to "the road plan supported by citizens" is an attempt to marginalize the preservation movement by implying that the Rinkōsen plan met the needs of most Otaru citizens. In other words, the wording of the Iida Plan implied that the redevelopment of the derelict canal district through the construction of a road long desired by ordinary citizens would be far more meaningful [than mere preservation]. "Preserving the canal in its entirety," at the expense of the Rinkōsen, was out of the question. The Rinkōsen road must be constructed on schedule. For the proponents of change, this was absolutely essential.

Yet the same passage in the Iida Plan reflects a new reality: the local authorities who had previously understood the canal merely as a "means of transportation" had been forced to acknowledge the canal district as part of Otaru's "historical legacy." Unlike the city's original blueprint, the Iida Plan includes frequent references to the canal and surrounding area as "a space with a unique ambience," and "an urban space that commemorates [the history of] Otaru," with "historical and cultural value" (Otaru City and Hokudai Iida Kenkyūshitsu 1979: 5–10).

Specifically, Iida responded to the burgeoning preservation movement by making two revisions to the city's original blueprint. First, instead of reclaiming the entire [southern end of the] canal for the Rinkōsen, Iida proposed narrowing this portion of the canal. The Rinkōsen would be constructed on land reclaimed from the canal's western side (the side farther from the port, and closer to the mountains). Second, a public footpath would separate the new road from what remained of the canal, which together with surrounding warehouses would be redeveloped as a "canal park" (Fig. 4.7). The plan also called for improvements to Ironaifutō Park and the narrow alleys (*denuki kōji*) that ran through the canal district, as well as the construction of a port tower. Thus, while the construction of the new road with government subsidies remained the chief object of the Iida Plan (Table 4.2), the revitalization of the tourism industry, urban manufacturing, and commerce through the new canal park were introduced as new goals (Otaru City and Hokudai Iida Kenkyūshitsu 1979: 11–24).

The Iida Plan was immediately put before the city's construction committee for deliberation:

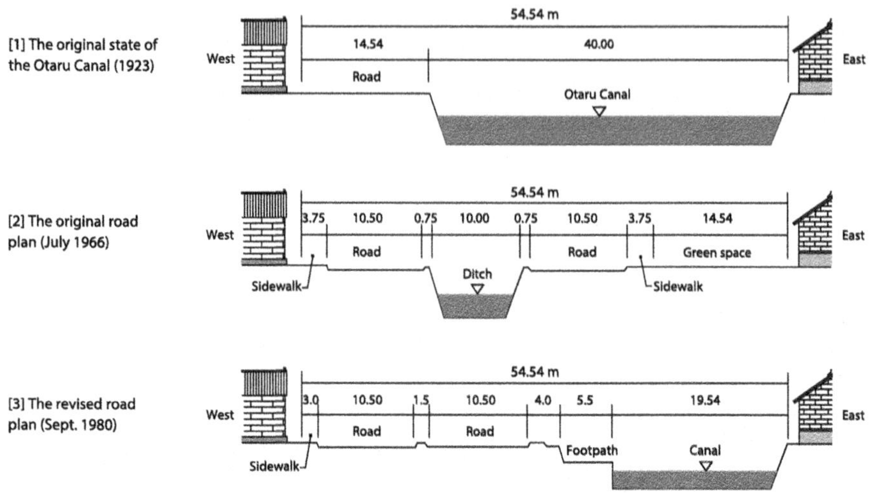

Fig. 4.7 A cross-sectional view of the revised blueprint for the Rinkōsen. Author's drawing, based upon Otarushi Dobokubu (c. 1986: 6) Unit: Meter (m)

Chairman: {…} An executive would like to report on the Environmental Improvement Plan for the Otaru Canal and Surrounding Area. Please go ahead.

Director of Civil Engineering: The plan we requested of Associate Professor Iida of Hokkaido University last July (1978) is now ready to present. Up until now this has been referred to as the Canal Park Plan, but this gives people the limited impression of a city-planned park, so from now on we will use the revised name. I'd like to use this plan as the springboard for discussion with all concerned parties, while focusing on a plan for implementation. The fiscal year of implementation and funding are future tasks [to discuss].

{…}

Chairman: We'll now take questions.

{…}

Director of Civil Engineering: We will construct the Rinkōsen according to the city plan. [The details can come later]. I'd like to take some time to focus on the Environmental Improvement Plan.

Takashina: The most problematic issue is building a six-lane road in this district. Given the sequence of events to date, isn't it a bit odd that [the city has] steadily moved ahead with the construction of the Rinkōsen?

Director of Civil Engineering: We looked into the possibility of resituating the Rinkōsen so that the canal could be left intact, and arrived at our [current] conclusion. While it's still imperfect as a road plan, we are working to meet the preservation camp halfway.

Takashina: What about the position of the Agency for Cultural Affairs?

Director of Civil Engineering: [The agency's position is that] it would be desirable to preserve the canal in as close to the current form as possible, but the will of the local region takes precedence.

Takashina: How about creating a city-wide commission to consider the issue?

Table 4.2 An itemization of operational expenses for the development of the canal district

Project	Source of funding			Total (Unit: million yen)
	National government	Hokkaido prefecture	Otaru City	
Rinkōsen road construction	3,200	1,900	0	5,100
Construction of footpath along Rinkōsen	0	0	93	93
Road modification at the north end of the canal	144	0	164	308
Repairs to Dehamabashi (bridge)	0	0	25	25
Renovation of *denuki kōji* (historic alleyways)	160	0	160	320
Construction of parking lots	0	0	18	18
Installation of settling basin	65	0	136	201
Dredging of settling basin	0	0	268	268
Canal dredging	407	0	813	1,220
Levee maintenance	1,060	0	441	1,501
New memorial hall	129	64	111	304
New municipal park	1,180	0	2,120	3,300
New Ironai Futō Park	214	0	379	593
Total expenditure by source	6,559	1,964	4,728	13,251

Source Created by the author using 1992 data from the Otaru Civil Engineering Department

Director of Civil Engineering: The road issue is of primary importance. If the commission first acknowledges the necessity of the new road and puts this issue aside to consider the question of how to handle the remainder [of the canal], I have no objection.

{…}

Director of Civil Engineering: We have concluded that we cannot avoid involving the canal in the construction of the road. I'd like to win acceptance of our decision through further explanation. In terms of achieving a consensus, we plan to explain our position directly to organizations with an interest in the issue, and use the [city flier] *Kōhō Otaru* to notify citizens. Ultimately, however, the decision will be made by the mayor and the city council.

Takeda: The Iida Plan may contain a few revisions, but fundamentally, this is still the city's original blueprint. Moreover, the special November edition of *Kōhō Otaru* carries just a one-sided opinion [on the issue]. Will this approach really give citizens a balanced view of the issue?

> Director of Civil Engineering: The Iida Plan retains as much of the canal as possible, so it is completely different from the city's original blueprint. We took care to avoid pushing an unbalanced position in the *Kōhō [Otaru]*, and included the opinions of the preservation movement and Professor Muramatsu.
>
> Takeda: You say that there is no way the new road can avoid [the canal], but given technological advances, aren't there other methods that could be considered? The preservation camp persists in saying that the only way to preserve the atmosphere of the canal is to keep it at the current width—that it's not as simple as saving half of the canal.
>
> Director of Civil Engineering: Even the current plan leaves quite a bit of the traditional scenery [untouched].
>
> Minutes of the Standing Committee on Construction for June 29, 1979. Record set: May 1979-April 1983 (1), Registration no. 432. Information in parentheses has been added by the author, information in brackets by the translator.

While the city dispensed with the title "Canal Park" to avoid confusion, the revised plan continued to be widely referred to as such, or as the "Iida Plan." City officials defended the Iida Plan as a sacrifice (the new proposal was "imperfect as a road plan") made to "meet the preservation camp halfway." They also stressed their conclusion that "we cannot avoid involving the canal in the construction of the road." The Iida Plan, which "retains as much of the canal as possible, making it completely different from the city's original blueprint," was the final version. Of course, this failed to satisfy preservation activists, who allowed that 'the Iida Plan may have made a few revisions" but maintained "the only way to preserve the atmosphere of the canal is to keep it at the current width—it's not as simple as saving half of the canal." The stalemate continued.

The November 13, 1979 meeting of the City Planning Committee ran from 1:30 to 6:40 p.m.—a lengthy session. The meeting involved a seemingly endless debate over the question of whether to append an opposing statement to the city's blueprint (the Iida Plan) when forwarded to the prefectural city planning committee. City executives dismissed the idea, while council members aligned with the preservation camp supported it. This was not a substantive debate so much as an attempt by all parties to use the council proceedings as a bargaining chip. Ultimately the Iida Plan was submitted to the prefecture in December and approved the following in September (1980) (Hokkaido Bulletin 2361, issued in September 1980, [compiled in] Hokkaido Jūtaku Toshibu Toshi Seibika 1989: 28–30, 65–70, 77–79). Various administrative steps followed: after the City General Affairs Committee deliberated the issue (May 1981), an application was filed for land reclamation in public waters (June 1981),[9] and the city council passed a resolution approving the land reclamation work (March 1982). Official permission for land reclamation work was issued by the Hokkaido prefectural government in September 1982.

As the bureaucratic approval process ground on, conflict shifted from the administrative to the political arena. In 1983, the preservation camp redoubled its efforts in overtly political form: the head of the Otaru Chamber of Commerce and Industry made an appeal to the city government to reconsider reclaiming the canal, activists

[9] Some mistakes in the blueprint were discovered after the initial application, and a revised version was reviewed and approved in 1982.

established the Committee of One Hundred for the Otaru Canal (September 12), the Otaru Youth Council approved a resolution supporting the canal's preservation (October 18), and the Committee of One Hundred submitted a petition, signed by 98,000 local citizens, demanding the canal be saved (November 19). 1984 would prove to be a watershed year, in which "five-party" political negotiations brokered by the Hokkaido governor attempted to break the deadlock on the issue. Hokkaido Governor Yokomichi Takahiro designated his deputy governor Araya Masaaki to act as an arbitrator during these negotiations. Governor Yokomichi brought Araya, an Otaru native, on an official visit to the city, where he consulted with Mayor Shimura in setting up the five-party talks. Shimura recalls:

> I conferred with the governor a number of times about setting up the 'five-party talks.' However, once the conference was underway, I never spoke privately with the governor, or with Mr. Araya. My remarks were confined to the conference venue itself, and the governor and Mr. Araya never spoke to me privately about what was going on there.
>
> Interview with former mayor Shimura Kazuo in Otaru, September 6, 1999.

Shimura's recollection of events underscores the point that negotiations to break the deadlock on the canal were no longer a purely municipal affair, but conducted at a higher level of politics.

The five-party talks failed to achieve a consensus, and the city moved forward with its plan to reclaim a portion of the canal.[10] However, Araya claims the Rinkōsen project received a special budget precisely because of the protracted controversy surrounding the canal:

> …Given the extent of both opposition and support for the project, the Ministry of Construction and the prefectural government wanted to make (the Rinkōsen) as high-quality as possible, and gave it an exceptional budget that wasn't typical at the time. They really went all out on the construction.
>
> Interview with Araya Masaaki in Otaru, August 31, 1999. Information in parentheses has been added by the author.

Special budgetary provisions for the "high-quality" construction of the Rinkōsen[11] and the new footpath alongside what remained of the canal gave the city of Otaru new life as a tourist destination:

> The port is facing extremely harsh conditions, and it's tough to attract businesses—of course this is the case anywhere in the country. So I guess you could say that tourism is where we are putting up the best fight. [But] when I took office, I don't think there was anyone (who could conceive of) Otaru as a "tourism city." After all, Otaru wasn't generally included in that category.
>
> Interview with Araya Masaaki in Otaru, August 31, 1999. Information in parentheses has been added by the author, information in brackets by the translator.

Here Araya summarizes the trajectory of Otaru, from a place that no one "could conceive of" as a tourist destination to one in which "tourism is where we are putting

[10] See Chap. 5 for a more detailed account of the breakdown of the five-party talks.

[11] Tamura Yoshiko (2009: 149–152) makes the same point.

up the best fight." More broadly, city officials had originally understood the canal district as "the most straightforward site for highway construction." The protracted dispute over the fate of the canal changed this: over time, officials came to regard the canal district as something more than the site of a new road: it was now also a "canal park."

4.4 The Continuity of Discontinuity: The Logic of Change

The Rinkōsen project was originally adopted without any opposition or politicization. Construction of the new highway duly began at the Sapporo end. After all, the plan had won easy passage in the city council and had been enacted following the completion of all appropriate procedures. Given this, city officials initially believed they could safely ignore the resident-led movement to save the canal, which began well after the budget for the project had been approved.

As we have seen, however, they eventually confronted a new reality, in which even Japan's Agency for Cultural Affairs was prepared to designate the Otaru canal and affiliated stone warehouses as a "Group of Traditional Buildings." It began to dawn on city officials that, if the oil shock continued to delay their construction schedule, the preservation movement could easily increase its clout, and potentially halt construction of the Rinkōsen altogether. Developments from 1975 onward could hardly have been welcomed by Otaru City Hall. It was thus in the city government's best interests to make certain concessions in order to reach an agreement that could move the project forward: in the process, a project that had originally been understood as a simple road construction project came to be viewed as something different—a plan for a city park.

Here we must stop and ask: why? Why did city officials feel the need to make certain concessions to preservationists? Conversely, why did they remain so stubbornly determined to construct the Rinkōsen road at all costs—even if this meant making disagreeable compromises that left the Rinkōsen "imperfect as a road plan?" To answer these questions, we must return to the June 1979 session of the Otaru City Standing Committee on Construction, and the words of the director of civil engineering:

> Director of Civil Engineering: My stance (on the Rinkōsen construction project) has been consistent from the very start. The [current] plan is a compromise that takes subsequent developments into account, and it is the very best we can do.
>
> Minutes of the Standing Committee on Construction for June 29, 1979. Record set: May 1979-April 1983 (1), Registration no. 432. Information in parentheses has been added by the author, information in brackets by the translator.

"Subsequent developments" refers to the sequence of events beginning with the amendment of the Law for the Protection of Cultural Properties—and to the gathering strength of the local preservation movement. At the same session, the director of civil engineering responded to another question:

Director of Civil Engineering: The city has made compromises in order to gain a consensus. I'd like to see the preservationists make some compromises, too. Any further demands for the city to compromise would actually represent [an effort at] compulsion by the preservation movement.

Minutes of the Standing Committee on Construction for June 29, 1979. Record set: May 1979-April 1983 (1), Registration no. 432. Information in brackets has been added by the translator.

In the above passages, the director of civil engineering makes it clear that the compromises reflected in the Iida Plan were political concessions made in order to "gain a consensus," but that his personal "stance" on the content of the highway construction plan had remained "consistent from the very start." Our next task is to ask why that was the case. While he could approve concessions aimed at "gaining consensus," why did the civil engineering director and other city officials remain so adamant on the underlying fundamentals of the plan—in other words, the necessity of paving over the canal to construct the new road?

It would seem that the disposition of Otaru Mayor Shimura Kazuo was of decisive importance. We already know that Shimura valued "how much effort" was put into reaching a consensus. In this case, we must consider yet another of the mayor's creeds:

All the officials involved said that I didn't waver, even when they accused (the city of unilaterally ending discussions with preservationists). {…} While the officials (who are actually accused of such things) will come and go, as the top [executive], the really important question for me is what the person at the top is working to achieve.

Now, you know, I went off as a sailor during the war. I had left Hokkaido University, and there was this reserve student system at the time. I took the test and enlisted… {…} And after just eight months [of training] I was made a second-lieutenant in the navy. I (was assigned to) Kaohsiung Air Base in Taiwan, where the Flying Corps were stationed. {…} I was [just a novice,] an accidental second-lieutenant—the farthest thing from a true navy lieutenant {…} and among my subordinates, there were these non-commissioned naval officers who were godlike (in their knowledge of) all kinds of things—artillery, semaphore flags… They were all like that, [they all acted like] these gods. {…} And below them were the soldiers, and those guys just looked at me as if they couldn't believe [I was their superior], they were [incredulous that I was in such a high-ranking position with no experience]. Well, when (a battle) broke out, the B29s would fly in and drop their bombs. And those non-commissioned officers and those soldiers… When things got dangerous, they would look at me. They didn't always take me very seriously, because I was (still) this very young officer—but in that instant they would still take a quick look at my face [to await their orders]. It wasn't just me. All the other guys [in my position] apparently had the same experience. All may be fine and well in normal times, but when you're confronted with difficulty I believe it's very important that you continue to behave in a way to avoid transferring your anxiety and agitation to your subordinates. This is the same everywhere—not just for soldiers, but in government offices, and in companies. When a company is facing liquidation or some life-or-death crisis, employees aren't going to stick behind a CEO who has been dallying [unproductively]. So, whatever the case may be, the mental attitude of the top person is crucial. First of all, you have to adopt a persona that won't convey your anxiety or agitation to your subordinates. You also have to clearly inform your subordinates of your intent and your way of thinking. You can't leave them wondering what you were talking about; you have to tell them 'these are my objectives and this is my way of thinking' in what I suppose you could call a 'top-down' way. {…} And there's the military expression 'the commanding officer out in front.'

> I felt very strongly that the commanding officer shouldn't hang back and simply urge others on—he has to take the lead.
>
> Interview with former Otaru Mayor Shimura Kazuo in Otaru, September 3, 1998. Information in parentheses has been added by the author, information in brackets by the translator.

In this passage, Shimura not only recalls his first-hand experiences in the war, but also explains the source of his deeply-held convictions on the subject of leadership. Shimura's self-deprecating description of his younger self as an "accidental second lieutenant" reflects the bitterness of his early experience commanding the "real" and more experienced non-commissioned officers. Shimura brought to his administration this understanding of the mental attitude of a leader ("top-down," "the commanding officer out in front," never "conveying anxiety or agitation" to his subordinates). Mayor Shimura's nicknames at Otaru City Hall were "that stubborn old man" and "the bear." It would seem that he had earned them.[12] Clearly, Shimura felt responsible, as Otaru's "commanding officer," to impart his decisions clearly to his subordinates and resolutely execute them.

One could also say that Shimura believed in "sticking to his policies:"

> And then there was one more thing—you know, the telegram. This was on February 4, 1984. It read: 'We're counting on you to stick to your public promises and policies. Keep at it, Mayor [Shimura]! From your young supporters in the prefectural government.' This [telegram] was sent by prefectural employees [who supported me].
>
> Interview with former mayor Shimura Kazuo in Otaru, September 6, 1999. Information in brackets has been added by the translator.

Shimura's reference to this message of support from Hokkaido prefectural employees reveals the pride he felt at "sticking" to his "public promises and policies." Moreover, Shimura believed he had been entrusted with the execution of the municipal government:

> The (Committee of One Hundred) were extremely energetic in stressing they had gathered the signatures of 98,000 people—the majority of Otaru citizens, residents, in support of [the canal's] preservation. They talked about it wherever they went. They said that's how they would announce it in the newspapers. Now, I couldn't help having some doubts about those signatures. I say this because it was in 1983 [that these signatures were collected]. And in the April of 1983, I ran for mayor on the promise of completing the Rinkōsen (project) according to the existing plan. People supported me and I was elected. I was elected with what I think was fifty thousand-odd votes (53,903). I really wondered whether citizens could have changed their minds on the issue in less than six months…
>
> Interview with former mayor Shimura Kazuo in Otaru, September 6, 1999. Information in parentheses has been added by the author, information in brackets has been added by the translator.

Here Shimura contends that he was entrusted with the final authority over the Rinkōsen issue because he ran "on the promise of completing the Rinkōsen according to the existing plan" and "people supported me and I was elected." According to Shimura's creed, once he had clearly laid out a plan and been entrusted with its

[12]NHK Sapporo Documentary *Kita no Chōsensha-tachi, No. 1.* Yomigaere Unga no Machi: Otaru o Nibunshita Daironsō (April 5, 2002: 20:00–20:45).

execution, it was his duty to act resolutely to avoid "conveying anxiety or agitation." For Shimura, who feared changing the road construction project halfway through its execution would negatively impact Otaru's eligibility for future subsidies, protecting the flow of subsidies to Otaru represented the most important task of local government: indeed, it represented the "fundamental principle of the administration."

> Horikawa: {…} Ms. Mineyama once told me that, at the very last stage of the canal dispute, she waited for you outside of city hall. When you arrived, she clasped her hands together and said: "This is my final plea—please [save the canal]." She told me that you replied: "I can't do that, because this matter involves a fundamental principle of my administration." Is this really what was said?
>
> Shimura: No, that's not right. But Ms. Mineyama did come to my house once.
>
> Horikawa: Is that so?
>
> Shimura: Yes. We couldn't talk at city hall or places where others were present, so I had her visit me at home. We talked all about the canal issue and about the city. I used the expression "fundamental principles of administration" on many occasions.
>
> Horikawa: And what did you mean by that?
>
> Shimura: That this wasn't just a passing fancy or a decision undertaken lightly. We had taken all sorts of steps up to this point, and we finally arrived [at this decision—i.e., the plan to situate the Rinkōsen over the canal]. We couldn't be careless and behave irresponsibly. Simply overturning this kind of decision [would be a rejection of] the fundamental principles of administration.
>
> Interview with former mayor Shimura Kazuo in Otaru, September 7, 1999. Information in brackets has been added by the translator.

For Shimura, the decision to promote the Rinkōsen construction project was not "just a passing fancy or a decision undertaken lightly;" "simply overturning" the plan would violate "the fundamental principle" of local administration. By this, Shimura meant the guarantee that a decision "finally" arrived at by the administration, after extensive deliberation, would be enacted as planned. Having successfully extracted subsidies for the Rinkōsen's construction, Shimura refused to countenance the idea of "simply overturning" the project. This was an article of faith for the mayor, and it determined his attitude to the dispute.[13] In order to protect this "fundamental principle," Shimura would only go so far in meeting the demands of the preservation movement:

> There may have been (individuals within the) APOC, or others, who privately felt that 'we got this far, so maybe that's enough.' But the APOC as a whole refused to accept anything short of the complete preservation of the entire canal—they wouldn't budge an inch. I thought we could meet each other halfway, and from my point of view, I felt as if 'well, I tried to meet them halfway.' But (the APOC) wouldn't concede a thing. That's how it was. That's how it turned out.

---

[13]It is important to note that Shimura viewed this principle (that the primary role of local government is to extract resources from the central government) *positively*. As we shall see in Chap. 5, local preservation activists such as Yamaguchi Tamotsu and Ogawara Tadashi expressed far more cynicism about Shimura's creed, which could also be understood as an indictment of Japan's national subsidy system, given its negative effect on local autonomy. They would argue that local executives like Shimura typically "looked to the east"—in other words, to Tokyo—instead of responding to local conditions and concerns.

Interview with Shimura Kazuo in Otaru, September 6, 1999. Information in parentheses has been added by the author, information in brackets by the translator.

Shimura had been willing to make certain concessions, but the APOC refused to return the favor ("I felt as if 'well, I tried to meet them halfway' but (the APOC) wouldn't concede a thing.") In short, Shimura Kazuo's beliefs informed the municipal administration's commitment to executing the Rinkōsen project—albeit with the concessions reflected in the Iida Plan.

Many other local actors stood to benefit from Shimura's commitment to his "public promises and policies." Some residents believed the new road was essential to resolving traffic congestion in Otaru. Members of the local business community saw the construction of the Rinkōsen as a necessary investment in distribution infrastructure that would complement other improvements to port facilities. Representatives of various port and construction-related industries assiduously represented their interests in the project. Within the city government, the project was supported most enthusiastically by the Port Bureau, the Civil Engineering Department (responsible for public works such as road construction across the entire city), and the City Planning Division, which had formulated the Rinkōsen plan.[14]

After the preservation movement took aim at the Rinkōsen project, these advocates of change made their positions clear. The construction industry was particularly vocal and organized petitions from private "construction promotion" organizations (in reality, vertical industry organizations). City offices voiced their interests in written opinions and reports, and through their vertical relationships with neighborhood associations, which were also encouraged to submit petitions in support of the Rinkōsen. In the summer of 1980, for example, the Otaru branch of the Sapporo district trucking association, the Otaru Warehouse Association, the "Rinkōsen Development Promotion Association" within the Otaru Chamber of Commerce and Industry, and the Otaru Port Promotion Association all submitted written opinions demanding the "early completion" of the Rinkōsen highway.

The various actors promoting change in Otaru shared certain perspectives on the issue.

First, they viewed the modernization of port facilities as a pressing policy issue. Indeed, throughout the postwar period, Otaru had grappled with the need to update its cargo handling and distribution systems in order to compete with other ports.

Second, proponents of change shared a fundamental approach to the canal as a "space." They saw the canal as an empty "vessel" to be filled and did not allow emotional attachment or old memories to complicate this understanding. Nor did they understand the canal as a physical repository of the city's history. Consequently, they viewed the reclamation project as a simple "repurposing" of physical space—a third shared perspective. The various petitions submitted in support of the Rinkōsen project over the course of the canal dispute repeatedly emphasized the need

[14]Securing national funding for such a project was understood as a great professional success among local bureaucrats. Their continued enthusiasm in the face of mounting local opposition to the Rinkōsen plan also speaks to the negative impact of national subsidies upon local autonomy (See preceding footnote).

to replace outmoded forms of use in order to meet the demands of different age.[15] Their evaluation criteria were based entirely on the economic efficacy of the space. In the words of an Otaru port official, "the Arihoro warehouses no longer served any economic purpose, so tearing them down was the right decision," and with regard to other warehouses still in active use during the long canal dispute, "while some (preservation activists) talk only about the physical appearance [of the warehouses], these warehouses were originally vessels of economic activity—and viewed from that perspective, they are useless."[16] Spaces that no longer fulfilled their original economic purpose should be repurposed. This recalls the words of the Otaru city publication cited earlier in the chapter: "Up until now, the canal itself had functioned as a method of transportation. *Its new mission was to become a road*."[17] No longer a viable means of transportation, the canal should be transformed into something of functional equivalency: a road.

A fourth distinctive theme was the general adherence to the "ripple effect paradigm" for public works projects. This paradigm was widely accepted across Japan in the postwar era, but necessity made particularly fervent believers out of the poorer local governments. Otaru's leaders were especially dependent on this paradigm.

Finally, there was a collective fear that failing to complete the Rinkōsen project would render Otaru ineligible for future subsidies. Local politicians' ability to extract subsidies from the central and prefectural governments was typically the most important criterion on which they were evaluated (Curtis 1971), and it would have been unrealistic for Otaru's leaders to ignore this concern. Regardless of the factual validity of this belief, this shared *assumption* made it a social reality. Indeed, sociologists have often observed how beliefs or assumptions can govern human behavior—and prompt forms of self-regulation—in the same way as established fact. In this case, the assumption that Otaru would lose out on future subsidies informed Shimura's obstinate insistence upon protecting the Rinkōsen project as "the fundamental principle" of his administration.

In summary, the proponents of change in Otaru believed that spaces should be constantly updated or improved—and when necessary, razed and rebuilt from scratch to accommodate entirely new uses. The need to continually improve the functionality of urban space leads to discontinuity and disruption, but they saw no reason to lament this *continual discontinuity*. This comprised their logic of change.

---

[15]For example, Sugawara Haruo, who headed the Committee to Redevelop Otaru's Canal District (Unga Chiku Saikaihatsu Iinkai) and previously led the Otaru Warehouse Association, wrote in his August 1982 report that "the current route of the Rinkōsen, which runs over a portion of the canal, was decided upon in 1966, and the port plan also [called for] the total reclamation of the *functionally obsolete* Otaru canal..." {...} "Rinkōsen no Sōki Kansei to Unga Chiku no Saikaihatsu ni tsuite," page seven of the August 1982 report from the Chairman of the Committee to Redevelop Otaru's Canal District. Emphasis added by the author, information in brackets by the translator.

[16]Interview with the director of the Otaru Port Bureau's planning division, March 18, 1993.

[17]Shisei no Ayumi Henshū Iinkai (1988: 200–201).

References

Curtis, G. L. (1971). *Election Campaigning, Japanese Style*. New York, NY: Columbia University Press [Yamaoka, Seiji, trans. *Daigishi no Tanjō: Nihonshiki Senkyo Undō no Kenkyū*. Saimaru Shuppankai].

Inoue, T. (1981). Tōkō, Otaru Unga Hozon ni Kan Suru Iken. *Yusō Tenbō 178*: 44. Nittsū Sōgō Kenkyūjo.

NHK Sapporo Documentary *Kita no Chōsensha-tachi, No. 1*. Yomigaere Unga no Machi: Otaru o Nibunshita Daironsō (April 5, 2002).

Niitani, Y., Ed. (1987). *Kōtsū Keikaku ni Okeru Yosoku no Jigo Hyōka ni Kan Suru Kenkyū* (Toyota Zaidan Josei Kenkyū Hōkokusho 1-022, #83-1-III-053). Toyota Zaidan.

Otarushi Dobokubu. (1986). *Otaru Keikaku Dōro Rinkōsen: Otaru Unga to Sono Shūhen Chiku Kankyō Seibi Keikaku*. Otaru City (color, 7 pp.).

Otaru Shigikai. (1966). *Shōwa 41-nendo Otaru Shigikai Kaigiroku (1): Dai-ikkai Teireikai*. Otaru City.

Otaru City and Hokudai Iida Kenkyūshitsu. (1979). *Otaru Unga to Sono Shūhen Chiku Kankyō Seibi Kōsō: Gaiyō*. Otaru City (B4, 24 pp.).

Seibika, H. J. T. T. (Ed.). (1989). *Machi no Saisei o Mezashite: Otaru Rinkōsen Kensetsu no Kiroku*. Sapporo: Hokkaido Doboku Kyōkai.

Shinozaki, T. (Ed.). (1989). *Chihō Toshi Saikaihatsu Seisaku no Sōgō Kagakuteki Kenkyū: Otarushi Unga Chiku Saikaihatsu o Megutte* (Shōwa 63-nendo Kagaku Kenkyūhi Hojokin [Ippan Kenkyū A60410012] Kenkyū Seika Hōkokusho). Otaru: Otaru Shōka Daigaku Shinozaki Tsuneo Kenkyūshitsu.

Shisei no Ayumi Henshū Iinkai. (1988). *12-nen: Sanki Shimura Shisei no Ashiato*. Otaru City.

Tamura, Y. (2009). *Otaru Unga Monogatari*. Kajima Shuppankai.

Chapter 5
The Logic of Preservation: The Preservationists' View of the "Canal Issue"

Abstract This chapter provides a detailed analysis of the canal preservation movement. Based on thirty-three years of fieldwork and extensive life history analysis of preservation activists, I illuminate the canal's significance to the preservation movement and explain how the canal came to symbolize the identity of the local community. I divide activists into four types (the "aesthetic school," the pure preservationist faction, the community development faction, and the traditional leftist faction), and describe their respective rationales for preservation and how they competed for leadership within the movement. Unlike local officials, members of the preservation movement saw the canal as a "place," but over time the movement's objectives evolved from static or "cryo" preservation to "community development." The ultimate goal of Otaru's preservationists was thus something very different from what the word "preservation" suggests. What activists actually sought was "social control over change." They recognized that their beloved hometown of Otaru would change and that this change was inevitable. What activists really wanted was to realize this change in the manner, and at the speed, of their own choosing.

Keywords Preservation movement · Social control of change · Life history

The city government's efforts to bring change to Otaru were resisted by those who rejected development and sought to preserve the canal district. This chapter focuses on the Association to Protect the Otaru Canal (APOC), the primary group behind the preservation effort, and the logic informing the actions of its members. The history of APOC activism can be divided into three periods: an initial phase from 1973–1976, a second phase from 1977–1984, and the "post-canal" period beginning in 1985.

S. Horikawa, *Why Place Matters*,
https://doi.org/10.1007/978-3-030-71600-4_5

5.1 The Canal as Cultural Property: The Logic of the Preservation Movement During Its First Phase (1973–1976)

In the early days of the movement, activists sought to save the canal by identifying it as an important cultural asset. Indeed, this period was characterized by activists' shared belief that the rare and valuable site of the canal was what defined their hometown, and thus could only be an important cultural property.

The story of the preservation effort begins with Fujimori Shigeo, an Otaru native who played a crucial role in the early days of the movement. Alarmed by the destruction of old warehouses in the Arihoro district of Otaru, Fujimori resolved to save the canal from a similar fate. The Arihoro district was a visually distinctive area of Otaru with numerous storehouses built with the local soft stone and characterized by their lined and gabled roofs. For Otaru's residents, however, the scenery of the Arihoro district was simply the familiar backdrop to everyday life. Schoolchildren would pass through the neighborhood on their way to play along the canal or at the port. The destruction of the Arihoro warehouses filled Fujimori with a profound sense of loss—and of urgency, as he realized the canal would become the next target for destruction. Fujimori, who had graduated from Tama Art University in Tokyo and made his living painting billboards that often depicted the scenery of Otaru, understood the disappearance of the familiar townscape of his hometown to mean that "Otaru would no longer be Otaru" (Ogasawara 1986: 48–65).

Fujimori was not alone. His concerns were shared by one of Otaru's most prominent citizens, the local historian, and businessman Koshizaki Sōichi. Koshizaki owned Koshizaki Shōten, a food wholesale business, and had chaired the Hokkaido Cultural Property Protection Association since 1961, as well as a local consultative committee on cultural properties since 1966. He viewed any drastic change to the canal district, which had once been so crucial to the city's growth, as an intolerable development. The destruction of the Arihoro district warehouses near the canal made the real implications of the city's Rinkōsen road plan painfully clear to Fujimori and Koshizaki. Fujimori started recruiting members for a new group, the Association to Protect the Otaru Canal, in October 1973. At the group's first meeting on December 4, Koshizaki was elected APOC chairman and Fujimori was chosen to head the secretariat. The two men would remain the driving force behind the movement for the next three years. Within the APOC leadership structure, they were supported by Yoneya Yūji (executive manager), Toyotomi Tomō (organization manager), Horii Toshio (finance), and Chiba Shichirō (publicity) (Table 5.1).

The accounts of early members put the APOC's membership during this period at approximately twenty-four[1]: sixteen men and eight women, ranging in age from their twenties to sixties. Their ranks included a school teacher, an artist, a carpenter, the

[1]This is the number given by Mineyama Fumi in a November 4, 1997 interview with the author, conducted in Otaru. Other documents refer to twenty-five, not twenty-four, key members but this discrepancy is unsurprising, given that membership was not clearly defined during the APOC's early days. Here twenty-four is used as the provisional number.

Table 5.1 The Organizational Structure of the APOC in 1973

Position	Name	Occupation/Organization
Chairman	Koshizaki Sōichi	Local historian, president of Koshizaki Shōten
Secretariat	Fujimori Shigeo	Billboard artist, Fujimori Co., Ltd
Executive manager	Yoneya Yūji	Writer, publisher of *Gekkan Otaru*
Publicity	Chiba Shichirō	Artist
Organizational manager	Toyotomi Tomō	Educator
Treasurer	Horii Toshio	Architect

manager of ironworks, a graduate student, a coffee shop proprietor, a newspaper sales office worker, and a full-time housewife. Significantly, membership in the APOC was not determined by political affiliation: the APOC included a member of the Hokkaido chapter of the left-wing Japanese teachers' union (Nikkyōso), as well as small business owners who supported the conservative Liberal Democratic Party. Rather, membership was primarily determined by personal connections between friends, acquaintances, and family members; founders' meetings were run "by whoever was able to attend."

By January 20, 1974, the APOC's official membership stood at 183. Of this number, eighty-two members were somehow involved in culture or education, while ordinary citizens comprised the remainder (Otaru Unga Mondai o Kangaeru Kai 1986a: 31–32). Most had simply agreed to lend their names to the movement without becoming active members. The real number of active members hovered somewhere between ten and twenty-five.

What were the goals of the movement during this period, and what message did activists emphasize? What ideas guided the movement? What was forefront in the minds of the movement's early leaders? The APOC's draft regulations and the *APOC Handbook*, created in the early days of the movement, offer some insight into these questions.

According to the draft version of the APOC's regulations, the group sought to: "preserve the Otaru Canal, which is subject to land reclamation under the city's plan, the neighboring stone warehouses, and other historic buildings as important cultural properties of historical significance." The APOC would "take necessary steps toward achieving this goal, including petitioning relevant authorities to change the city's plan." (APOC, 1974: 8). Activists understood these "necessary steps" to mean the petitioning of all relevant parties and educating the general public on the need for preservation.

The guiding spirit of the movement is evident from the *APOC Handbook*:

> (2) The Spirit of the Otaru Canal Preservation Movement
>
> {…}
>
> "We appeal to all Otaru citizens to join us as individuals united solely by the desire to 'protect the irreplaceable canal,' without regard to personal ideologies or beliefs, occupation, age, gender, *connection* to the canal, awareness of the canal [issue], or position on redevelopment."
>
> APOC, 1974:1; emphasis in original.

"Connection to the canal" presumably refers to those with commercial interests in the canal district, or contractors who stood to benefit from the Rinkōsen project. In other words, the APOC sought to craft an apolitical citizen's movement with "participation on an individual basis," motivated "solely by the desire to protect the irreplaceable canal."[2]

The unusually emotional tone of this passage is striking, and all the more evident in the following excerpt from the *APOC Handbook*, regarding the "origin" of the preservation movement:

> There is more than a bit of *old-fashioned emotion* in the hearts of the citizens who are attempting to save the "canal." Actually, they are almost entirely filled with this emotion.
>
> In some undefinable way, we love the canal and feel that paving it over would be a pitiful [fate]. This is an *irrational* sentiment... {...} Our movement stems from the energy that lurks deep in the hearts of each *Otarukko* [Otaru native], from their *love of their hometown*.
>
> APOC, 1974:1; emphasis in the original.

It was this love of hometown characterized by "irrational" and "old-fashioned" sentiment that animated APOC members. Neither Koshizaki nor Fujimori wanted the APOC to become "a political tool," and this desire likely informed their elaboration of the association's "spirit." (Ogasawara 1986).[3] Indeed, the use of emotionally-charged language allowed the APOC to distance itself from any perceived connection to political parties or movements, and present itself as a moderate cultural movement focused exclusively on the canal.

The APOC immediately began petitioning Mayor Inagaki, the chairman and vice-chairman of the city council, local party heads, local representatives to the Hokkaido Prefectural Assembly, and the Hokkaido Civil Engineering Agency. On December 10, the APOC presented Inagaki and his assistants with a proposal to alter the route of the Rinkōsen in order to avoid paving over the canal. The same proposal was presented to the Civil Engineering Agency on December 13. The nature of these early petitions is eloquently put forth in the following abstract:

> Petitioner: Koshizaki Sōichi, The Association to Protect the Otaru Canal (*address omitted*)
>
> Overview
>
> {...}
>
> With the canal at its heart, the Otaru port imbues the lives and feelings of each and every of the many generations of citizens who live and were raised in Otaru with vitality. Both before and after the war, [the port] has been crucially significant to the historical formation of our hometown, Otaru, and remains so as we resolutely pursue regeneration. Not only is the "canal" itself of obvious historical value, but the distinctive architecture of the stone

---

[2]Given the APOC's limited capacity, it seems clear that, at least during this early period, APOC members did not recognize the need to specify precise bylaws to govern their decision-making process or voting processes. Matters were thus "decided upon by the people who showed up" to meetings. Although there was some consideration of members' factional interests (the *APOC Handbook* refers to participants "as individuals," "without regard for...[their] connection to the canal)," the lack of clear decision-making procedures became only too apparent in the final days of the preservation effort.

[3]Interview with Naka Kazuo in Otaru, March 16, 1993.

> warehouses in its vicinity are without equal anywhere in Japan. In recent years, there has been growing recognition of this district's cultural significance, to which numerous specialists can attest. {...}
>
> Not only the physical attributes of the canal, but also its spiritual, cultural, and touristic value are unique to Otaru. The canal is a major presence that should remain as an encouraging guide to the future for Otaru's next generation, who will assume responsibility for the city's development.
>
> It would be extremely regrettable if even a portion of the canal were to vanish due to the changing times or demands for development. The enormity of this loss to citizens should not be dismissed as mere sentimental argument. We have never opposed development plans that are essential to the development of our hometown of Otaru. We have high hopes for port-based regional development and to that end, we support the construction of the Rinkōsen in the most effective and appropriate form.
>
> The problem lies with the conclusion that the "Otaru canal" is a useless object that hinders the city's plans, and with a plan that stipulates that development is impossible unless the canal is reclaimed. We strongly suspect that this conclusion will not elicit much sympathy or support from the generous souls who comprise the majority of Otaru's citizens.
>
> When the Rinkōsen road plan was adopted (in 1967), the city did not actively convey its intentions, but we believe there is enormous significance to reconfirming—even belatedly—the meaning behind the existence of the "Otaru canal and its environs" and considering whether it would be possible to alter the [Rinkōsen] plan. We do not oppose the city's plan or the Rinkōsen road. In the hopes of contributing to the next century of development in our hometown, we are working to mobilize the citizens of Otaru: young and old, men and women, with no regard to ideology or beliefs. From the standpoint of [Otaru's] citizens, the APOC is prepared to conduct further study of such issues as devising an ingenious plan to route the Rinkōsen so that it circumvents the canal and the subsequent redevelopment of the preserved [canal and surrounding area], and we will passionately take up the campaign [toward this end]. With regard to changing the plan [of the Rinkōsen], for now we have submitted a diagram with our counter-proposals, and request the opportunity and venue for deliberation.
>
> Petition 161, Submitted to the Otaru City Standing Committee on Construction. Received December 17, 1973. Replicated in Otaru Unga Mondai o Kangaeru Kai, ed., 1986a: 25–26. Information in parentheses has been added by the author, information in brackets by the translator.

Here the APOC identifies not only "the physical attributes of the canal, but also its spiritual, cultural, and touristic value" as "unique to Otaru," and the canal as a "major presence" that "should be handed down" to future generations. The effort to save the canal was thus a cultural movement, not a political one. The problem, as the APOC saw it, was the city government's "conclusion that the canal is a useless object that hinders the city's plans" and "a plan that stipulates that development is impossible unless the canal is reclaimed." The APOC cast doubt on whether "this conclusion" would "elicit much sympathy or support from the generous souls who comprise the majority of Otaru's citizens."

In short, the APOC understood the canal and the surrounding area as a single entity with "spiritual, cultural, and touristic value." Consequently, the APOC appealed to municipal authorities to reconsider the Rinkōsen plan, which identified the canal, so "unique to Otaru," as "useless." Significantly, the APOC did not oppose the Rinkōsen plan itself, but merely suggested revising the route of the new road. Koshizaki's explanation as he submitted Petition 161 illuminates this point:

> Explanation of Petition 161
>
> Speaker: APOC Chairman Koshizaki Sōichi
>
> "The canal, which extends from Minamihama to Kitahama, has contributed to Otaru's economy for half a century. Such a large canal cannot be found elsewhere in Hokkaido or mainland Japan. The canal was still clean when we were children, and we would swim or fish there. It was wonderful. And so a few of us wondered whether there might be some way to avoid destroying the canal, and got together to form this association.
>
> In the early December of last year (1973), we met with the mayor and submitted a petition. The mayor told us that while he understood our position, the matter had already been decided at a meeting of the port council in 1966 and there was no way of changing course now, because construction was already underway. The mayor asked us at the time if by asking [the city] not to reclaim the canal we were also rejecting the construction of the bypass. This may sound greedy, but *we want to avoid reclaiming the canal and move the bypass to a different location.* The mayor told us this was unthinkable. {…}
>
> Minutes of the Standing Committee on Construction, February 15, 1974. Record set: 1971–1974, Registration no. 157. Emphasis and information within parentheses has been added by the author.

Koshizaki's statement confirms that both the city government and the APOC recognized the *necessity of development* and the *necessity of the bypass road.* Thus, from the very beginning, the APOC supported an alternate road plan that would avoid the need for land reclamation at the canal (APOC, 1974: 5–6). In fact, the APOC's alternate plan included three different proposals (Fig. 5.1).[4]

The *Asahi Shinbun* reported Mayor Inagaki's response to the APOC's petition and alternate proposals:

> The mayor responded (to the petition) by discussing the difficulty of saving the canal, saying: "Personally, I would like to save the canal if that were possible, but we have already adopted a plan to reclaim land from the canal for the construction of the Rinkōsen road, and the expense of preserving the canal together with the neighboring stone warehouses would be colossal. The construction of this road is a matter of life and death for Otaru, and we have received forceful demands from people in port-related businesses, who tell me that 'the canal is obstructing Otaru's development, and should be paved over.'"
>
> "Unga Mondai Jyūnen no Ayumi," Part 1, No. 8, *Asahi Shinbun*, Hokkaido edition, December 13, 1983. Information in parentheses has been added by the author.

It seems clear that Inagaki viewed the Rinkōsen construction project as a settled matter and did not give serious consideration to the APOC's counterproposals. The APOC submitted another petition during the fourth session of the Otaru city council, where it was tabled for further deliberation. City officials did not yet view the APOC as a formidable rival. At this point in time, it would seem they simply decided to keep a watchful eye on the nascent preservation movement.

---

[4]There are two variations to the third proposal, so technically speaking, there are five alternative plans. Here we will follow the APOC literature and refer to three alternate plans.

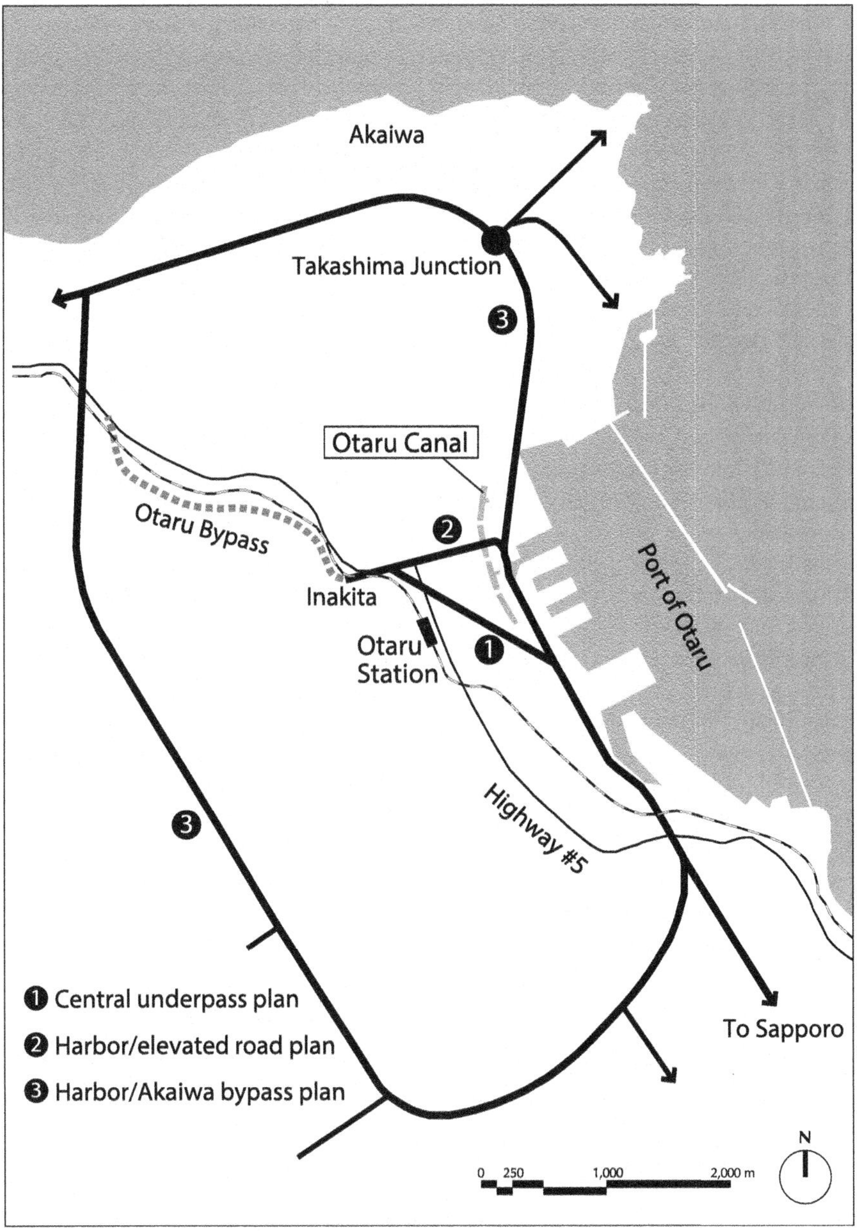

Fig. 5.1 An Overview of the APOC's Proposals for Relocating the Rinkōsen.
Source Replicated and enlarged from Otaru Unga Mondai o Kangaeru Kai, ed. (1986a)
1. Build a tunnel through the city center (Central Underpass Proposal)
2. Devise a route that would pass through the base of the wharf area and link to the Inakita intersection. An elevated road would comprise one portion of this route (the Elevated Road/Harbor Proposal)
3. Plot a route that would pass through the base of the wharf area before making a large detour to the Takashima/Akaiwa district (Harbor/Akaiwa Bypass Proposal)

Shortly after the submission of Petition 161, the Otaru Warehouse Association and the Otaru Chamber of Commerce launched a counteroffensive, submitting their own petitions urging the "early completion of the Rinkōsen." Thus began the rival lobbying campaigns of the "canal preservation faction" and the "road faction."[5] Meanwhile, the Otaru city government proceeded with the construction of the Rinkōsen road from the Sapporo end. However, the oil shock of 1973 led to substantial delays in construction. This gave activists a window of opportunity: between February 23–26, 1974, they organized public speeches and an "Otaru Canal Exhibition" in the event hall of a local department store.

The Otaru Junior Chamber (JC) responded to these developments by organizing a public debate on the canal's fate as an opportunity to educate all Otaru residents on the issue. The April 16 debate was attended by more than one hundred people, including Otaru's deputy mayor, director of civil engineering, and members of the APOC leadership, but no headway was made in the course of the two and a half hour debate ("Unga Mondai Jyūnen no Ayumi" 1984; Otaru Unga Mondai o Kangaeru Kai ed. 1986b: 274). For its part, the Otaru JC maintained a strict neutrality on the issue until the very end.

In May, the APOC's Fujimori set off across Japan to visit other towns, such as Kurashiki and Tsumago, known for their pioneering efforts to preserve historic townscapes, and met with local preservation activists. Fujimori's "townscape pilgrimage" allowed him to witness the developments and driving forces behind preservation efforts at a time when the national cultural preservation authorities were shifting focus, from "individual buildings to groups of buildings, and from elite architecture to vernacular architecture" (Kihara 1992: 1–58). The three-week pilgrimage was an important initiative, intended to create opportunities for exchanges between local preservation activists.[6] On June 1, Fujimori's journey culminated in a visit to the Agency for Cultural Affairs in Tokyo. There Fujimori met with agency officials overseeing the preservation of historic buildings and elicited an important opinion from these authorities: the canal and surrounding stone warehouses were excellent examples of Japan's national heritage (Ogasawara 1986: 59, 101–103).

> We weren't making any headway (no matter how much we stressed the canal's value to local citizens in our petitions to the city government). So I went to the Agency for Cultural Affairs. I carried all the [relevant] documents on my back. And lo and behold, [I met with] the most influential person at the agency, Itō Nobuo—the section chief responsible for the buildings

---

[5]Between January and March 1974, the Otaru Warehouse Association and the Otaru Chamber of Commerce petitioned the city council for the early completion of the Rinkōsen highway, as did the neighborhood associations of Sakaimachi and Ironaimachi, which are close to the canal and the wharves (Miyano, ed. 1975: 27–29). Thereafter, the "road faction" and the "preservation faction" made repeated petitions on behalf of their cause. The *Asahi Shinbun* reported that the Otaru city council secretariat had received more than seventy-seven petitions on the canal issue during the period between the APOC's establishment in 1973 and March 1984. "Unga Mondai Jyūnen no Ayumi," Part 2, No. 1, *Asahi Shinbun*, Hokkaido edition, May 8, 1984.

[6]In chronological order, Fujimori's destinations were Osaka, Kobe, Kurashiki, Okayama, Kyoto, Nara, Kameyama (Mie Prefecture), Nagoya, Takayama (Gifu Prefecture), Tsumago (Nagano Prefecture), Nagoya, Yokohama, and Tokyo. "Unga Mondai Jyūnen no Ayumi" Part 1, No. 22–28, *Asahi Shinbun*, Hokkaido edition, January 18–26, 1984; Ogasawara (1986: 101–103).

> section of the Cultural Property Preservation [Division]. {...} [He told me] that this year they had secured a budget for conducting full-scale surveys of groups [of buildings] for the first time. [He told me] he wanted to conduct a survey of townscape preservation. And that they were pursuing the amendment of regulations right now, through Diet legislation. And that [agency officials] had gone around Japan, making assessments. They had visited Otaru, and I asked [Itō] what his impression had been. He told me that he thought [the canal district] was of exceedingly high merit. I was thunderstruck. Even the Agency for Cultural Affairs was willing to say that plainly.
>
> Interview with Fujimori Shigeo, conducted and transcribed by Ishizuka Masaaki in Ishizuka 1975: 80. Information in brackets added by the translator.

Here Fujimori expressed his irritation with the fact that "even" someone at the Agency for Cultural Affairs in faraway Tokyo could recognize the value of the canal, while local officials could not.

Miyano's (1975) account of Fujimori's "pilgrimage" supports Fujimori's own recollections:

> With a rucksack on his back, APOC Secretary Fujimori Shigeo visited the Agency for Cultural Affairs on May 31 ((1974)) to appeal for the canal's preservation. Itō Nobuo, section chief for [historic] buildings within the Cultural Property Preservation Division, told Fujimori: "[the canal's] merit as a cultural heritage [site] is extremely high. The one hundred-odd stone warehouses along the canal are the best in Japan. I imagine they will be destroyed if a road is constructed, so I'd like to conduct a survey immediately, without waiting for revisions to the law (from individual buildings to groups of buildings). If the city requests such a survey, I'll allocate a budget for this straight away."
>
> Miyano, ed. 1975: 30. Information in single parentheses appears in the original, information in double parentheses has been added by Horikawa, information in brackets has been added by the translator.[7]

Fujimori arrived at the agency wearing a rucksack because "I carried all the materials on my back." Indeed, he had managed to cram various documents pertaining to Otaru, as well as signed petitions in support of the canal's preservation, into his rucksack.[8]

The Agency for Cultural Affairs' assessment of the "extremely high" value of the Otaru canal was the highlight of Fujimori's pilgrimage across Japan. The fact that national authorities were both ready and willing to conduct a survey of Otaru's canal district was also significant: that the national government would go so far while the municipal government obstinately refused to acknowledge the canal's cultural value gave powerful ammunition to local preservation activists. At the same time, the

[7] Miyano puts the date of Fujimori's visit to the Agency for Cultural Affairs as May 31. Ogasawara (1986: 103) gives the date as June 1, but writes elsewhere (1986: 59) that "in May 1974 [APOC] secretary Fujimori Shigeo made his appeal to the Agency for Cultural Affairs." Based on the *Asahi Shinbun* reporter Shin'chi Mitsuo's interview with Fujimori himself, this book uses the date of June 1, 1974. "Unga Mondai Jyūnen no Ayumi," Part 1 No. 28, *Asahi Shinbun,* Hokkaido edition, January 26, 1984.

[8] This account corresponds with the one given in the *Asahi Shinbun*: "Fujimori says that he used the floor of the Agency for Cultural Affairs to spread out all the materials he had packed into his rucksack or brought to the meeting wrapped up in cloth." "Unga Mondai Jyūnen no Ayumi" Part 1, No. 28 *Asahi Shinbun*, Hokkaido edition, January 26, 1984.

central government could not automatically involve itself in local affairs, and agency officials made it clear they would not undertake a survey unless one was requested by the Otaru municipal government. It was thus up to the preservation movement to create the opportunity for national intervention, by convincing local officials of the need for such a survey. In this sense, the "canal problem" remained very much a local issue.

News reports of the central government's position did give momentum to their cause. The newly public position of the Agency for Cultural Affairs meant that the Otaru city government could no longer simply keep a watchful eye on the preservation movement. At the December 1974 meeting of the city construction committee, members rejected the APOC's Petition 161 and adopted another petition in favor of the Rinkōsen's construction. At the meeting, opposition party members supported the preservation effort and argued that the petition ought to be carried over to subsequent sessions for further deliberation. They were opposed by pro-construction LDP council members, who pushed for a speedy conclusion to the matter. In an extraordinary deviation from normal proceedings, the committee remained deadlocked in negotiations for hours before reconvening shortly before midnight and, in under a minute, voting on the various petitions. The vote, which some have characterized as a case of steamrollering,[9] is notable as an effort by Mayor Inagaki and his supporters to bring the canal issue to an early close before their opponents grew more powerful.

Undeterred, the APOC submitted a petition to the Hokkaido prefectural assembly in March 1975. Although this petition was adopted unanimously, it did not have a significant political effect. While the Rinkōsen construction project was primarily funded by the prefecture, the Otaru municipal government had drafted the original road plan and submitted it to the prefectural government for deliberation and approval. The prefectural government wanted the matter to be settled locally. While the APOC was able to leverage the opinions of the Agency for Cultural Affairs and the prefectural assembly to a certain point, ultimately the matter continued to be referred back to Otaru.

A mayoral election was held in April 1975. The candidates to replace the outgoing Inagaki were Shimura Kazuo, a deputy mayor in the Inagaki administration running with the endorsement of the LDP, and the JCP and JSP-endorsed Toyotomi Tomō, who also belonged to the APOC leadership. The election resulted in a resounding victory for Shimura, who viewed the Rinkōsen project as a settled matter and avoided any reference to "the canal issue" during his campaign. Toyotomi had stressed the need to save the canal, but in keeping with the APOC's principle of "not being used as a political tool," ran without the association's explicit endorsement or support. The election results, which followed a campaign lacking in substantive debate, suggest the canal's fate was not yet seen as an issue of paramount political significance.

What is interesting, from the perspective of understanding the evolving "logic of preservation," is that at this point in time, the justification for saving the canal was understood as an extension of its status as a "cultural property." The canal district

[9]See Ogasawara (1986: 86); "Unga Mondai Jyūnen no Ayumi" Part 2, No. 3, *Asahi Shinbun*, Hokkaido edition, May 10, 1984.

evoked "old-fashioned emotions" and should not be tampered with—this was the logic for preservation during the initial phase of the preservation movement, and preservation activists looked to earlier campaigns to protect cultural properties for their frame of reference. This was only natural: APOC Chairman Koshizaki had served successive terms as the chairman of Otaru's committee on cultural properties, and at the time the APOC was established, the Law for the Preservation of Cultural Properties was the logical framework for any effort to save something old and precious. Given this, there was no question that the final stop on Fujimori's "pilgrimage" would be the Agency for Cultural Affairs and that the goal of his visit would be the commissioning of a survey, conducted under the guidance of authorities responsible for Japan's cultural properties.

On June 24, two months after the mayoral election of 1975, the APOC held an inaugural meeting as a formal entity with a steadily growing membership. According to the APOC's report, one hundred people attended the inaugural meeting, while overall membership stood at 1,200 (Otaru Unga Mondai o Kangaeru Kai, ed. 1986a: 47-51). To be sure, many had simply lent their name to the preservation effort, but the inflated number was a tactic to make the movement appear larger than it actually was to gain influence.[10]

The newly inaugurated APOC went on to propose a comprehensive survey of the canal district. In August, APOC representatives met with city officials to discuss the possibility of a survey. However, they could not agree on who should undertake the survey or how it should be conducted—indeed, the very premise of such a survey remained a point of contention between the two sides. Over the course of seven meetings, both sides attempted to establish advantageous conditions for the survey, to which the other party could not agree. These talks were finally aborted in September 1976.

Through petitions and informal meetings with city officials, the APOC had attempted to resolve the canal's fate through discussion. Having exhausted these methods, the preservation campaign reached an impasse. Given the mounting "malicious pressure" on his family business, Fujimori Shigeo resigned from his position as APOC secretary in late June [1976] ("Unga Mondai Jyūnen no Ayumi," Part 1, No. 34, *Asahi Shinbun,* Hokkaido edition, February 3, 1984). On October 19 of the same year, APOC Chairman Koshizaki committed suicide.[11] According to some accounts,

[10]In the words of local activist Yamaguchi Tamotsu: "You know, in that sense, citizens' movements are political. Strategically, it was important to represent a ten-person organization as something bigger." Interview with Yamaguchi Tamotsu in Otaru, September 16, 1988.

[11]Ogasawara Masaru, who supported the preservation movement from Sapporo and led the Otaru Unga o Kangaeru Kai, was well-acquainted with the internal circumstances of the APOC, and wrote: "in yet another blow, in the October of that year (1976), [APOC] chairman Koshizaki took his own life. There were reports that he had suffered a breakdown due to his failed management of a hotel, but the news came as a shock to the APOC" (Ogasawara 1986: 105). The *Asahi Shinbun* series on the Otaru canal issue also reported Koshizaki's death as a "suicide" ("Unga Jyūnen no Ayumi, Part 2, No. 30, *Asahi Shinbun,* Hokkaido edition, June 19, 1984). Meanwhile, Tamura writes that "Fujimori died shortly after resigning as APOC secretary, and Koshizaki died soon after that. It was rumored among those who knew them that they had not died from illness, but had taken their own lives" (Tamura Nobuko 2009: 96). Tamura's account mistakes both the cause and timing of

Koshizaki had become increasingly neurotic after being harassed for his involvement in the preservation effort. Other reports cite his despondence over a failed business venture. Yet the true circumstances surrounding Koshizaki's suicide remain unclear to this day.

Koshizaki's death and Fujimori's resignation had an enormous effect on the preservation effort. At this point, almost all the original members of the APOC leadership had defected. Not only had activists confronted the difficulty of opposing a fully funded construction plan seven years in the making, but having based their activities on the example of the cultural property campaign, they found themselves unable to forge a new approach to the cause. The movement to save the Otaru canal had stalled and seemed to have exhausted all options.

During this initial phase of the movement, preservation activists approached the canal as a precious "cultural property" and an essential part of Otaru's identity. They argued that the Rinkōsen should be rerouted so that the canal could be preserved in its entirety. Their resistance to change was based on one simple explanation: "the canal is a cultural property." As such, they could do little more than identifying the problem at hand and failed to develop the countervailing political force capable of eliciting a compromise from the city administration. For their part, city officials refused to engage in any substantive discussion of the validity of protecting the precious canal district townscape. Their response to the emotional appeals of activists began and ended with references to procedural correctness: "The [canal] matter has already been settled." The two sides debated the issue in entirely different terms, making their continued disagreement inevitable.

Meanwhile, the construction of the Rinkōsen road proceeded quietly in the direction of Otaru.

5.2 The Canal as Tourism Resource: The Logic of the Preservation Movement in Its Second Phase (1977–1984)

The preservation movement that had stalled following Koshizaki's death and Fujimori's resignation would soon be revived by new supporters with a different ideology. This section examines the dramatic comeback of the preservation effort during the second phase in the "canal war," which unfolded between 1977 and 1984.

First, however, it is worth asking how the average resident of Otaru understood the "canal problem." Initially, the residents of Otaru were either indifferent to the canal's fate, or viewed the canal as a smelly nuisance. To be sure, as the battle over the canal wore on, people grew more familiar with the place in question. Yet how a resident

---

the two men's deaths. Fujimori resigned as APOC secretary in 1976, but while he battled chronic health problems, he went on to run a gallery dedicated to the canal in Otaru. He died thirteen years after Koshizaki, on January 13, 1987. These factual errors call into question the overall credibility of Tamura's (2009) documentary, *Otaru Unga Saisei no Dokyumentarī*.

felt about the canal's fate depended on his or her personal experience of the place. The canal as a physical space had lost much of its importance to Otaru's citizens: it no longer fulfilled an economic function, and barges rarely passed through its waters. Abandoned boats lay submerged in the waters of the canal, and many residents had taken to throwing their garbage into the canal under cover of darkness. At some point, the city had stopped dredging sludge from the canal. The only visitors to the canal were a handful of amateur painters and fishermen. It was because of this general indifference that the smell of the canal had grown so oppressive. For those living or working in the vicinity of the canal, however, the smell of methane gas emitted by the accumulated sludge in the canal was an intense and immediate problem. Let us turn to the testimony of a man who worked in the Kyōwa Hama Building, a four-story building of reinforced concrete constructed on the banks of the canal in 1933, and later designated as a historic structure by the city of Otaru:

> Interviewer: …What is your reaction to talk of the canal being destroyed?
>
> Respondent: I'm convinced [it's the right thing to do]. For us, and for the canal, this is the beginning. All the rivers carry garbage (into the canal). The city used to dig up two or three meters of [sludge] at a time and carry it out to dump in open water. The city can't do that anymore, now that it's out of money. And so, in summer, methane gas oozes up from the bottom. People living in the area are driven crazy by the smell.
>
> Interviewer: It's quite smelly, you say?
>
> Respondent: "Quite" doesn't begin to describe it! You don't see many private homes along this street, do you? That's why no one has made a fuss. But if these buildings had all been private homes, [their owners] would have protested and had the canal destroyed or covered over a long time ago. The flies and the stench are really unbearable. In midsummer, for instance, it gets into your throat when you're asleep at night. That's why [paving over] the canal is such a popular idea. There's not a single person on this street who's against it. On the other hand, I guess this "Save the Canal" group and the painters oppose the idea.
>
> Interview conducted by Ishizuka Masaaki 1975: 193. Quoted matter has been transcribed exactly as it appears in the source text. Parentheses have been added by the author, brackets by the translator.

The canal was so polluted that residents of the area "were driven crazy by the smell," and "convinced" the canal should be destroyed. Yet people living in the hilly areas of Otaru, at a remove from the canal, had no idea the smell of the canal was such a problem. Their ignorance—coupled with the ready assent of the canal district's immediate residents—explains why the city's Rinkōsen plan remained unopposed for seven years. It also explains the duality of Otaru residents' approach to the canal issue: the urgent demands by some to resolve a pressing pollution issue were countered by those who, viewing the canal from a greater distance, longed to preserve a distinctive landscape from their past—in many cases, the backdrop to their earliest childhood memories.

The advent of the APOC, and growing newspaper coverage of the group's activities, began to raise the awareness of Otaru residents. Unfortunately, there is little data from this period that can be used to ascertain the extent to which Otaru residents took an interest in the canal's fate, or the strength of popular support for the city's reclamation plan. Our only source is a June 1977 survey undertaken by a group of students at the Otaru University of Commerce who had formed a study group on Otaru's economic history. Seven third-year students set out to ascertain how the average Otaru citizen perceived the canal, now that it no longer functioned as the economic backbone of the city.

Survey period: June 5–19, 1977

Survey method: Door-to-door interviews

Sample size: 800 households (there is no data on the response rate)

Criteria for selection: Students divided Otaru into fifteen districts and took a random sampling of 40–80 households in each district, based on population.

Respondent profile: 60% of respondents were female. 70% of respondents were over the age of forty.

Survey structure: Five-question survey

Survey content:

Question 1: Do you know about the Otaru canal?

Question 2: Have you ever been to the canal?

Question 3: Do you know about the plan to turn the canal into a road?

Question 4: What do you think about the canal becoming a road?

Question 5: Please share your impressions of Otaru and any comments.

Toshi Isan Kenkyūjo ed, 1977: 69–75; "Otaru Unga 46% ga 'Nokoshite,'" *Asahi Shinbun*, Otaru city morning edition, September 9, 1977; "Hatsu no 'Unga Ishiki Chōsa' Matomaru," *Hokkaido Shinbun* (Otaru city morning edition), September 9, 1977.

This survey has several fatal defects. For instance, there is no information on the response rate or the number of valid responses. Thus, the survey's reliability cannot be ascertained, nor can the data be subjected to reanalysis. However, given that the survey represents our only insight into the attitude of Otaru citizens at that time, let us briefly turn to the results.

According to the survey, 95% of respondents knew about the canal, and 80% had visited the canal at some point. The fact that 80% of respondents knew about the road project indicates that the canal issue was growing in visibility (Table 5.2). When asked whether they supported the city's plan to pave over the canal to build the new road, 46% replied that they would prefer to save the canal, 27% supported the city's road plan, and another 27% either did not know or were indifferent to the canal's fate (Toshi Isan Kenkyūjo ed, 1977: 72). Notwithstanding the survey's dubious reliability, the fact that the results were widely reported in local newspapers suggests that its impact on Otaru at the time was significant ("'Unga Ronsō' ni Mondai Teiki," morning edition of the *Hokkai Times*, September 9, 1977).

In October 1977, Mayor Shimura announced the city government would conduct its own survey of the canal district, premised on the plan to pave over the canal. This prompted a backlash from the APOC (Otaru Unga Mondai o Kangaeru Kai ed., 1986b: 275), which published an interim report on its own canal survey, conducted with the backing of the Hokkaido Chapter of the Japan Scientists' Association, and Hokkaido University's Architectural History Laboratory and Institute for Cityscape

Table 5.2 Results of the otaru economic history research group survey on the otaru canal

Question 1: Do you know about the Otaru canal?	
Yes	95
No	5
Question 2: Have you ever been to the canal?	
Yes	80
No	20
Question 3: Do you know about the plan to turn the canal into a road?	
Yes	80
No	20

Source Otaru Unga o Kangaeru Kai, ed. (1977: 69–75)

Preservation (Toshi Isan Kenkyūjo) (APOC ed., 1977).[12] In the same month, the Otaru Chamber of Commerce countered the APOC's efforts by organizing a pro-Rinkōsen development association. Nevertheless, public support for the APOC continued to grow.

The Japan National Trust initiated its own survey of the Otaru canal in May 1978. The following month, the Association to Think About the Otaru Canal Issue (Otaru Unga Mondai o Kangaeru Kai) was established in Sapporo. At this meeting, cultural authorities from across Hokkaido voted to support the APOC's efforts. In July, residents of the Inakita shopping district of Otaru, who faced the prospect of eviction to make way for the Rinkōsen, formed their own association to "unconditionally oppose" the planned route of the new road. The APOC had moved beyond mere petitions, and now used surveys—and the published results of these surveys—to launch a broader public appeal. The preservation campaign had outgrown the simple framework of "the canal as cultural property." What accounts for this transformation?

The defining characteristic of this second phase in the preservation movement was the influx of new blood. The fortunes of the APOC were revived by a change in leadership, as the sixty-year old housewife Mineyama Fumi took over the APOC chairmanship, and the arrival of many new members from the outside world—primarily young people in their twenties and thirties, including many Otaru natives (sometimes referred to as the "U-turn group").

The motivations of these young members were quite different from those of the APOC's founders. Sasaki Kyojirō belonged to a prominent local family and had traveled throughout Europe before returning to Otaru. Ogawara Tadashi was a leading student activist at Tokyo University before returning to Otaru to take over his

[12]This is a 185-page handwritten document issued by the APOC, but the imprint attributes the editing solely to the Toshi Isan Kenkyūjo (Hokkaido University Institute for Cityscape Preservation). Prominent figures in the architecture world contributed to the report's preface: Matsumura Teijirō (then of Tokyo University), Yoshizaka Takamasa (Waseda University), Koshino Takeshi (Hokkaido University), the architecture critic Hasegawa Takashi, and Itō Teiji (Kōgakuin University). Their names demonstrate the APOC's growing strength and reach during this phase of the preservation campaign.

father's soba noodle restaurant. Disillusioned with the radical student movement that consumed Japanese universities in the 1960s and early 1970s, Yamaguchi Tamotsu traveled across Europe and Japan before making an impromptu stop in Otaru, where he decided to remain. Despite their different backgrounds, all three young men deliberately chose Otaru as the place where they would settle and make their living. And if they were to live in Otaru, they felt the city needed to offer its young people employment, along with "something to feel passionate about." All three made a conscious decision to involve themselves in the local community. The decision of these young men to defend the canal had little to do with an interest in "the preservation of a cultural property."

Moreover, given their unhappy experiences in the Japanese student movement, Ogawara and Yamaguchi were reluctant to reengage with national politics. They gravitated toward a different approach: addressing issues specific to their local community, where the smaller population size made the viability of their causes easier to ascertain (Horikawa 1994). Yamaguchi, for example, describes his own involvement with Otaru as an attempt to "try and achieve something at the regional city level, rather than the national level."[13]

Three young men from the architecture department of Hokkaido University—popularly referred to as the "Hokudai[14] trio"—also wielded a decisive influence over this period of the preservation movement. These three young architects and urban planners (Yanagida Ryōzo, Ishizuka Masaaki, and Morishita Mitsuru) formed a powerful APOC "think tank" that put the preservation movement on equal footing with the city government. By scrutinizing the road construction ordinance and the Rinkōsen plan, the three men identified subtle flaws that could be used to critique the city's plans. The three could also be described as the APOC's "translators," who rendered the technical terminology of city planning authorities into talking points for the preservation campaign. The Hokudai trio were also a vital source of new ammunition, as they continued to provide the APOC with the latest information on townscape preservation efforts overseas and across Japan. Moreover, dissemination of their publications and research notes gave the APOC some powerful new allies among academic heavyweights in the fields of architecture and engineering, including Matsumura Teijirō (the University of Tokyo), Yoshizaka Takamasa (Waseda University), architecture critic Hasegawa Takashi, Itō Teiji (Kōgakuin University), Miyamoto Ken'ichi (Osaka City University), Shinozuka Shōji (Waseda University), and other national experts, who publicly supported the preservation of the canal. Their efforts, together with those of Mineyama and the APOC's new members, dramatically increased popular support for the canal's preservation, both in Otaru and across Japan.

The activism of this phase in the preservation effort, during which the APOC coordinated its efforts with those of other townscape movements across Japan and developed a new logic of preservation, was qualitatively different from that of the

[13]Joint interview with Yamaguchi Tamotsu and Ogawara Tadashi in Otaru, August 31, 1996.

[14]Hokudai is shorthand for Hokkaido Daigaku, or Hokkaido University.

first phase. Inspired by the Hokudai trio, three young members of the APOC's "U-turn" contingent organized a new kind of local event in July 1978, known as the Port Festival. This festival was a homegrown affair, organized entirely by the young people of Otaru. High school and university student volunteers planned the events, solicited donations, and handled every aspect of the festival, from the construction of stages to the post-event clean-up. Initially, little economic support was forthcoming for a festival organized by a group of anonymous local youth, and organizers almost gave up on their plan. After a midnight conversation along the canal, however, they vowed to redouble their efforts. Gradually they found their way: they applied for a license to use the road along the canal as the festival site and used barges floating on the canal as open-air stages. They gathered after school or work to share ideas, and slowly built up local support for the event. The organizers, who were still hard at work on the morning of the festival, were overwhelmed by the success of the event: more than 100,000 people attended the first Port Festival (Ogasawara 1986).[15]

It is worth pausing to examine the foundational principles of the Port Festival. One idea expressed by every member of the festival staff was that of "pure amateurism:"

> This had nothing to do with saving the canal… {…} The Port [Festival] was about treasuring friendships and had nothing to do with [personal] gain. Our only shared interest was in holding the Port Festival; we connected simply through a desire to hold the festival. Our work had nothing to do with it either. Interests had nothing to do with it. I made a lot of friends who felt that way.
>
> Interview with Ōta Masaaki in Otaru, September 9, 1997. Information in parentheses has been added by the author, information in brackets by the translator.
>
> After all, there was absolutely no profit motive, and there was no way that we could run [the festival] as a business, where profits come first, so there were no real interests at stake. That's why it was [all about] genuine personal relationships. Everyone's [motivations] were really pure, since they had nothing to [lose or gain]… {…} There are some things that won't last long without a [profit motive,] but in this case, it was precisely because [our motives were] so pure that there could be no lies…you know, even if I said something grandiose it didn't sound like a lie. {…}] Human relationships (during the preparation period leading up to the festival) were completely honest, and we did everything together.
>
> Interview with Ōhashi Akira in Otaru, September 10, 2011. Information in parentheses has been added by the author, information in brackets by the translator.

Two things are readily apparent from these recollections. The first is the apolitical nature of the Port Festival. As Ōta's immediate disavowal ("this had nothing to do with saving the canal") makes clear, the Port Festival was not explicitly aimed at saving the canal. To be sure, the young "U-turn" contingent was inspired to organize the festival at the suggestion of the Hokudai trio, and they certainly conceived of the event as part of their effort to save the canal. However, in order to mobilize the youth of Otaru, they maintained that the festival was an event for young people that simply happened to be set against the backdrop of the canal, where there was ample

---

[15]Ogawara Tadashi, who spearheaded the first Port Festival, told reporters from the Sapporo branch of NHK television news that the festival organizers themselves were surprised by the turnout. "Kita no Chōsensha-tachi Dai-ikkai: Yomigaeru Unga no Machi," NHK Sapporo, April 5, 2002.

space for rock music performances.[16] The apolitical character of the Port Festival also derived from the philosophy of certain key organizers, who were determined not to replicate their earlier experiences as student activists. We shall return to this point in the life histories recorded later in the chapter.

Second, the event was characterized by "pure volunteerism." This is evident in Ōhashi's description of the "completely honest human relationships" that bound the festival organizers together, and his observation that "there was no way that we could run [the festival] as a business, where profits come first." A "pure" desire to put on the Port Festival brought people together to work "honestly" for the success of the event. This apolitical approach was rooted in an anti-business mentality and a conscious rejection of an earlier model of student activism. Here one can identify the layered nature of Japan's postwar social activism, in which the experiences of previous movements have shaped—both consciously and unconsciously—the development of subsequent movements.

A seemingly straightforward regional youth event was thus shaped by the organizers' rejection of the tactics employed by the student movement of the 1960s and early 1970s, and their refusal to become enmeshed in ideological confrontations or party politics. One should also note the ingenious way in which the activists' goal of preserving the canal as a means of achieving Otaru's rebirth was woven into the fabric of the festival itself. The festival's organizers sought to inspire the citizens of Otaru to reassess their hometown by experiencing the space of the canal firsthand. Indeed, the true essence of the Port Festival lay in the grand design of this event, which enabled thousands of people to personally experience one potential future for Otaru (Horikawa 2009, 2010). The booths that lined the canal during the Port Festival were not computer simulations, but actual businesses. The liveliness of their trade demonstrated how Otaru could be reinvigorated. And by organizing a nonpolitical event that was completely unrelated to the conventional political divide between progressives and conservatives, and attracting ordinary people to the canal to experience for themselves the latent potential of the space, organizers hoped to bolster support and sympathy for the cause of preservation. The organizers themselves described this as an "environment-based event" (Sasaki, ed. 1983). This was an entirely different approach from the one employed during the preservation movement's first phase, when the canal was positioned as a cultural property in successive petitions.

Following the grand success of the first Port Festival, members of the festival's executive committeespearheaded the creation of a new group, the Yume no Machizukuri Committee.[17] An excerpt from the first issue of *Fiesta Otaru*, the group's newsletter, aptly summarizes the second phase of the canal preservation effort:

> Our point of departure is to move beyond the image of the destruction of our city's character, culture, and historic environment in the name of 'growth' and 'development.' This is because the first thing one thinks of being destroyed are the historic structures and heritage sites that

---

[16]In the very final stage of the battle over the canal's fate, they abandoned this neutral stance and made a clear and direct appeal for the canal's preservation on August 28, 1983.

[17]There is no concise translation of the committee name, but it can be loosely translated as the Committee for Vision-based Community Development in Otaru.

> have silently molded the character, culture, and historic environment of our city. It is essential to ask how we, the people of Otaru, relate to these aspects of our cultural heritage and historic buildings. {...} We are not 'preservationists.' We do not want to see the historic and cultural environment of our city, Otaru, preserved in the conventional manner of cultural properties: as mere museum pieces, "frozen" in time. Rather, we seek to restore its value by reviving it as the lived environment of generations [of residents]. In other words, we want to fill this old vessel with a new energy. {....} We hereby announce our campaign to fill this good old vessel (the city) with something new (community and human development).
>
> *Fiesta Otaru*, inaugural edition, December 1978. Quoted in Otaru Unga Mondai o Kangaeru Kai ed, 1986a: 222–223. Information in brackets has been added by the translator.

In this excerpt, activists decry "destruction" carried out in the name of "development," and stress that "we, the people of Otaru" must decide how to address the city's cultural heritage. Naturally, they conclude that this heritage should be saved from destruction, but refuse to identify themselves as "preservationists" willing to preserve "the historical and cultural environment of our city, Otaru" as "cultural properties"—in other words, as "museum pieces, 'frozen' in time." On the contrary, they sought to restore life to a historic city through "community and human development." Saving the canal district—that "good old vessel"—simply by freezing it in time was not good enough. The preservation of the canal district's townscape was not the ultimate goal but rather the means to an even more important end: community development. According to this logic of preservation, neither total demolition nor 'cryopreservation' could achieve Otaru's rebirth. This could only be realized through the renewal of Otaru's historic buildings and infrastructure. The ideal form of "community development" was to seek development in concert with preservation.

What did this evolution in the logic of preservation achieve? For one, the goal of saving the entire canal district in its current form—in other words, freezing it in time—could only bring activists into full-out confrontation with city officials. However, when activists began emphasizing "the canal as a tourism resource" and suggesting that there need not be any contradiction between preserving the canal and redeveloping the canal district, they gave the local business community new space to join a growing number of people outside of Otaru[18] in supporting their cause.

A 1982 decision by the Seibu Group illustrates this point. In February 1982, leading members of the Otaru Chamber of Commerce began to seek out investment,

---

[18]In September 1978 a group dedicated to the Otaru canal was launched in the Hokkaido city of Asahikawa, and in November of the same year, in Tokyo. These new groups joined the Otaru APOC and the Sapporo group, bringing the total number of organizations dedicated to saving the canal to four. The activities of these four groups were supported by the annual Port Festival, the Yume-machi executive committee, and their publication, the *Fiesta Otaru*. Meanwhile, on November 17, 1978 the Otaru city government distributed a special edition of the city newsletter, the *Kōhō Otaru*, on the subject of "The Canal and the Rinkōsen." In June 1979, the city submitted a revised blueprint for the Rinkōsen that reduced by one-third the area of the canal to be paved over for the new road (Otaru Unga to Sono Shūhen Chiku Kankyō Seibi Kōsō). The revised plan was adopted on September 20, 1980 (Hokkaido bulletin #2361), and in March 1981 the Otaru mayor announced the city would purchase and preserve [some of] Otaru's historic warehouses. During the period in which the city's revised plan was open for public review, 1,300 written opinions were submitted urging the city government to reconsider the plan and preserve the entire canal.

support, and cooperation from major companies for the redevelopment of the canal district. They approached the Seibu Group and reached an informal agreement to support Seibu's bid to launch a new department store in a group of renovated warehouses in the canal district. The 1974 launch of Kurashiki Ivy Square in Okayama had inspired a wave of similar projects, in which the interiors of historic structures were renovated for new use, and Seibu was interested in adopting this approach in Otaru.

While early discussions seemed promising, in September 1982, Seibu's chairman, Tsutsumi Seiji, told the chamber that Seibu could not participate in the venture if the Rinkōsen road project was completed as planned. Tsutsumi noted that "preserving the canal in its entirety would be of far greater value," and "inserting Seibu into the controversy would be bad for our corporate image" ("Hozonshite Koso Kachi Aru—Tsutsumi Seibu Daihyō Kataru," *Asahi Shinbun,* Hokkaido edition, October 1, 1982). For the Seibu Group, the more attractive option would be to establish a branch of its department store chain in a repurposed historic building along the original canal. Out of concern for its corporate image, Seibu refused to come to Otaru in any form that might convey its complicity in the canal's destruction.

Given Mayor Shimura's resolve to execute the Rinkōsen road plan, the Otaru Chamber of Commerce had little choice but to continue to publicly support the project. However, the Seibu Group's decision gave many of the chamber's "businessmen"[19] pause. Many began to wonder if it would be better to collaborate with Seibu in the redevelopment of the preserved canal district. A secret "pro-preservation camp" emerged within the chamber. Between late 1982 and the summer of 1983, the chairman and vice-chairman of the Otaru Chamber of Commerce held a series of closed-door discussions with Shimura, at which they urged the mayor to reverse course and leave the canal intact. Citing the decision of the LDP party council and the potential damage to "the essence of his administration,"[20] Shimura steadfastly refused to consider any change to the plan (Miyamaru 1983).

On August 17, 1983, the chairman of the Chamber of Commerce reversed his position and issued a public request to "save the canal," explaining that "as things stand now, we will miss our window of opportunity for outside investment."[21] His

---

[19]The word "businessman" is not merely descriptive. It underscores a significant departure from the first phase of the canal preservation effort, which was a self-defined "cultural movement." As an apolitical cultural movement, the preservation effort was typically perceived as the pastime of a few "dilettantes" with no real understanding of the political and economic factors informing the new road plan. The involvement and support of "businessmen" conveyed a new message: pragmatic people could also reasonably support the canal's preservation.

[20]Joint interview with Yamaguchi Tamotsu and Sasaki Kyojirō in Otaru, June 5, 1990.

[21]Apparently, the pro-preservation members of the Chamber of Commerce had hoped to convey a more moderate stance through the timing and wording of their announcement. Once a newspaper published an exclusive account of their secret negotiations, however, they had no choice but to immediately change their position. Ōno Tomo, then the vice-chairman of the Chamber of Commerce, recalls these events as follows: "The chairman and I decided together that we had to discuss the matter with the city, and explain that we wanted to leave historic [places] intact—that this was the only way to achieve Otaru's rebirth. But doing so publicly would leave the mayor in a difficult position, so we approached him confidentially. That was on August 6. On August 16 [our confidential

decision sent shock waves through the local business world, and many Otaru businesses followed his lead. A group of local businessmen added their names to another petition submitted by a group of local citizens, who had voluntarily joined the head of the Chamber of Commerce to make yet another appeal to Shimura (August 24, 1983). This new group, later referred to as the "Voluntary League" (*Katteren*), had no connection to the APOC and was headed by Murakami Katsutoshi, the thirty-nine-year old manager of a local accessory and precious metals business.

Murakami believed the sudden reversal by individual members of the Chamber of Commerce's leadership did not carry the same weight as that of the chamber as a whole. He thus sought to mobilize the entire local business community behind the cause of preservation ("Otaru no Keizaijin Yūshi Unga Hozon e Shomei Undō," *Hokkaido Shinbun*, Otaru city edition, August 25, 1983). Murakami's petition sought to restrain the city government and transform an isolated rebellion within the Chamber of Commerce into a broader movement by the entire Otaru business community.

APOC members were energized by these new developments. Not only had one of the mayor's most powerful support bases endorsed the canal preservation movement, but many prominent members of the local business community had voluntarily taken a public stand in support of the canal's preservation. They hastened to form an umbrella organization to bring together the different groups working to save the canal: this became known as the Committee of One Hundred for the Otaru Canal (Otaru Unga Hyakunin Iinkai), and was comprised of Otaru citizens, financiers, local business owners, and political party members (September 12, 1983). It was also a deliberate attempt to incorporate a broad range of conservative business leaders. The overriding goal of the new organization was to dispel any notion that the canal was the pet cause of an eccentric few and to present the Otaru city government with a united front—extending from Otaru's leading cultural figures to its business leaders—in support of the canal's preservation.

To facilitate the participation of the "secret preservation camp" within the Chamber of Commerce, the Voluntary League's Murakami was selected to join the APOC's Mineyama Fumi and Yamaguchi Tamotsu among the five representatives of the Committee of One Hundred. Including Murakami was a symbolic move, intended to convey the Otaru business community's support for the cause. Other committee members included many local business owners, key members of the APOC and the Yume-machi executive committee, and professors from the Otaru University of Commerce.

---

meeting] was reported in the newspaper, and all hell broke loose. Of course there had been no sudden change in our opinions. I had been listening to everyone in the APOC for a long time, and couldn't help feeling they had a legitimate claim. {...} But when we talked to the mayor on August 6, he told us that the matter had already been decided, and that it was the best possible plan. If the plan changed now, he would be forced to resign. We told him that he needn't be so rigid, that times were changing and we hoped he could be more flexible in his decision-making...We told him we would bring him any good proposals we could come up with, and left." Interview with Otaru Chamber of Commerce vice-chairman Ōno Tomo, conducted by Miyamaru Yoshie, then the president of the Japan Environment and Culture Research Institute. Quoted in Miyamaru, 1983: 55.

The Committee of One Hundred immediately launched a signature-gathering campaign. Members canvassed every corner of the city, eventually collecting signatures from 98,000 residents who supported the canal's preservation. In a city with a population of just 180,000, this achievement bolstered the Committee's political strength immeasurably. To be sure, the city government had no legal obligation to respond to a voluntary signature-gathering campaign. Indeed, the petition would have lacked legal force even if each and every one of Otaru's 180,000 citizens had signed. Nevertheless, the response from the city government was immediate: officials began to compare each signature on the petition with the local voter registry in an effort to ascertain how many signatures belonged to Otaru residents of legal voting age. Indeed, the city government employed part-time workers to complete this task at top speed and hastened to announce that the number of registered voters who had signed the petition was well below 98,000. Yet such a sensitive reaction to a petition without legal bearing underscores the growing impatience of the Otaru city government. In fact, the city's heavy-handed response and use of public funds to undermine the signature collection effort ended up increasing the public's sympathy for the preservation movement.

Emboldened by the success of their petition, members of the preservation movement began hinting at a mayoral recall effort (January 1984). The spirit of the preservation movement had long been entirely apolitical, yet some activists were now prepared to plunge into local politics by launching a recall effort. How do we explain this reversal?

In fact, at this point in their campaign, activists were left with little choice. Throughout the protracted confrontation between the Otaru city government and preservation activists, the construction of the Rinkōsen had quietly continued. Activists needed to find a way to halt construction or secure some type of compromise with the city government as soon as possible, otherwise the new road would become an accomplished fact: in December 1982, work to solidify the sludge along the bottom of the canal had begun, and pile-driving work had commenced in November 1983. Moreover, Mayor Shimura appeared undeterred by the success of the Committee of One Hundred's signature-gathering campaign. Activists' willingness to contemplate a recall effort must be understood within the context of the steady advance of the Rinkōsen road toward Otaru.[22]

On February 15, 1984, the Committee of One Hundred convened an emergency meeting to formally approve the formation of a "recall committee." Given the broad range of its membership, the Committee of One Hundred could not directly involve itself in a recall effort without forcing many of its members to leave out of concern for their public position. A separate committee was thus established and dedicated to the recall effort.[23] The formerly apolitical preservation campaign was now a formidable political enemy of the city administration. Put differently, the effort to save the

[22] Interview with Mineyama Fumi in Otaru, December 3, 1989.

[23] Records show that the Committee of One Hundred had only approved the establishment of a recall committee. All other matters remained open for discussion. Minutes of the February 15, 1984 emergency meeting of the Committee of One Hundred for the Otaru Canal. Page four of seven handwritten pages of minutes, issued at the April 10, 1984 meeting. In other words, the

Otaru canal was no longer a "cultural movement" motivated by the "old-fashioned" sentiment of a handful of Otaru residents. The preservation campaign was now a powerful new actor in the regional political process, capable of mobilizing local business interests behind its cause. The local chapter of the LDP immediately formed a group to protect Shimura from a potential recall campaign. In March 1984, yet another group, known as the "Committee to Achieve the Speedy Completion of the Otaru Rinkōsen," emerged to defend the Shimura administration.

Local and national officials were alarmed by the growing strength of the preservation movement. The Otaru Expo, an event which received central government funding, was scheduled to run between June 10 and August 24. The canal controversy showed no sign of dying down before the start of the Expo, compelling the national and prefectural governments to intervene in the hopes of achieving some type of resolution. Everyone in the hometown host of the Expo typically banded together to ensure the success of the exposition. The Otaru Expo could not be held in a city divided against itself and mired in controversy. On March 24, Minister of Construction Mizuno Kiyoshi sought to bring the situation under control, suggesting "temporarily suspending construction during the exposition period so that [both parties] can try discussing the matter again." Acting as the designated intermediary between the two sides, Hokkaido Governor Yokomichi initiated discussions with representatives from both the preservation and the road factions (May 24, 1984). The pro-Rinkōsen faction was represented at these "five-party talks" by Mayor Shimura and Sugawara Haruo, chairman of the Committee to Achieve the Speedy Completion of the Otaru Rinkōsen. The pro-preservation faction was represented by APOC chairwoman Mineyama Fumi,[24] the Committee of One Hundred for the Otaru Canal's Murakami Katsutoshi, and Otaru Chamber of Commerce President Kawai Issei. The meetings were organized and mediated by Yokomichi and Araya Masaaki, supervisor of Hokkaido public works.[25]

Within the preservation camp, however, a debate raged over whether to participate in the talks at all. Leading members of the APOC feared they would be blamed if the launch of a recall effort during the Otaru Expo period affected the success of the event. Moreover, preservation activists had always sought dialog; APOC activists feared that the general public would withdraw their support if they rejected this new opportunity for dialog, offered by Governor Yokomichi himself. Meanwhile, Murakami and other members of the Committee of One Hundred stubbornly insisted on proceeding with a recall effort, and began attacking Mineyama, Yamaguchi, and Ishizuka as "opportunists." Murakami complained that people like Mineyama and other central figures in the APOC were willing to hint at a recall effort, but refused to proceed in earnest when the time came. Mineyama and Yamaguchi countered by arguing that rejecting an opportunity for dialog was wholly inconsistent with the approach of a citizens' campaign. A true citizens' movement would negotiate up

establishment of a recall committee had been approved, but the specific timing and announcement of a recall effort remained undecided and subject to future discussion within the committee.

[24]In fact, Yamaguchi and Ishizuka also attended the meetings as APOC representatives.

[25]Araya went on to serve three terms as mayor of Otaru between 1987–1999.

until the last possible minute, and only initiate a recall effort after every opportunity for negotiation had been exhausted. To do otherwise would reduce a genuine citizens' campaign into a crude power struggle.

A series of meetings failed to break the impasse between the Murakami and Mineyama factions. On June 28, 1984, Murakami, Morimoto Mitsuko, and a few others with ties to the JCP bypassed the Committee of One Hundred's formal decision-making process and held a press conference to announce the launch of a recall effort against Mayor Shimura. ("Otaru Unga Hyakunin'i Tsui ni Shichō Rikōru," *Hokkaido Times*, June 29, 1984). According to Murakami, "we couldn't reach an agreement on whether or not to immediately begin the recall effort, but because five out of eight members of the recall committee pushed for an 'immediate' start we have decided to move ahead with the recall" ("Otaru Shichō Rikōru Isshukan Inai ni Undō Kaishi," *Asahi Shinbun*, Otaru edition, June 29, 1984). In other words, Murakami had unilaterally announced a recall effort without the consensus of either the Committee of One Hundred *or* the recall committee. And while the recall committee was ostensibly separate from the Committee of One Hundred, given the circumstances surrounding its establishment and the overall organizational structure of the two groups, the approval of the latter should have been an essential prerequisite to any announcement. Moreover, Murakami's actions contravened a February 15 resolution by the Committee of One Hundred to continue deliberation on all agenda items. The announcement of a recall was the action of a small group with no regard for official protocol. As noted in the same *Asahi Shinbun* report, "Mineyama and others who have long campaigned to save the canal were nowhere to be seen at the press conference, which was held by Murakami and others pushing for 'immediate' [action]" ("Otaru Shichō Rikōru Isshukan Inai ni Undō Kaishi," *Asahi Shinbun*, Otaru edition, June 29, 1984).

Strategically, Murakami's promise to "initiate a recall effort within the week" changed everything. The recall effort was no longer a hypothetical last resort, but a concrete attack on Shimura. Moreover, the potency of the recall threat stemmed largely from a certain opacity surrounding timing and execution. The preservation camp's bargaining power dissipated as soon as Murakami made his announcement, as the focus immediately shifted to the actual mechanics of the recall—and the financial resources available for such an effort. Exposing the paltry state of the movement's finances immediately undermined the clout of a preservation campaign that had always tried to present itself as more powerful than it really was. Mineyama, Yamaguchi, and Ogawara had anticipated as much and pushed to postpone the recall effort in order to retain their bargaining power. When asked about these events, Mineyama replied:

> We said we would '*have no choice but to launch a recall* effort.' We never said we '*would* launch a recall.' When we met with the mayor, we told him: 'You'll leave us no choice but to start a recall campaign. If you won't acknowledge a petition with 100,000 signatures, we'll have no choice in the matter. If you refuse to consider the wishes of [Otaru] citizens, we have no choice.' Do you see?' That's how we put it to the mayor. Do you understand?
>
> Interview with Mineyama Fumi in Otaru, December 3, 1989. Emphasis is Mineyama's.

In this interview, Mineyama took great care to emphasize the specific nuance of her wording ("We said we would '*have no choice but to launch a recall* effort.' We never said we '*would* launch a recall") and impress this upon her audience (repeatedly confirming: "do you understand?"). Mineyama knew that the source of her bargaining power was the *suggestion* of a recall, not its actual implementation.

Yamaguchi and others subjected Murakami's actions to a close inquiry, forcing the latter to admit that his press conference had not been approved by the Committee of One Hundred. However, Murakami's attempt to force a recall effort had already thrown the preservation camp into disarray, aggravated the antagonism between the APOC and the Committee of One Hundred, and highlighted the lack of sufficient decision-making protocol within these organizations. A formal recall effort was never launched. The preservation movement's internal divisions had destroyed their final chance to save the canal and eroded local support for the cause—particularly among conservatives.

Meanwhile, the five-party talks had made little headway after four sessions, and [the governor concluded] that the Rinkōsen project would proceed as planned (August 18, 1984). Opposing a fully funded road plan that was already underway had been a daunting task from the start, and internal discord had done nothing to help the cause of preservation. Nor had the personal mediation of the Hokkaido governor produced a new consensus. At the conclusion of the talks, Governor Yokomichi suggested the city establish a new "revitalization committee" to include preservation activists in the development of a blueprint for the local economy. Yokomichi also offered the prefecture's financial support for the preservation or relocation of the stone warehouses that lined the canal (From a reproduction of the original proposal, "Chiji Teian," included in a file of documents collected for the city government's book on Otaru history, *Otarushi-shi*).

Murakami refused to give an answer on the spot but promised to raise the matter at the Committee of One Hundred. A general meeting of the Committee of One Hundred was convened on August 22 to decide how to proceed, now that the five-party talks were at an end. "Various arguments" were laid out, but ultimately there was little to do but decide among "three proposals for the future course of the committee put forth by members" (Document signed by Chairman Shinozaki Tsuneo and Yūki Yōichirō of the Committee of One Hundred for the Otaru Canal, dated August 26, 1984):

1. Dissolve the Committee of One Hundred to allow members to respond to circumstances however they choose.
2. Maintain the Committee of One Hundred in its current form and
    a. Seek the support of Otaru citizens through such measures as a recall campaign
    b. Collaborate with the Revitalization Committee [proposed by the governor]

Notice of a General Meeting on Continuing with the Committee of One Hundred, August 26, 1984. Signed by Committee of One Hundred Chairmen Shinozaki Tsuneo and Yūki Yōichirō. Cited in Otaru Unga o Kangaeru Kai, ed., 1986a: 520.

Founding members of the Committee of One Hundred were told to submit their vote by August 31. With talks at an end, how would the Committee of One Hundred decide to proceed? Otaru citizens and the local media awaited the decision with bated breath.

Meanwhile, six days after the conclusion of the five-party talks, the leadership of the APOC convened in a civic hall conference room. Mineyama took "responsibility for failing to protect the canal" by offering her resignation as chairwoman. She also withdrew her membership in the APOC ("Otaru Unga o Mamorukai Mineyama Kaichō Jinin," *Asahi Shinbun,* Otaru edition, August 25, 1984). A devout Christian who was widely trusted in the local community, Mineyama had come to symbolize the preservation movement in its second phase. Her resignation confirmed suspicions that the rupture within the preservation campaign would prove decisive.

On the evening of September 1, 1984, the members of the Committee of One Hundred gathered to decide how they would proceed.[26] Although committee members had convened to decide the very fate of the organization, the meeting was immediately derailed by a routine financial report:

> Murakami: You shouldn't submit the financial report without informing the secretariat and executive leaders.
>
> *Morimoto and Iida agree.*
>
> Yamaguchi (K): {...} [I agree]. We shouldn't be discussing matters that haven't yet been shared with the secretariat.
>
> Iida (*also agreeing*): The financial report should be submitted only after it has been reviewed by the secretariat.
>
> Yamaguchi (Tamotsu): The financial report was put together by Hirata, myself, Ogasawara, and Uchiyama.
>
> Yūki: This is the same procedure as at all previous meetings. I have no idea why this procedure was acceptable in the past but is suddenly up for debate now.
>
> Kasai: Isn't this just an attempt to get right to the subject of the committee's infighting? We ought to be having a serious (debate) about the future of this organization, not arguing over operational issues. Let's acknowledge as much and stop complaining about procedure.
>
> Iida: I'm not here to create confusion for the committee.
>
> From the first page of the handwritten minutes of the September 1, 1984 meeting of the Committee of One Hundred. Information in parentheses added by the author, information in brackets by the translator.

The exchange begins with Murakami casting doubt on the committee's financial reporting practices, and receiving support from several other members. The typical procedure at general meetings of the committee was for the treasurer to report on accounts and open the matter up for discussion. Murakami had suddenly found fault with this approach. This explains the dismay expressed by other members, who had "no idea why this procedure was acceptable in the past but is up for debate now." The

[26]Preservation activists had shared information on their gatherings with the local media. As their internal divisions deepened, however, a growing number of meetings were held behind closed doors. The final meeting of the Committee of One Hundred was closed to the public. Just two outsiders received permission to attend: this author and Shin'chi Mitsuo, a reporter from the Otaru branch of the *Asahi Shinbun.* The account of this final meeting is based on my own field notes and the handwritten transcription of the meeting, which I later obtained.

key to understanding this odd episode is the failed recall effort. Murakami and the other members who had unilaterally announced a recall effort were the same ones now censuring the treasurer's report. Those who had advocated a more cautious approach to the recall defended the existing practices at committee meetings. The failed recall had opened up a deep division between different factions within the Committee of One Hundred: Murakami and his supporters, who still clung to the idea of a recall, and Yamaguchi's faction, who blamed Murakami's group for imperiling the entire preservation movement with their misguided actions.[27] The former's criticism of the treasurer's report was a concrete expression of this division: "Isn't this just an attempt to get right to the subject of the committee's infighting?".

Once this procedural issue was finally resolved, the committee members turned to the question of the future. As the results of the vote were read aloud, members realized that a sizable majority had voted to disband.

> Of the one hundred founding members (three members had already left), 59 cast votes:
>
> In favor of dissolution: 41
>
> Opposing dissolution: 8
>
> Undecided: 7
>
> Invalid responses: 3
>
> From the second page of the handwritten minutes of the September 1, 1984 meeting of the Committee of One Hundred.[28]

The results of the vote threw the meeting into even greater chaos. Still reeling from the results, individual members began explaining why they had voted as they did:

> Morita: I voted against disbanding, but I also wrote that if opinion was divided here today, we should go ahead with the dissolution. We originally had a lot of support from citizens, but we lost it after the June 28 recall announcement, and the canal (issue) has been settled. I don't think this movement was in vain. Members of the preservation camp should join the [Otaru] revitalization committee, and I will support their efforts. We should proclaim that any future efforts are up to each individual, and disband.
>
> Yamaguchi (Tamotsu): I did not mail in my vote, but I will give my answer [here and now]. I urge the committee to disband. This group expanded our support base, but our lack of internal unity and mishandling of operations has cost us the support of citizens. The five-party talks failed to achieve a consensus on the canal, but I will continue to push for the total preservation of the canal by participating in the revitalization committee. Even if the Committee does not disband, I will withdraw my membership.
>
> Shimozawa: The committee should be dissolved.
>
> Yarimizu: I think the Committee of One Hundred still has a role to play, and should continue. The governor has said the canal will be filled in and has ordered construction [to resume],

[27] The specific complaint of this latter faction was that "after more than ten years of activism, the most critical juncture for the [preservation] movement had been tragically wasted by *the usual factional interests*." "Futatabi 'Machinami Hozon no Kyōkasho' to Yobareru 'Machizukuri Undō' no Saikōchiku o Mezashite," cited in Otaru Unga Mondai o Kangaeru Kai ed, 1986: 500. Information in parentheses and emphasis added by the author.

[28] Three votes (two for dissolution, one opposed to dissolution) were deemed invalid: one arrived after the deadline, one was missing a seal, and one did not use the designated reply form.

but I think there's a way to keep up the fight. We could petition the governor directly, for example. We should make an appeal to the governor for the total preservation of the canal.

Morita: The reason the governor finally handed down the judgement that he did is because we couldn't come up with anything different during the period that construction was suspended for the Otaru Expo. The situation would have been different if it hadn't been for that pointless recall announcement. There's no point in pestering the governor further. Leave that to some political party.

Ishizuka (?): That's a strange thing to say. Wasn't Morita pushing for a recall early on?

Kasai: I'm not one of the committee's original members so I came here today without an established position. I doubt that even those who voted for dissolution (on their mail-in ballot) arrived (here) with the weight off their shoulders. But given the debate at the start of today's meeting (regarding the treasurer's report) and the kind of opinions that are being expressed, I don't think we have any choice but to disband. We can't appeal to the governor in our current state—without any popularity, and mired in that kind of [stupid] debate about an audit report.

Sasaki Kyojirō: When the committee was first formed there was [a real sense of] community-building, but now there's only loathing. We have to disband.

Tani: There were three proposals to choose from. I don't think the Committee of One Hundred should join the revitalization committee. People who want to leave should leave. I will remain. A citizens' movement should not be dissolved.

Iida: The revitalization committee is just a compromise solution. Our citizens' movement ended up being used for political purposes. I also have to question the neutrality of the mass media. When it came to a recall, there were some remarkable changes (to certain people's stances on the subject) midway through the process. The people who were serious [about conducting a recall] somehow ended up [being cast] as the bad guys. Given my sincere desire to save the entire canal, I would like to continue [with the committee].

Yamaguchi (Tamotsu): We can't even agree on what happened with the recall. On the recall committee, Ms. Mineyama, Mr. Morita, and I opposed [the recall effort launched by Murakami], but ever since May I had thought a recall would be inevitable. But things ended up as they did [as a result of] our internal problems. Governor Yokomichi's conclusion is disappointing, but fundamentally it is a consequence of our own actions. I'd like to reflect on that and then dedicate all of our efforts to developing a new way forward. After spending two whole months rehashing the recall, we still can't reach a common understanding [of what went wrong]. We have no choice but to disband.

Mineyama: I won't offer any kind of explanation for past events. But I'm surprised by the results of the survey. The question is how to make sense of this harsh reality, but given the course of events, attempting to continue would be impossible. A new start is essential.

Yamaguchi: I didn't mail in a vote, either. As someone working in the secretariat, I felt I couldn't make a decision without listening to what other people, who haven't been able to attend every meeting, had to say. I held out some hope for today's meeting, but like Mr. Kasai said, after we wasted all that time at the very beginning, I've given up all hope. I vote to dissolve.

Morimoto: I sense a hatred of politics. I think the Committee of One Hundred should continue—it doesn't matter if only a few remain .

Shimozawa: The Committee of One Hundred should disband for a while. Those who would like to continue should start a new organization.

Yarimizu: Dissolving [the committee] now would be the same as consenting to the governor's land reclamation [plan for the canal]. Anyone who agrees with that should resign. Calling for a recall only to crush it, and falling into step with the governor once the five-party talks were held… I still believe the whole thing was some kind of plot.

Kakimoto: After listening to that declaration, we really have no choice but to disband.

Shimozawa: Even if the Committee of One Hundred continues, can we really expect people to remain? A few people will just end up using [the committee] for their own purposes. We should place our hopes in the Revitalization Committee.

Yarimizu: You may think as much, but [remember that] the Revitalization Committee is predicated on the reclamation of the canal.

Murakami: With regard to the survey results, three of the people who voted to dissolve the committee have notified me that they have since changed their mind.

Yamaguchi (K): A few people tried to bring down Murakami in order to replace him at the five-party talks. I hear they promised to suppress the movement once they were at the talks. I really cannot comprehend why the APOC's representative at the talks gave her consent [to the city's reclamation of the canal] there.

Yamaguchi (Tamotsu): Wasn't that after the June 28 recall announcement? If it happened after June 28, that was completely reasonable.

Yamaguchi (K): There is something very suspicious about the [APOC] representative making concessions at the five-party talks.

Uchiyama: One year ago, at our inaugural meeting, we talked about how we didn't want to engage in these kinds of disputes. These ugly quarrels are a betrayal [of our agreement]. I don't want this kind of group to use the name of the "Committee of One Hundred." I insist we disband.

Murakami: Disbanding before we have any idea what this Revitalization Committee will look like makes a mockery of Otaru citizens.

Shimozawa: That's an odd argument. There is no connection between the Committee of One Hundred and the Revitalization Committee.

Tsukamoto: Bidding (for the canal reclamation work) isn't even over yet. At this [early] stage, it would be wrong to break up the committee and announce that we lost to the administration.

Yūki: I deferred my vote (on the ballot I sent in). [I didn't know how I felt about] simply disbanding because we failed to save the whole canal, after all that we did. I don't hold out high hopes for the Revitalization Committee either. But coming here today has made me wonder: what can a group that gets into these kinds of arguments really hope to achieve? Everything changed when [some members decided to] disregard [proper] procedures and launch a recall effort. But I persevered because there was no other option. Until I arrived here [tonight], I thought there was a chance we could work something out. But what do ten-odd people bearing the name of the Committee of One Hundred really think they can do? Getting together to attack the people who want to disband won't achieve anything. It's the people who aren't capable of starting a new group that are insisting we should carry on. That's all.

Kakimoto: Instead of trying to reach a conclusion here today, I suggest we entrust the decision to the [committee] leadership.

Yūki: The matter is to be decided here today, so (anyone who objects should) submit a motion.

(Name cannot be deciphered from the handwritten transcript): This committee is full of distrust and should be dissolved. Please take a vote.

Yūki: Is there a motion?

X-ta (name is undecipherable), Tani, and Morimoto oppose resolving the matter through a vote.

Yūki: I'll listen to that as your opinion. If there is no motion, we will take a vote on the dissolution.

XX (illegible): I move to take a vote here, and then have the leadership decide.

Motion is adopted by the majority.

In support of the motion: 17

Opposing the motion: 0

The remainder abstain.

Stenographic record of the September 1, 1984 meeting of the Committee of One Hundred, pp. 2–4 of the four (B4 size) pages. Information in parentheses has been added by the author. The author has also clarified clear instances of abbreviation (the use of the letter "R" for recall, for example).

The minutes of the September 1 meeting offer a vivid depiction of how the internal division produced by Murakami's doomed recall announcement had only deepened with time: "We originally had a lot of support from citizens, but we lost it after the June 28 recall announcement;" "This group expanded our support base, but the lack of internal unity and mismanagement of operations have cost us the support of [Otaru] citizens;" "The situation would have been different if it hadn't been for that pointless recall announcement." Committee members who had opposed Murakami's recall announcement blamed the defeat of the preservation effort on Murakami's actions, and the rift he created. Yamaguchi Tamotsu's observation that "even after spending two whole months going over the recall, we can't reach a common understanding [of what went wrong]" speaks to the impossibility of bridging the growing gap between committee members, and suggests the citizens' movement was "crushed" by the "usual factional interests" ("Futatabi 'Machinami Hozon Undō no Kyōkasho' to Yobareru 'Machizukuri Undō' no Saikōchiku o Mezashite," cited in Otaru Unga o Kangaeru Kai ed, 1986a: 500). Yamaguchi also recognized that while "Governor Yokomichi's conclusion is disappointing... *fundamentally it is a consequence of our own actions*." Yamaguchi's point is that it would be impossible to "develop a new way forward" without grappling with the reasons for their inability to build a genuine citizens' campaign that could overcome political interests. The only remaining choice was to disband. In other words, a fresh start would not be possible without self-reflection—not only by those behind the misguided recall announcement, but also by those who failed to prevent it. The dissolution of the Committee of One Hundred was the specific expression of this reflection and repentance.

Members of the group who had pushed forward with the recall effort now fought to maintain the Committee of One Hundred: "A citizen's movement (shouldn't disband);" "I sense a hatred of politics;" "I think the Committee of One Hundred should remain, even with just a handful [of people]." They also attacked the people who voted to disband—the same people who had opposed their hasty recall announcement: "When it came to a recall, there were some remarkable changes (to certain people's stances on the subject) midway through the process, and the people who were serious [about conducting a recall] somehow ended up [being cast] as the bad guys;" "Calling for a recall only to crush it, and falling into step with the governor during the five-party talks... I still believe the whole thing was some kind of plot;" "A few people tried to bring down Murakami in order to replace him at the five-party talks." In other words, the people "calling for a recall" subsequently made a "remarkable" change of stance, and ended up "crushing" the recall effort. Consequently, not only were those who had been "serious" about a recall branded "the bad guys" of the story, but Murakami was almost "brought down" by "a few people" who wanted to "replace him at the five-party talks." This could only have been "some kind of plot," in which the anti-recall faction made a secret pact in order to gain access to the five-party talks by agreeing to the [partial] reclamation of the canal. They countered the accusation that they had been guided by factional interests by suggesting that the internal cohesion of the preservation campaign had actually been destroyed by the anti-recall group and this secret pact.

No common ground was found in the course of the September 1 meeting. Those who had opposed Murakami's recall announcement maintained that the Committee of One Hundred had been dissolved as a result of their majority vote on September 1, while Murakami and his supporters refused to disband. The two sides went their separate ways and made no further attempt at dialog. But the Committee of One Hundred was not the only casualty of the stymied preservation effort. The APOC was also forced to confront a new reality: having lost the final opportunity to save the canal, the group's reason for existence disappeared.

One contingent within the APOC claimed that participation in the Otaru Revitalization Committee was tantamount to condoning the canal's destruction. After expelling Mineyama and Ishizuka on these grounds, they announced they would "return once more to the starting point: the preservation of the entire canal," and launch a new movement (Otaru Unga o Kangaeru Kai, 1986a: 524ff). This contingent of the APOC also issued a statement of protest when the Otaru Revitalization Committee[29] was convened in November 1984:

> Statement Opposing the Launch of the Otaru Revitalization Committee
>
> {…}
>
> The second point of the governor's arbitration was his proposal of a "Otaru Revitalization Committee." We see no value in such an undertaking. There is a fundamental contradiction within any discussion of Otaru's revitalization that deliberately avoids the most crucial issue: whether or not the canal will be reclaimed. {…} This sham Revitalization Committee should be shut down immediately. The APOC will develop a tenacious citizens' campaign to pursue Otaru's revitalization in the real sense of the word, *insisting, as it has from the very start, that the canal be preserved in its entirety.*

---

[29]The fifteen members of the Otaru Revitalization Committee were Fujii Eiichi (President of Otaru University of Commerce), Mayor Shimura Kazuo, Kuno Mitsurō (Otaru University of Commerce professor), Professor Shinozaki Tsuneo (Otaru University of Commerce), Kawai Kazunari (president of the Otaru Chamber of Commerce), Ōno Tomonobu (vice-president of the Otaru Chamber of Commerce), Yamabuki Seiichi (head of the Otaru chapter of the Liberal Democratic Party), Sugawara Haruo (president of the Committee to Achieve the Speedy Completion of the Otaru Rinkōsen), Mineyama Fumi (former chairwoman of the APOC), Yamaguchi Tamotsu (secretary of the Committee of One Hundred for the Otaru Canal), Ishizuka Masaaki (former planning director of the Otaru Unga Mondai o Kangaeru Kai), Sasaki Kyōjirō (chairman of the *Yume-machi* executive committee), Ishii Nobukazu (chairman of the Port Festival executive committee), Satō Hiroaki (president of the Otaru Tourism Association), and Yamamoto Kazuhiro (incoming president of the Otaru Junior Chamber). A representative of the Hokkaido prefectural government participated as an observer. Fujii was elected chairman of the Revitalization Committee. The office of the Revitalization Committee was established within the Otaru Chamber of Commerce. Otaru Unga Mondai o Kangaeru Kai ed. (1986: 533).

November 19, 1984

The Association to Protect the Otaru Canal, Acting Director Kitamura Toshiko

Acting Secretary Satō Jun'ichi

Quoted in *Otaru Unga Mondai o Kangaeru Kai* ed, 1986a: 532–533. Information in parentheses and emphasis have been added by the author.[30]

Following the dissolution of the Committee of One Hundred, Mineyama, Ishizuka, and Yamaguchi had joined the Otaru Revitalization Committee and adopted the keyword of "community development." By contrast, the APOC's new leaders, Kitamura and Satō, together with longtime APOC member Morimoto Mitsuko, clung to the catchphrase "save the entire canal." This latter group insisted that anyone who had participated in the five-party talks or the Otaru Revitalization Committee had surrendered to the pro-Rinkōsen interests and accepted the partial reclamation of the canal as inevitable. However, the claims of this latter group are not borne out by the following communication between the Hokkaido Government Office and the Ministry of Construction's City Bureau, dated August 27, 1984:

To the Director of the City Bureau, Ministry of Construction

From the Director of Housing, Hokkaido Government Office City Bureau

Regarding the 3.2.4 Prefectural Road Otaru Rinkōsen Project

We would like to express our gratitude for the exceptional consideration you have given to the establishment of prefectural road projects. Thanks to the good offices of yourself and other related parties, we have decided to continue with the above-named project.

The course of events leading up to this point are as follows:

1. In order to avoid any disruption to the Otaru Expo and develop a local consensus on the issue, the Minister of Construction and the Hokkaido Governor agreed to suspend construction on the Rinkōsen during the exposition period.

2. The Governor postponed construction and requested that the various parties on opposing sides of the road project sit down around the same table to discuss the issue, in the hopes of reaching a consensus. The prefecture acted as a coordinator at these five-party talks, which were attended by the mayor of Otaru, the chairman of the Committee to Achieve the Speedy Completion of the Rinkōsen, the chairman of the Otaru Chamber of Commerce, the chairwoman of the Association to Protect the Otaru Canal, and a representative from the Council of One Hundred for the Otaru Canal.

3. Talks were held four times, on May 24, June 15, August 17, and August 18. Participants shared their reasons for opposing or supporting the project, and also discussed the revitalization of Otaru. *They did not reach a consensus regarding the partial reclamation of the Otaru canal stipulated by this plan.* However, the need to develop the port and the canal district as the focal point of Otaru's rebirth and revitalization was acknowledged by all parties. The governor brought the five-party talks to a close by proposing the establishment of an Otaru Revitalization Committee (provisional name), which would include the five parties to the talks, and examine issues such as the preservation and maintenance of historic buildings and the redevelopment of the port and canal district. The governor promised to provide active support (including financial support) for any concrete plans developed by the committee.

[30]Members of the Committee of One Hundred who opposed disbanding returned to the APOC and attempted to make a fresh start within a new leadership structure. However, this November 19 statement was the last action taken by the group.

4. The mayor of Otaru, the chairman of the Committee to Achieve the Speedy Completion of the Rinkōsen, and the chairman of the Otaru Chamber of Commerce all agreed to the governor's proposal. *The chairwoman of the Association to Protect the Otaru Canal and the representative of the Committee of One Hundred for the Otaru Canal reserved judgement on the matter and said they would "discuss [the issue] within their respective organizations."* The Committee of One Hundred held a general meeting on (August) 22, and the APOC board of governors convened on August 24, but *neither group has reached a conclusion on the matter.*

5. Given that the five-party talks ended on August 18 and the Otaru Expo concluded on August 24, the prefecture will order the start to levee protection works on September 7, and will continue to issue sequential orders for other scheduled work in order to complete work on the canal by the 1986 deadline and remain on-schedule to complete other projects by fiscal year-end.

From a six-page handwritten communication on prefectural government stationery (B5 size) between the Hokkaido Prefectural Government and the Ministry of Construction. Dated August 27, 1984. Information in parentheses and emphasis added by the author.

This communication confirms the fact that the canal's fate was contested until the very end. Indeed, a final consensus was never reached: not only did the parties represented at the talks fail to "gain a consensus on the partial reclamation of the Otaru canal," but Mineyama and Ishizuka even reserved judgment on the "Otaru Revitalization Committee" proposed by the governor. Thus, while Mineyama and Ishizuka were accused of "compromising" and "betraying" the preservation movement, this is not borne out by the evidence. On the contrary, both appear to have consistently advocated for the canal's total preservation throughout the duration of the five-party talks. Araya Masaaki, who represented Hokkaido prefecture at the five-party talks as a moderator (and went on to become mayor of Otaru) supports this conclusion: "Ms. Mineyama was a tough-spirited person *who never wavered in her insistence that the entire canal be saved*. The way the matter was decided must have been really hard on her" (*Hokkaido Shinbun*, Otaru morning edition, December 30, 2010. Emphasis added.).

Nevertheless, if we read between the lines, we can identify two wholly different logics of preservation at work in the fatal split within the preservation movement: "community development" and "total preservation at any cost." Activists taking the latter approach sought to leave the canal intact—to preserve it in its current form, "frozen in time." It would be no exaggeration to describe their actions as the repudiation of change. By contrast, activists who followed the "community development" logic to preservation showed more flexibility—but also emphasized the importance of giving residents primary control over changes to their city. "Community development" activists saw preservation as a means to this end: Otaru would not lose its distinctive scenery and history if a mechanism for controlling change could be developed. While their acceptance of change was far from unconditional, their theoretical tolerance for some limited forms of change stands in stark contrast to those who remained fixated on the total preservation of the canal. Once the five-party talks broke down, the die-hard proponents of "total preservation at any cost" saw no reason to participate in the Otaru Revitalization Committee. By contrast, community development-oriented activists recognized that while they may have lost the battle to save the canal, they had acquired an influential voice in local affairs. They saw the Otaru Revitalization Committee proposed by Yokomichi as an important venue for exercising this newfound influence.

The implosion of the preservation movement unleashed recriminations and accusations between former friends and allies in the fight to save the canal. Public support for the preservation movement soon plummeted. Activists did not waste time making a formal announcement of their resignation, but simply hastened to escape further controversy by returning to their daily lives as quickly as possible. The protracted "canal war" had finally come to an end.

5.3 Community Development in the "Post-Canal" Period: The Third Phase (1985–2013)

The Otaru Revitalization Committee held twenty meetings, and six subcommittee meetings, before issuing a final report (Otaru Revitalization Committee 1987). The report proposed two focal points for Otaru's redevelopment: a waterfront district that showcased the city's historic townscape, and the traditional craftsmanship of Otaru artisans, including the local glasswork industry. The report identified a new plan for making the port a top priority and laid out detailed proposals for differentiating the Otaru Port from the nearby Ishihikari Bay New Port, drawing on case studies of port redevelopment projects overseas. Ishizuka Masaaki of the Hokudai trio and the APOC, who had also served as the Committee of One Hundred's planning director, proposed a waterfront-based form of "community development" and used computer graphics to give the proposal a concrete, visual form.

The proposals developed by the Otaru Revitalization Committee were largely ignored. This was partially the result of bad timing. The committee issued its final report in January 1987, during the final months of the Shimura administration. Three months later, Araya Masaaki was elected mayor of Otaru (Shimura, whose tenure as mayor had been almost entirely consumed by the fight over the canal, did not run in the election and retired from political life). The Araya administration never gave serious consideration to the report of the Revitalization Committee, which had been established by the Hokkaido governor, not the mayor, and carried no legal bearing.

The Otaru Rebirth Forum, established by community development activists in October 1985, subsequently petitioned Araya to implement the Revitalization Committee's proposals, but their requests went unheeded. Yet one can still detect certain changes in Otaru's approach to city planning and redevelopment. In November 1987, the city government established a deliberative council to formulate a new comprehensive plan for Otaru, and included a citizen's colloquium on community development. On April 1, 1992, the city promulgated a "Landscape Ordinance for Community Development through Otaru's History and Natural Environment."

For Otaru's former preservation activists, however, the ten years following the end of the "canal war" are rightly described as "the silent decade." Many activists rededicated themselves to their occupations during this period, and often agonized over the irreparable damage done to personal relationships following the breakdown of the preservation campaign. The die-hard "total preservation" contingent of the movement had issued their final salvo in November 1984, when they protested the establishment of the Otaru Revitalization Committee. The community development contingent of the preservation movement, including Mineyama, went on to establish the Otaru Rebirth Forum, but the activities of this new organization were limited to symposiums, panel discussions on the canal issue, and walking tours of Otaru.

However, this "silent decade" was hardly a quiet time for the city of Otaru. Indeed, the decade witnessed the extensive redevelopment of the city center and a tourism boom. As former activists fell silent, tourism-related businesses struck up a lively trade throughout Sakaimachi and along the banks of what remained of the canal. As detailed in Chapter 6, a number of historic buildings were destroyed, or torn down to make room for parking lots. The very scenery that so many had fought to preserve began to disappear (Myōen et al. 1993; Horikawa 2000; 2010. Horikawa and Fukaya, ed. 2013). Despite the economic promise of the tourism boom, the Otaru city government had no tourism policy to speak of, nor any system for assessing or guiding merchants involved in the local tourism trade. Former preservation activists bemoaned the state of local tourism as "a cheap sort of souvenir-based tourism," or "Dejima tourism."[31]

Despite their misgivings, former activists remained spectators as Otaru was transformed into a major tourist destination, eventually welcoming more than nine million visitors each year. Yet even from the sidelines, they exerted an influence on a city government newly sensitive to citizen watchdogs: in order to avoid another "canal war," local officials grew more scrupulous in their response to local issues, and carefully selected representatives to citizen councils. As one former activist put it, the protracted battle over the fate of the canal remained "like a sharp splinter lodged in the throat."[32] Neither former activists nor city officials could easily forget such a painful experience.

Meanwhile, redevelopment in the city center transformed Otaru's townscape. Controversy returned to Otaru with the advent of the commercial complex Mycal Otaru, which took over the former site of the Otaru Chikkō railyard in March 1999. While the national retail giant was welcomed by some, others protested Mycal's presence in Otaru, citing the need to protect local businesses and Otaru's traditional townscape (Morihisa 2002).

It is worth asking why the "Mycal dispute" never took on the force or proportions of the canal controversy. To be sure, this was a different issue, arising at a different time in Otaru's history. Yet one can detect the legacy of the "canal wars"

[31]The tiny man-made island of Dejima, just off the coast of the southern Japanese city of Nagasaki, was once a Dutch trading post and the only point of exchange between Tokugawa-era Japan and the outside world. Dejima has since been transformed into a popular tourist attraction, yet it has very little connection to the modern city of Nagasaki or its residents. Describing Otaru's tourism scene as "Dejima tourism" is thus a critique of a tourism model that is concentrated in a narrow sliver of a city and completely detached from the everyday life of residents.

[32]This is the expression used by Yamaguchi Tamotsu and Ogawara Tadashi.

in the Mycal dispute. During their campaign to save the canal, activists canvassed every neighborhood in the city to drum up support for their cause. They received no support from the merchants in the city center (proprietors of businesses along the city's main shopping streets), who feared that taking a clear stand for or against the canal's reclamation would cost them customers and jeopardize their business. By and large, Otaru's merchants kept a safe distance from the canal controversy. Preservation activists and their supporters resented this indifference—in their words, the merchants "let us die alone." Because these merchants never came to the aid of their fellow citizens in the fight against the city government, they received no help in return when, years later, they took up their own battle against Mycal. Community development activists remained silent, offering neither support nor advice. Former activists likely found it too difficult to overcome their sense of indignation that the merchants would seek their help after refusing to join the fight to save the canal. Moreover, in the eyes of former preservation activists within the community development camp, the local merchants' campaign to protect themselves from the retail giant smacked of narrow self-interest, not concern for the greater public good. They watched with cool indifference as the merchants' efforts faltered, and Mycal Otaru opened for business in 1999. One can thus conclude that the "canal war" cast a long shadow over the fight to keep Mycal out of Otaru.

Following their decade of silence, many former preservation activists felt compelled to revive community development efforts—in part to address what they saw as the sorry state of tourism in Otaru. However, they struggled to overcome the deep rifts created during the breakdown of the canal preservation movement. Nevertheless, several former activists went on to join semi-governmental organizations related to tourism and used their positions to promote community development efforts. Yamaguchi Tamotsu and Ogawara Tadashi became active members of the Otaru Tourism Promotion Council, while Sasaki Kyōjirō served as secretary of the Otaru Tourism Association.[33] They discovered, however, that these new positions were limited: the state of Otaru could only be discussed in relation to tourism, and their message was often lost on the local tourism industry. It was a far cry from their former activism when they were free to adopt a more ad-hoc approach.

Throughout the canal preservation campaign, Yamaguchi Tamotsu had demonstrated his abilities as an adviser and strategist. Later, having witnessed the limitations of the local referendum system revealed by referenda in places such as Makimachi (Nakazawa 2005), Yamaguchi and other former activists identified the city council as a venue for pursuing change and issue resolution.[34] After a fiercely-fought campaign, Yamaguchi was elected on April 28, 2003. The canal preservation movement that

[33] All three left these positions upon completion of their terms.

[34] In August 1996 the residents of Makimachi, Niigata Prefecture successfully used a referendum to block the construction of a nuclear power plant. Yet referenda remain costly and limited (voters are asked for a yes or no vote on a single issue). Given this, Yamaguchi saw the city council as the proper site for mediation and resolution of local issues, and ran for office accordingly. This is yet another example of the layered nature of citizen activism in Japan. Activists in Makimachi had watched earlier developments in communities throughout Japan, including Otaru. Makimachi in turn demonstrated the inherent limitations of local referenda to Otaru activists like Yamaguchi.

began in 1973 had finally produced a legislator of its very own, who worked to expand community development efforts from within the city council.[35] Yamaguchi was reelected in 2007 and 2011, and retired upon completing three terms in office.

Today, the city of Otaru must decide how it will pass on the lessons of the "canal wars" in order to meet the challenges posed by a dwindling population and the loss of its traditional townscape. The "beginning of the end" is no longer just a warning—it has arrived, and the lessons of the canal war have become more necessary than ever.

5.4 The Activists: Life Histories and Patterns of Activism

Up until now, the subject of this chapter has been the "preservation movement" as a whole. But the preservation movement was far from monolithic. In the final section of this chapter, we turn to the individual men and women who mobilized in defense of the canal. What led them to the cause of preservation, and how did they take up the fight? What motivated them to embrace a cause with no opportunity for profit or compensation?

We cannot hope to answer such questions if our unit of analysis remains the "preservation movement" as a whole. We have already seen how the movement to save the canal incorporated two broad factions, whose shared goal of preservation stemmed from quite different ideals. One faction clung to the goal of "saving the entire canal," while the other came to focus upon developing a form of canal-based tourism in Otaru. The two factions did not adhere to the traditional cleavage within Japanese politics: both factions included progressives and conservatives. How did a movement that straddled the conservative-progressive divide survive for over a decade? How did people of such different political stripes manage to work together within the movement? How and why did the leadership of the preservation campaign eventually transition from one faction to the other? Questions such as these make it clear that the "preservation movement" as a whole cannot remain the unit of analysis if we are to understand the reality of the preservation campaign. It is time to move our level of analysis from the "movement" to the individual activists.

The following section examines the preservation movement from the individual level, as we review the life histories and events that led some to join the effort to save the Otaru canal.[36] Classifying these life histories according to certain patterns allows

---

[35]Certain members of the city council had previously voiced support for the canal's preservation and the preservation movement, but Yamaguchi was the first preservation activist elected to local office.

[36]This issue is not limited to the unit of analysis (in this case, the movement or the individual activist). It is equally applicable to the *mode* and *style* of analysis. This is a narrative analysis aimed at understanding a one-time phenomenon, not a paradigmatic analysis aimed at deriving a fixed theory of a causal relationship. Suzuki Tomoyuki's (2013) succinct explanation gets to the heart of this difference: "It is possible to explain the causes behind actions in the 'logical-scientific mode.' However, the result of such an exercise is qualitatively different from one that seeks to understand the motivations for action during a single event. There is a huge difference between stating that 'Four

us to explain the evolution of the movement from its early to later phases. Interviews with activists led to the identification of four significant variables for classifying activists within the canal preservation movement (Horikawa 2007, 2010):

Locality-oriented (LOC)

Nostalgia-oriented (NOS)

Tolerance of change (CNG)

Political attitude (POA)

The locality orientation describes the importance one attaches to the distinctive features of the local culture and landscape. For example, an activist with a positive locality orientation (LOC +) believes that it is precisely the distinctive scenery and resources of Otaru that should be protected, and is either indifferent or opposed to homogenizing forces in the environment, such as the entrance of national chain stores. This locality orientation is indicated by a positive (+); its absence is designated with a negative (−). Someone with a consistently positive locality orientation (LOC +) would naturally work to preserve the distinctive features and landscapes of his or her local society. Conversely, people lacking this locality orientation (LOC −) would only participate in such an effort if they could meaningfully connect it to a different logic for preservation. The locality orientation variable is an extremely useful tool for distinguishing between the two camps within the preservation movement.

The nostalgia orientation (NOS) indicates whether or not an individual's logic for preservation is rooted in a nostalgic yearning for the past. Those who demonstrate a strong desire to preserve the beloved objects of their past at any cost can be said to manifest a positive nostalgia orientation (NOS +). Those who do not are NOS −. This variable is essential to understand the political processes at the interface of the preservation movement and the administration. Preservation movements are almost always tinged with nostalgia (Barthel, 1984, 1996; Horikawa 1989, 1990, 1994). However, nostalgia is entirely incompatible with the administrative and bureaucratic system. Indeed, there is probably nothing more difficult for an administrative organization to deal with than opposition rooted in a nostalgia for the past. We need not consult Max Weber to conclude that bureaucracies are guided by laws and regulations, not sentiment. Consequently, the success or failure of a preservation movement rests largely upon activists' ability to translate their nostalgia into a form capable of being addressed by bureaucratic organizations. Nostalgia is thus one crucial factor in assessing the success of a preservation movement.

The CNG variable is an indication of how much change an individual is inclined to permit—in other words, his or her tolerance of change. An individual with a positive CNG orientation (CNG +) tolerates change, while a CNG—individual is likely to reject change, or only tolerate it to a very small degree. A detailed analysis of the inner workings of the preservation movement is only possible when this variable is

out of ten women who have been cheated on decide to leave their partner,' and 'I decided to break up with my boyfriend because he cheated on me.' In most cases, our approach to understanding individual events in our everyday life takes the latter form—in other words, a narrative mode" (Suzuki 2013: 144).

factored into our analysis. All activists used the same word to describe their goal: "preservation." It is only by assessing their relative tolerance of change that we can begin to identify the differences in their approaches to the issue and untangle the various logics that informed their efforts to save the canal. After all, there was no single rationale for saving the canal. The CNG variable illuminates the range of arguments for the canal's preservation, from "static" or "cryo-preservation" to "dynamic preservation."

Finally, the POA variable indicates whether or not an individual embraces overt confrontation (POA -) or adopts a more conciliatory, cooperative approach (POA +). To be clear, this variable does not indicate an individual's political *affiliation* or *membership* in a political party. Rather, it describes a negotiation *style*. POA − activists did not shirk from outright confrontation during negotiations with city leaders, and attempted to achieve their goals at any cost. POA + activists, meanwhile, were more likely to avoid direct confrontation and attempt to negotiate with city officials through alternate means: they used personal connections to sound out solutions after the formal end of a committee meeting, over a midnight drink at a bar in Hanazonochō, or during an informal home visit. In the small, provincial society of Otaru, the attitude displayed during negotiations has a huge influence on how much support a movement can garner. Similarly, the decisive factors affecting the city council's response to a petition often have less to do with the petition's content than with the sequence of events leading up to its submission, and the manner in which the petition is presented and explained. The same petition will be received very differently by city council members and local bureaucrats depending upon whether or not it has been preceded by informal consultations. In yet another frequently observed phenomenon, activists' deployment of arguments that utilize the standard vocabulary of modern democracy—including words such as "rights" and "obligations"—can elicit a negative reaction, and prompt the administration to assume a defensive posture (Akashi 2002). While the ideal–typical bureaucracy should be incapable of such a response, in the real world it is a commonly observed phenomenon. For all of these reasons, political behavior is an essential final variable in our analysis.

Following Manuel Castells' (1983) method of analysis, these four variables can be used to establish four categories of activist within the canal preservation movement:

1. The "Aesthetic School"
2. The Pure Preservationists
3. The Community Development Activists
4. The Traditional Leftists.

A more detailed discussion of these four categories is forthcoming in subsequent sections. First, however, let us define the "two broad groups with contrasting ideals for preservation." The camp that sought to save the canal in its entirety included the "aesthetic school" and the "traditional leftists." In the opposing camp were the "community development activists," who proposed a tourism promotion plan centered upon the canal. Conflict between these two camps would eventually split the movement, but for a long time, Mineyama Fumi and her fellow "pure preservationists" managed to bridge the divide between these two opposing camps.

In the sections that follow, the special characteristics of each category are described with reference to the life history of a representative activist (Kikkawa 2001a). The goal of this approach is to depict the canal preservation movement as the lived activism of real, flesh-and-blood human beings, not an abstract group of nameless participants (Arisue 1992; Kito 1996; Tamano 1997; Kito ed. 1999). This book considers the "world of subjective meaning." It is thus crucial that we understand the thoughts, feelings, and memories of these subjects. The canal—that single body of water—was the subject of many different "gazes." Unless we can grasp the meaning ascribed by these various gazes, it will be impossible to fully understand the Otaru canal preservation campaign, which struggled for years and had a profound influence on the course of national preservation policy. Let us now turn to the different gazes cast upon the canal, and seek to understand the meaning that real people assigned to each.[37]

5.4.1 The Aesthetic School

The first Otaru citizens to mount an effort to save the canal were members of the "aesthetic school" within the preservation movement. These men and women saw the canal as an essential part of their identity. To them, preserving the canal was not a political issue, but rather a matter of designating and protecting an important cultural property. These activists referred to earlier movements across Japan aimed at designating cultural properties, and consequently relied primarily upon petitions to further their cause. They also genuinely loved the landscape of the canal district and refused to consider any change to the canal or surrounding environment. Indeed, they detected a certain beauty in the decay and decline of the canal district. While they also loved their hometown and saw the canal as an affirmation of their identity as

[37]Even Kikkawa Toru, the foremost researcher in the so-called "quantitative method," noted when discussing his book *Gakureki Shakai no Rōkaru Torakku* [The Local Track in an Education Society] (Kikkawa 2001a) that: "I realized the limits of the paper questionnaire survey when it comes to the exploration of micro-level and dynamic social relations. In such cases, quantitative data can offer only a single cross-sectional view" (Kikkawa 2001b) and "quality can weed out quantity" (Kikkawa 2001b). Kikkawa's comments should not be taken as an argument for the superiority of qualitative data, but rather as part of his larger review of the special characteristics, and pros and cons, of both quantitative and qualitative data. The crucial point here is that even an expert in quantitative data like Kikkawa recognized that qualitative data, such as life histories, are essential to making sense of the momentary "cross-sectional views" captured at different times through quantitative data. If we apply Kikkawa's discussion to this study of the Otaru preservation movement, we could say that in order to understand the "micro and dynamic social relations of the preservation movement," we require qualitative data to explain why the "cross-sectional views" of a particular instance changed over time. This explains this book's emphasis upon the life histories of Otaru's preservation activists.

Otaru residents, it is this particular aesthetic sensibility that justifies their description as "the aesthetic school" of the preservation campaign.

Yet activists of the "aesthetic school" proved unable to fully express their attachment to the scenery of the canal, speaking to others only of their "love for their hometown" and their fear that "Otaru would no longer be Otaru" without the "irreplaceable" canal. There is no need to assess the degree of emotion, or the level of significance, they attached to their choice of words. On the contrary, it was precisely because these activists experienced the canal space at such a primal and physical level that they proved unable to speak of it in more objective terms. As a result, however, their primacy within the preservation movement was ultimately supplanted by members of the "community development" group. Members of the "aesthetic school," who continued to discuss their beloved canal and its environs in terms of its distinctiveness, included Koshizaki Sōichi, the first chairman of the APOC, Fujimori Shigeo, and the storytelling duo of Naka Kazuo and Matsuoka Tsutomu, who used *kamishibai* (a traditional form of paper storytelling) to appeal for the canal's preservation. While this group certainly saw culture as a battlefield, they did not remain entirely uninvolved with politics—unlike the "pure preservationists" discussed in the following section. However, their political activity was characterized by its moderation: they limited themselves to petitions, appeals to city council members, and signature-gathering campaigns.

Let us turn to one of the primary figures within the preservation movement's aesthetic school: Fujimori Shigeo.

A Life History of Fujimori Shigeo: The Pioneer of the Preservation Movement

Fujimori Shigeo[38] was born in Otaru, in 1936. His family ran the Fujimori Benkyōdō, a venerable local sign-making business founded in 1897. Fujimori showed artistic talent from a young age, and in 1954, he enrolled at Tama Art University in Tokyo. After graduating in 1958, Fujimori returned to Otaru and joined his elder brother in the family business. More of a designer than a painter, Fujimori distinguished himself in commercial art and advertising campaigns. In the year of Fujimori's graduation from university, one of his works was used as the poster for the Hokkaido Exposition, and he was repeatedly recognized for the tourism posters he submitted to the Japan Advertising Art Association (Horii 2009: 136–137). Soon after his graduation, the Otaru-born, Otaru-raised Fujimori submitted two winning posters to the association: *Otaru in Winter* and *Otaru Snow*. Fujimori, who was also involved in the planning of the 1958 Otaru Exhibition and *Ushio Matsuri*, the local summer festival, was clearly bound inseparably to his hometown.

One day, Fujimori witnessed the demolition of the Arihoro warehouses, and decided that something must be done to save the canal from a similar fate:

[38]The author was unable to personally interview Fujimori Shigeo, who died in 1987 at the age of forty-nine. This life history is based upon other interviews, including the author's interview with Fujimori's widow, Shigeko, and other secondary sources. Of the latter, the graduation thesis (1975) of Ishizuka Masaaki, a member of the Hokudai trio who would later play an active role in the Otaru life, and Horii (2009) are particularly valuable sources of information on Fujimori.

> You know, last summer (in 1972) the amateur artists of Otaru put on a "canal exhibition." They probably intended this as a kind of "farewell to the canal." It was their way of saying goodbye to a canal that would soon disappear. That made me so angry! Then the senseless destruction of the Arihoro warehouses prompted me to take action. {...} Ten of us gathered on November 21, (1973) with the feeling that we could no longer remain silent.
>
> Quoted in Ishizuka Masaaki, 1975: 76. Information in parentheses has been added by the author.

Fujimori's narrative explains the factors behind the formation of the APOC. First came the demolition of the Arihoro warehouses—once a subject of Fujimori's painting. Then came Fujimori's disgust at the assumption widely shared among the artists ("amateurs") who painted along the canal: that this particular landscape was destined to disappear ("they probably intended this as a kind of 'farewell to the canal'"). He rebelled against this assumption. Thus, began the "brave fight led by a young sign-maker as commander" (Ogasawara 1986: 99). The APOC was established in 1973, with local historian Koshizaki Sōichi as chairman, the poet Yoneya Yūji as executive manager, painter Chiba Shichirō as publicity manager, the educator Toyotomi Tomō as organization manager, and the architect Horii Toshio as finance director. Fujimori himself headed the secretariat.

The new association's members immediately began to campaign vigorously, submitting petitions to the mayor and city council. Fujimori was central to these early efforts. The most vivid demonstration of Fujimori's dedication to the cause was his self-funded two-week "pilgrimage" across Japan to learn from earlier townscape preservation efforts (Ogasawara 1986: 101–103). As he visited Tsumago, Kobe, Kurashiki, Okayama, Kyoto, Nara, Kameyama, and Takayama, Fujimori began to comprehend the significance behind the effort to save the canal and learned concrete tactics for achieving this goal. On June 1, 1974, Fujimori made the final stop on his pilgrimage: the Agency for Cultural Affairs in Tokyo. There he succeeded in eliciting an important judgment from agency officials, who acknowledged the "high value" of the canal and nearby warehouses. Yamaguchi Tamotsu, who would later play a central role in the preservation effort, recalled his first encounter with Fujimori and Fujimori's lasting influence:

> Thirty-three years ago, in October 1976, I first encountered the strong personality and passion of Fujimori Shigeo. He believed that the Otaru canal *was* Otaru. Paving over the canal would destroy the town. It needed to be saved at all costs. He was driven [by this belief]. {...} Today there has been a reappraisal of post-Meiji era modern architecture, and the need to preserve it is widely accepted. But back then [these buildings] were only beginning to draw attention. As [APOC] secretary, Fujimori fought hard—negotiating with the Agency for Cultural Affairs and going to the Hokkaido government offices to file a complaint. {...} Fujimori thought that young people should play a central role, and remained the unsung hero. {...} [One could say that] I'm still infected by the "Fujimori virus.
>
> Yamaguchi Tamotsu, "A Pioneer in Canal Preservation" (Horii 2009: 91). Information in parentheses has been added by the author, information in brackets by the translator.

Yamaguchi identifies the "driven" Fujimori as the "pioneer of canal preservation." By using the analogy of "the Fujimori virus," he also conveys Fujimori's lasting impact.

Fujimori's high school classmate Satō Kimiaki agreed: "(Fujimori) had extremely intense feelings about saving the canal. It laid bare his fighting spirit. He painted so many pictures of the canal area and felt that the area had to be saved somehow. I suppose he should be called the leading figure of the canal preservation movement" (Horii 2009: 94; Information in parentheses has been added by this author).

The canal must be saved "at all costs." "Driven" by this belief, Fujimori pushed forward, "laying bare" his "fighting spirit." This is the image of Fujimori Shigeo that emerges from the accounts of Yamaguchi and Satō. But how do we explain the extent to which Fujimori felt saving the canal was imperative? Fujimori gave the young Ishizuka Masaaki the following answer:

> ...The APOC is about just one thing: saving the irreplaceable canal. Any understanding of the canal in relation to the port [of Otaru] is entirely beside the point—it's simply that we love the canal, and want to save it.
>
> Ishizuka (1975:75). Information in brackets has been added by the translator.

Fujimori's reference to "the port [of Otaru]" is likely an allusion to the city's port redevelopment plan and the current status of the port. His point is that his desire to save the canal was completely unrelated to political or economic issues—it stemmed solely from the fact that "we love the canal." Fujimori demonstrates an extremely nostalgic orientation: NOS + + . Moreover, his stubborn determination to save the canal indicates a negative CNG orientation.

In conversation with Ishizuka, Fujimori continued:

> As far as I'm concerned, it doesn't matter if [the canal becomes a] nationally-designated cultural property, or a cultural property designated by the Hokkaido [prefectural government]. The citizens of Otaru truly love this place. Otaru natives love this place. These citizens have identified the cultural value [of the canal]. That's what is essential. And that is the spirit that lies at the very heart of our organization.
>
> Ishizuka (1975:80). Information in brackets has been added by the translator.
>
> Now is the time for the mayor to take action on development in our hometown. But he should try to avoid doing so in a common, ill-mannered way—it should be a cultural approach. I don't want to describe a cultural type of citizen's movement as a 'cultural movement.' (I'd make a point of referring to it as) a cultural citizen's movement. I want to demonstrate the character [of our movement].
>
> Ishizuka (1975): 87. Information in parentheses has been added by the author, information in brackets by the translator.

From these two excerpts from Ishizuka's conversation with Fujimori, it is clear that Fujimori believed the canal was a "cultural property" precisely because it was beloved by Otaru natives. Their love for the canal justified his efforts to save it. Also, Fujimori believed it was important to avoid a "common" approach to development in favor of a "cultural citizen's movement." This sentiment was shared by Koshizaki, who wanted to "avoid being used as a tool in a political battle." Given this, Fujimori's political orientation is positive (POA +).

Fujimori's comments correspond perfectly with the description of the preservation movement included in the APOC's "Campaign Guide," which was drafted primarily by Fujimori. Clearly, Fujimori was far more than the APOC's administrative head; he also played a pivotal role in developing the objectives and principles of the preservation movement. Significantly, Fujimori's actions stemmed from his "*designare* spirit." Horii Kiyotaka (2009) explains:

> [Fujimori] Shigeo came across the Latin word *designare* in a book. The word *designare*, from which the word 'design' stems, refers to the anticipation and development of a "specific material response to the shape and requirements of everyday life." Shigeo says that when he learned this word, "it felt as if everything before me immediately became brighter" {...} Design was not a matter of circles, triangles or squares, but of "anticipating and dealing very directly with one's daily existence." That is what design was about. This is what Shigeo decided when he made his life in Otaru as a designer, with the "spirit of *designare*."
>
> Horii (2009: 137).

If we accept Horii's explanation, Fujimori rejected the armchair definition of design as a matter of "circles, triangles or squares." He identified the true meaning of design as "dealing directly" with the proposed reclamation of the canal, a place so [deeply] rooted in his daily life. In 1974, Fujimori's wife told Horii that "recently Shigeo thinks about more than just himself—he's started thinking about how to make all the residents of Otaru happy" (Horii 2009: 138). Taking both of these comments into account, it seems clear that Fujimori began his work to save the canal precisely because, as a designer, he could "anticipate" that the consequences of the city's road plan for the "citizens of Otaru" would not be "happy" ones. And, because Fujimori believed that *designare* summed up the ideal spirit of design, he aspired to pursue a form of design embedded within the local community. As a designer living in Otaru, Fujimori believed it was his mission to launch the canal preservation campaign, and revealed his "fighting spirit" and a determination to save the canal "at all costs." For these reasons, Fujimori demonstrates a strong locality orientation (LOC +).

Fujimori's *designare* spirit also made him a target of the local supporters of the new road plan, who forced him to leave the preservation movement. After the formal establishment of the APOC, he began to cross swords with the city administration (through petitions and surveys). Fujimori was eventually summoned before leading members of Otaru's business community. According to his wife, "he was summoned to the lobby of the Hokkaido Hotel [in downtown Otaru], and told that 'unless you promise to quit the movement within the next five minutes, we'll stop patronizing your business.' After he returned home, he burst into tears. I cried, too."[39] Fujimori had no choice but to quit the movement: the family business of billboard-making and event design and preparation would go bankrupt without the support of local businesses. In June 1976, Fujimori reluctantly stepped down from his position as APOC secretary.[40] The observation that "economic pressures forced key members

[39]Interview with Fujimori Shigeko in Otaru, September 5, 2000.

[40]Fujimori's illness was another factor in his resignation, but Ogasawara (1986: 104–105) records the same incident at the Hokkaido Hotel, and we also have the testimony of Mineyama Fumi: "Fujimori resigned in part due to health-related reasons, but the pressure from the city administration and the business community made it difficult for him as a self-employed business owner to run his business... I don't know all the details, but I couldn't believe that minor health considerations would

of the preservation movement to drop out" (Sasaki ed. 1983: 11) is a clear reference to Fujimori. Thereafter, Fujimori relinquished his central role within the preservation movement and devoted himself to the family business and his own medical treatment.

Yet Fujimori's feelings for the canal remained unchanged. One episode vividly demonstrates the consistency of Fujimori's position. On the day that pile-driving began in the canal, APOC chairwoman Mineyama Fumi tearfully watched from a bridge. Fujimori was not at the scene, but "witnessed" the pile-driving on the evening news. Fujimori's eldest daughter, Satsuki, recorded her father's reaction at the time:

> On November 12, 1983 the news reported that pile-driving work had begun for the reclamation of the canal. My father, who had watched the news report, was so infuriated that he wept tears of frustration, and hurriedly began scrawling on a large canvas. What our family witnessed that night was our father's implacable creation—in bright red—of the "Red Canal." Of all his emotion-filled paintings, in no other piece did he lay bare his own emotions as in this one.
>
> Fujimori Satsuki, "My Father, Fujimori Shigeo" (Horii 2009: 154).

In this author's interview with Fujimori's widow, she also recalled that "when he heard the news of the destruction (of the Otaru canal), he painted it in a single day," and "he poured his sadness and frustration into the red [of the painting]."[41] This color choice was not aimed at achieving a particular artistic effect but rather revealed Fujimori's true feelings. By painting the canal deep red, Fujimori was likely expressing his fury with the city administration and his deep chagrin at the loss of the canal.

Despite his poor health, Fujimori's intense feelings for the canal prompted him to reengage with the canal preservation movement in 1983. At this point, key members of the Otaru Chamber of Commerce were revisiting the planned reclamation of the canal and beginning to publicly support the canal's preservation. The Committee of One Hundred for the Otaru Canal had also begun preparing for a recall effort against Mayor Shimura. Fujimori was openly critical of how the movement, which he himself had founded, had changed during his absence and pressed for a reorganization of the APOC and a reassessment of its relationship to the Committee of One Hundred. The APOC's second leader, Mineyama Fumi, was often in the public spotlight, and Fujimori also reportedly resented the fact that his own achievements and contributions to the preservation effort garnered less notice.

Fujimori immediately embarked on a reorganization of the preservation campaign: he established an executive committee without approval from a general meeting at APOC members and appointed a proxy leader of the secretariat. Fujimori's attempted reforms ignored the objectives and achievements of the people who had led the preservation campaign during his absence, and elicited a strong backlash. No one could dismiss Fujimori's early efforts to mount a preservation campaign and remind Otaru

be enough for him to abandon the movement. When I remember his passion, I find this impossible to accept" (Mineyama 1995: 48). Given that this viewpoint is supported by Sasaki ed. (1983) and the author's other interviews in Otaru, it seems clear that the financial threat to the Fujimori family business was a primary reason for his resignation.

[41] Interview with Fujimori Shigeko in Otaru, September 5, 2000.

residents of the canal's true value. After his return to the preservation movement, however, Fujimori's actions and remarks were not welcomed by Mineyama and others.[42]

Enraged by the loss of the canal—the beloved landscape of his childhood—and also by his treatment within the movement that he himself had founded, Fujimori began to harbor a hatred for his opponents in the movement:

> Shigeo was the one who started the movement, but [they] talked as if they were the ones who had. Nobody would call to [tell him when] the (canal preservation movement) attracted attention or received some form of recognition. Shigeo always said that 'people are assets,' but he once told me that in his lifetime, there was 'just one person I can't bring myself to forgive, and that person is Mrs. Mineyama.'
>
> Interview with Fujimori Shigeko at the Canal Gallery in Otaru, September 5, 2000. Information in parentheses has been added by the author, information in brackets by the translator.

The intensity of Fujimori's feelings can likely be explained by his passionate conviction that the canal be saved—and the fact that he understood the preservation movement as his own creation.

While he was forced to abandon the cause for some years and found his goals stymied upon his return, Fujimori remains a pioneer of the canal preservation movement. He was the first person to demand that the canal be saved, and the one who established a movement toward that end. Fujimori eventually gravitated toward the "community development" faction of the movement and indeed, one of his greatest contributions to the preservation effort was his education and support of Yamaguchi Tamotsu and other community development-oriented activists. However, Fujimori himself is classified as a member of the aesthetic school because he was unable to offer a clear "reason" for why the canal should be saved. Fujimori himself would have acknowledged that his championship of preservation was rooted in his nostalgic orientation, and characterized by an "irrational" "love for his hometown" and "old-fashioned emotions." Fujimori's life history and mode of activism is thus rightly positioned within the initial period of the preservation movement.

5.4.2 *The Pure Preservationists*

A second group within the movement was characterized by their exclusive focus upon preserving the canal as an end in and of itself. By carefully distancing themselves from politics and economic concerns, and focusing exclusively on a shared sense of attachment to the canal, this group consistently received a certain degree of support from many older and middle-aged Otaru natives. While their resolutely apolitical stance exposed them to criticism from both within and outside the movement, their

---

[42]For an account of Fujimori's actions and declarations after rejoining the preservation movement, see "Unga Mondai Jyūnen no Ayumi," Part 3: Nos. 31–44, published between November 29 and December 22, 1984 in the Hokkaido edition of the *Asahi Shinbun*.

single-minded focus upon preservation also appealed to political conservatives. The most representative example of this second group of activists is Mineyama Fumi, who became chairwoman of the APOC following Koshizaki's untimely death. A devout Christian, Mineyama was a highly respected church elder. Her wholesome public image helped differentiate the canal preservation movement from other forms of ostensibly "civic-minded" activism with hidden profit motives. Mineyama's statements consistently transcended specific political and business interests, thereby helping to unify and sustain a movement riven by conflicting motivations. It was precisely the transparent innocence of her goals that facilitated the convergence of such diverse interests into a single movement to save the canal. Indeed, it was the presence of Mineyama and other "pure preservationists" that enabled otherwise bitter enemies to pursue the common cause of saving the canal.

A Life History of Mineyama Fumi: "One's hometown is close at hand, and to be protected."

Given her deep involvement with the issue, the name Mineyama Fumi has become almost synonymous with the Otaru canal. Already sixty when she began her efforts to save the canal, Mineyama consistently supported a form of community development rooted in the canal and other historic buildings in the district.

Mineyama Fumi was born in 1914, in the village of Makkari, in the Abuta district of Hokkaido. In 1924, Mineyama moved with her parents and younger sister to Otaru, and was immediately dazzled by the energy of her new home—an energy that had been unimaginable in quiet, rural Makkari. After graduating from the prefectural girls high school in 1931, she became the first female employee at the local branch of Mitsubishi. She later married Mineyama Iwao, a high school teacher and archeologist who went on to head the Hokkaido Cultural Properties Research Association and teach at the Hokkaido University. In 1944, the Mineyamas left Otaru for Date after Iwao received a teaching assignment in that city. It was in Date that Mineyama converted to Christianity. After returning to Otaru eleven years later, in 1955,[43] Mineyama briefly taught at a municipal elementary school. In 1973, she joined the APOC and went on to lead the organization following the sudden death of the APOC's first chairman, Koshizaki Sōichi, in 1978. She stepped down from the position in 1984 to "take responsibility for failing to save the canal." She later served as an advisor to the Otaru Revitalization Forum and was recognized by the Architectural Institute of Japan for her years of work in community development *(machizukuri)* in 2008. Mineyama died on December 28, 2010, at the age of 96. At the time of her death, Mineyama was in the care of many of the old friends who had fought alongside her to save the canal.[44]

[43]In her interviews, Mineyama speaks of a twelve-year absence from Otaru. Elsewhere (Mineyama 1995: 23), however, she writes that she lived in Date between 1944 and 1955—an absence of eleven years. Although eleven years would seem to be the actual duration of her absence, the transcript of her interview has not been altered.

[44]Mineyama's national reputation as a preservation activist was confirmed at the time of her death. According to one passenger, her death was reported on the scrolling news screen aboard the Tōkaidō *shinkansen*, as well as by all the national newspapers the following day.

Born in the peaceful village of Makkari and raised by her teacher father, Mineyama described her younger self as a quiet "honor student" (Mineyama 1995: 20). Her former students at Kitatenomiya Elementary School remember Mineyama as "such a mild-mannered teacher" (Mineyama 1995: 77). However, this "mild-mannered" woman devoted a decade of her life to the preservation movement—on one occasion flying to Tokyo in the middle of a blizzard for direct talks with the Minister of Construction and the Minister of Internal Affairs. In order to make sense of Mineyama's transformation from quiet teacher to dynamic activist, we must remember that Mineyama knew *two Otarus* in the course of her lifetime. Mineyama had personally experienced both the prosperous Otaru of old and the more recent, declining Otaru, and this fact was of crucial importance to her later activism:

> I first set foot in Otaru in 1924… the canal had just been completed in the previous year. I was born in a small farming town, so when I first set foot in Otaru as a second grader, in 1924, the vitality of the place—the liveliness of it—took me by surprise. The people on the street seemed so busy and full of energy. Things were already really moving. There was a bus stop, cars… I thought: this must be what a vibrant city is like—like Otaru. I was a child at the time, but later, when I thought back to that time, I realized that the Otaru of those days was exactly the type of city that people would describe as 'vibrant,' or 'bustling.' And the reason I could look back and see that living, vibrant Otaru was because I later saw another Otaru—a city that had died. That's why I could look back and see how vibrant and bustling it once was.
>
> After I graduated from school here, I went to work for Mitsubishi Corporation. Did you know that when I was hired I was the first female employee at Mitsubishi's local office? And there were so many men working there who had been sent from headquarters—university graduates, all of them.[45] You had Mitsubishi, the company, and then [below that], Mitsubishi Commercial Company, which sold grain, coal, and oil in the prefecture. The phone would start ringing in the morning and wouldn't stop ringing all day. Orders came in from Abashiri and from all over the prefecture—and I had to respond to all of them. And it just so happens that [the Mitsubishi Building] was next to the canal. {…} There was a breakwater at the port, and when I would leave the building—say at lunchtime—I would see a long line of ships within the breakwater, their prows bumping together. Their cargo would be transferred onto a big barge, and then this barge would enter the canal. The barge would unload cargo at the stone warehouses on the side away from the street. {….} That was the kind of place Otaru was, back then. And as the port grew busier, with more and more ships coming through, commerce grew too—a number of shops started opening up. People would come to the Irifune warehouse district from all over to Hokkaido for provisions. Irifune was packed with people. {…} Otaru was really extraordinary back then. But then, I suppose Tokyo was as well. Perhaps it was because I didn't know anything but Otaru that I thought Otaru was so wonderful, but when I think back on it now, that's how I remember it. It had a strong economy and a refined quality of life. It was a sophisticated place, full of singing [to the *shamisen*, and other cultural pursuits].
>
> I lived there for ten years after finishing school, but then I had to move to a city called Date, although I really didn't want to leave. I loved Otaru and hated the idea of leaving, but I didn't have a choice. I always thought about making my way back to Otaru somehow, and twelve years later, in 1955, I returned. After my return I went to the port and there was not a single ship in the harbor. When I first came to Otaru it had been full of ships—within the breakwater as well. [But when I came back to Otaru] there was not a single ship in the

[45] Mineyama's reference to the educational background of her colleagues is important, as it suggests that Otaru's prosperity was attracting some of the country's best and brightest to the city.

port. I was shocked. In twelve years, all the ships had disappeared. And all the banks were gone—the old bank buildings stood vacant. Just three or four remained. All the shipbuilding companies had closed up shop, and the businesses in the warehouse district were shuttered. I was constantly surprised by what I saw. {…} I had wanted so badly to return to Otaru, but I was so shocked by what had become of it.

Interview with Mineyama Fumi in Otaru, September 1, 1997.

In the above passage, Mineyama describes her early fascination with the energy of old Otaru, explaining that "I loved Otaru and hated the idea of leaving." Clearly, Mineyama can be characterized as LOC + and NOS +. Upon her return, however, she was shocked to find not the prosperous Otaru she had left, but a city in decline. Her shock was all the greater for not having experienced this decline gradually in the course of her daily life, but suddenly, after a long absence. Therein lies the significance of Mineyama's *second* encounter with Otaru. As Mineyama herself says:

(Otaru had declined to the point that) you wondered what had happened to that city, which had once been so full of energy? I have seen Otaru twice: once when it was thriving, and once when things seemed hopeless.

Interview with Mineyama Fumi in Otaru on September 2, 1999.

Mineyama's second encounter with Otaru made her painfully aware of changes to the city, and it was this second encounter that propelled Mineyama to join the preservation movement. As she once reflected:

As someone who grew up here in Otaru, I just love this place. When I returned after twelve years, I was deeply moved by the changes to the city. I felt more than just a nostalgic yearning for the past—my sense of shock was much stronger than that. There wasn't a soul around the canal. Weeds were growing out of deserted boats. Rusted warehouses cast their shadow over the sludge-filled canal. I couldn't bear it.

This is when there was first talk of paving over the canal to make a road. I was just one ordinary citizen of Otaru, but this news shook me to the core. My reaction wasn't logical or rational—all I felt was a simple desire to protect the canal.

Mineyama (1995: 167–168).

Given Mineyama's shock at Otaru's transformation, and her "simple desire to protect the canal," her decision to become involved with the canal preservation movement is easy to understand.

Why should the canal remain? What was Mineyama's "logic" for preservation? Mineyama did not feel the canal should be saved because of its age:

…For ten years I fought to save the canal. Now why did I do that? After all, I fought for the canal even though all the procedures had already been taken [to pave it over and build a road].

For one thing, the canal followed the arc of [the coastline of] the Otaru port. And the sight of the dark grey warehouses lined up along the canal was really beautiful—it was a scene that had been painted, written about in books, and shown in movies and on television. But that's not why I wanted to save it—and it wasn't because the Architectural Institute of Japan's subcommittee on Meiji architecture had written that there were three Meiji-era landscapes they wanted to preserve at all costs: the Glover Garden in Nagasaki, the Kobe

> foreign settlement, and the Otaru canal and warehouses. No, these weren't the reasons why I took up the fight to save the canal. The decisive factor for me was the significance of the place to Otaru citizens.
>
> Otaru Shinpojiumu Jikkō Iinkai, ed. (2009: 32).

The decisive factor for Mineyama was not the fact that the canal was the subject of paintings, or had been recognized by the Architectural Institute of Japan, but rather the significance of the canal to Otaru citizens.[46] The canal should be preserved because it was valued by the city's residents, not simply because it was old.

Mineyama also repeatedly referred to a second logic, that of "living in the region" (*chiiki ni ikiru*). Mineyama explained this concept (also the title of her 1995 book), as follows:

> So it was for ten years that we—that I—fought to save the canal. And it's been ten years since that fight ended (as of the interview date in 1997). [But] we're still involved with the canal. We're still focused on it. Let me talk a little bit about why I'm so focused on the canal. In Otaru's most prosperous period, three thousand people—laborers—worked themselves to the bone there, and because of [the canal], because of [the canal], three thousand people picked over beans [on the upper floors of the warehouses] over there.
>
> The canal was where the work was—not the town center. The canal was the starting point for Otaru as a city. The starting point. It was the heart of Otaru—the heart. It was the heart of *Otaru-machi*. And this heartbeat reverberated throughout the whole city. That is how the city of Otaru came into existence. {…} There would be no city of Otaru if it weren't for that canal and those warehouses. They were the starting point. {…} Those of us alive today need to fully understand the work of people back then. {…} My generation needs to add to this new culture, and transmit this—impart this, hand this down—to younger generations. That is what I mean by "living in the town" [*machi*].[47] Well, actually I call it the region (*chiiki*). To me, that is the meaning of "living in the region." That's why I [continue with the movement].
>
> Interview with Mineyama Fumi on September 1, 1997. Information in parentheses has been added by the author, information in brackets by the translator.

In this passionate speech, Mineyama emphasized that we cannot ignore the work of "people back then."[48] In later years, she articulated this perspective more clearly:

> Those of us alive today really have to grasp the experiences of the people who worked so hard back in the old days, and understand what kind of place Otaru was back then, in order to mobilize [to save the canal and historic buildings]. We need to know what kind of place Otaru was, and pass this conviction on to the next generation, and to the people of the future. A town is the joint creation of the past, the present, and the future. That is what a town is.
>
> Interview with Fumi Mineyama in Otaru, September 7, 2007. Information in parentheses has been added by the translator.

---

[46]Elsewhere Mineyama stated: "we wanted to preserve the canal because it was an irreplaceable part of this city, not because it was 'old' or 'unusual,' as townscape preservationists would stress" (Mineyama 1995: 171).

[47]Here Mineyama's use of the term *machi* for the word city or town conveys a sense of warmth and familiarity.

[48]It is because Mineyama felt that there was still insufficient recognition of the fact that the present exists as a result of "the work of people back then" that she says "we're still involved with the canal, we're still fixated on it," and "that's why I continue with the movement".

If a town is, as Mineyama proclaimed, the joint creation of "the past, the present, and the future,"[49] then the present generation should not arbitrarily transform the town solely according to its own preferences. This is Mineyama's second point: that the current generation should not meddle with the city according to its own short-sighted interests. Moreover, Mineyama felt it was imperative that the "emotions" and culture of previous generations be "accepted," learned from, and "passed on to future generations." In other words, "living in the region" means protecting its origins and, "with conviction," passing it "on to future generations."

Activists became more deeply acquainted with their own town as they fought to preserve it. Gradually they realized that saving the canal was actually an effort to save the entire town of Otaru. The movement was not a nostalgia-fueled pastime, but rather an attempt to invest in the future by building upon both the present and the past:

> Looking to the past is often seen as a passive act, motivated by nostalgia. I disagree. Understanding the work of people in the past gives roots to our visions for the future. In that sense, the past is a guide to the future.
>
> Mineyama (1995: 170–171).

Mineyama believed that residents of the region should play the principal role in such preservation movements. This belief is reflected in her signature catchphrase: "One's hometown is close at hand, and to be protected." This is a play on words based on the famous phrase, "I think of my far-away hometown," in which Mineyama draws a clear contrast between the words "far-away/think" in the original phrase and the words "close/protect" in her version.[50] It is a succinct declaration that residents should be the central actors [in preservation].

It was precisely for this reason that Mineyama felt the movement should not rely upon political parties, but rather stick to a platform of citizen-based activism.

> Citizen activism cannot align itself with a certain political party. In a citizens' movement, don't people with connections to different political parties come together to support a particular cause? And also people with absolutely no connection to any political party whatsoever. If the movement became associated with a single political party—the JSP or the JCP or the LDP—we would lose the character of a citizens' movement. You know, to tell you the truth, people in the JCP really looked into the canal issue at the time—they really looked into it. And they truly involved themselves with the movement. So, in theory, I should have been sympathetic to the JCP's passion for the issue—but I wasn't. I felt that aligning myself with the JCP would change the character of the movement, and so I maintained the same distance from them as from all the other political parties. I stayed neutral. One of my husband's students became a JSP member, and he also really worked hard for our cause. So people sometimes asked me if I disliked the JCP because I was a Socialist myself, but that wasn't the case. I didn't want the movement to become tinged by a particular political association, so I maintained the same stance toward all the different parties.
>
> Interview with Mineyama Fumi in Otaru, September 3, (1998).

---

[49] A similar assertion can be found in Nishimura & Rachi, eds (2007: 34).

[50] Mineyama writes "a hometown is a far-away place to be thought of and a hometown is close at hand and to be protected" (Mineyama 1995: 177).

Precisely because the preservation movement included people from all walks of life, Mineyama believed it was essential to avoid close association with any political party and participate in the movement as residents and citizens.[51] As Mineyama recalled in the above interview, she did not join forces with the JCP despite the party's zealous defense of the Otaru canal.[52]

Mineyama's ideology and activism would not have been possible without the understanding and support of her husband and mother:

> Horikawa: You were over sixty when you joined the preservation movement. At that point, your husband Iwao was still [alive and] healthy. Was he supportive of your activism?
>
> Mineyama: He was very supportive. (After Koshizaki died and Mineyama was nominated to replace him) I had no interest in becoming chairwoman—I didn't think I could do it. I kept trying to get out of it. But he told me: "You can't act this way if you really want to save the canal. You need to do this, even if it's just until they can find someone else." You see? And then, it would cost between 120,000 and 130,000 yen to attend the *Machinami Zemi* [an annual convention of the Japan Association for MACHI-NAMI Townscape Conservation and Regeneration], because airplane tickets were expensive back then, and you needed a hotel room and so on. And I wasn't working at all then. I wasn't earning so much as a yen so I decided I wouldn't go again the next year. But my husband told me I was being silly. Every year he paid for me to attend the seminar. I'd go to Tokyo and people there would say: "You come here each year, Mineyama-san, but your husband is the one pulling the strings" (*laughter*).
>
> Interview with Mineyama Fumi, September 11, 2007. Information in parentheses has been added by the author, information in brackets by the translator.

The support of Mineyama's mother was also hugely important. The following excerpt from a pamphlet written by Mineyama Fumi, entitled *Memories of My Mother*, offers a glimpse of how central activism was to her daily life.

> I worried about my eighty-five year-old mother living alone, so I decided to have her come and live with me [in Otaru]. {...}
>
> Eventually her life in Otaru began. At the time I was completely consumed with the canal issue. For my mother, who had been living a quiet life alone in Date, my home life must have felt far too hectic.
>
> Her first shock as a mother came when she discovered that her quiet and mild-mannered daughter was making appeals out in the street, and lodging protests with the city, the prefecture, and the national government. She was amazed at my transformation, and distressed at the sight of her daughter [in constant motion]. But over the course of her first year with us, her views changed.
>
> She came to accept the idea that it was up to those of us living in Otaru today to save the canal from being paved over, and pass it on to the next generation.
>
> My mother felt that there had to be something she could do to help me, in her own way. Before she had tried to live passively so as not to inconvenience me. Now she became more proactive, thinking of ways in which she could make herself useful.

---

[51] Mineyama also once said that "citizen activism can't be a matter of winning or losing, or of crushing an opponent" (Interview with Mineyama Fumi, March 30, 1984).

[52] As we have seen, despite Mineyama's good intentions, the movement was brought down by internal strife.

> My eighty-five year-old mother had found a new purpose in life. She helped me with the things I couldn't get to because I was out almost every day, and paid scrupulous attention to my husband's daily needs. {...}
>
> My mother read the newspaper in painstaking detail every day.
>
> I would look for large-print articles on the canal issue to save [for a scrapbook]. But my mother would find the small-print articles, or opinion pieces by local citizens. This was a task that couldn't be overlooked, and she was of indescribable assistance to me. {...}
>
> On January 15 of last year, it was suddenly decided that we would deliver a petition signed by 100,000 people to the Minister of Construction, Mizuno [Kiyoshi], in Tokyo. We decided to leave on the first flight out on January 16. My mother woke up at 4 am that day to help me in various ways and to see me off, just as she would do when my husband traveled on business.
>
> When I went to the station that day to buy my ticket, I discovered that my mother had wrapped up a month's worth of her spending money in tissue paper and smuggled it into my purse.
>
> Trying to save the canal was no easy task. It involved fierce arguments, both internally and externally. I would return home worn out, both physically and mentally, and my exhaustion would suddenly overwhelm me.
>
> No matter how late I got in, my mother would wait up for me. She would take one look at my expression and sense what I had gone through that day.
>
> Silently she would pour me a cup of hot tea and set out my favorite sweets. No actual words were needed during these conversations between a ninety year-old mother and her seventy year-old daughter.
>
> As the warm tea coated my throat and I gazed into my mother's face, I would grow calmer, and summon up the courage to keep going.
>
> This is the mother that I lost suddenly.... {...}
>
> My mother, like all mothers, died having given me everything she had to give.
>
> Mineyama (1985: 3–6).

In this passage, one can sense the extent of Mineyama's transformation from the "quiet, mild-mannered" daughter to the woman who was "out every day," and the constant pressures of a daily life that involved "lodging protests with the city, the prefecture, and the national government," "making public appeals in the street," and hunting down both "large print articles" and "small print letters from citizens" in the newspapers.

Despite Mineyama's valiant efforts, the movement to save the Otaru canal eventually broke down. Ultimately, only a portion of the canal was preserved. How did Mineyama evaluate the "fruits" of her efforts in light of this outcome? In response to a straightforward question from a student of the author's, Mineyama answered frankly.

> Student: "Excuse me, you just talked about the fact that half of the canal ended up being paved over. Mrs. Mineyama, were you happy with this outcome, or had you hoped to save the entire canal?"

> Mineyama: "All of it. I wanted to save the entire canal. By no means was I satisfied that only half of the canal was preserved. After all, that [original] width of forty meters was part of what made the canal what it was. I'll say as much frankly."[53]
>
> Student: "Do you still feel it would have been better if the entire canal had been preserved, even now that half has been saved and turned into a tourist attraction?"
>
> Mineyama: "Saving the canal means saving all of it. That would have been the right thing to do. I won't back down and say that saving half was enough. But you know, over the course of this movement I experienced both the bitterness of losing half the canal and, wonderfully enough, the [realization] that 98,000 citizens wanted to save the canal. *Ninety-eight thousand*. And for me, that is more important. That is more important. Not the width of the canal, but the fact that citizens [became] aware of what our city once was. They discovered that they had been living here in this city without realizing what a fine place it was. And that is more important to me."
>
> Interview with Mineyama Fumi in Otaru, September 11, 2007. Information in brackets has been added by the translator.

It was not "the width of the canal," but the growing awareness of Otaru citizens that mattered most to Mineyama. This was confirmed in yet another interview:

> The fact that students from Hokkaido Otaru Ōyō High School asked to conduct an inspection of water quality, for example... That students from different schools, all in their own way, involved themselves with the movement to save the canal... That they set their minds to this at a time when the administration wouldn't even consider the canal [issue]. This was so gratifying to me, and I felt that it had been worth it; that this is what a movement is really all about... {...} The best part about working to save the canal has been the change in people's values. To the extent that the people of Otaru or the mayor had thought about the issue, they simply thought that filling in the smelly, sludge-filled canal would be an improvement. But their attitudes began to change—they started to realize: 'That's not the case. This place is an essential part of the city of Otaru.' I think you could say this was my greatest success—the biggest accomplishment of the movement. Because while this might not be in the *Otaru City History* [*Otarushi-shi*, published by the Otaru municipal government], the fact that [the canal] is part of the consciousness of Otaru citizens is of enormous value. It's worth a lot.
>
> Interview with Mineyama Fumi in Otaru, September 2, 1999. Information in brackets has been added by the translator.

The "greatest accomplishment" of the preservation movement was the growing public perception of the canal as "an essential part of the city of Otaru." This cognitional victory was "so gratifying" to Mineyama, a true and devoted defender of the canal.

Mineyama's concept of "living in the region" (*chiiki ni ikiru*) means living according to the environmental conditions imposed by a specific region. It also means addressing the demands of "the present" by consciously positioning oneself upon

---

[53] On another occasion, Mineyama said: "When I say it's not a canal, what I mean is, if [there isn't room] for barges to pull up together lengthwise, like this, it's not a [real] canal. Isn't that right? Now that it's half the [original] width, we call it (*sarcastically*) the 'new waterfront,' not the 'preserved canal' or the 'restored canal.' That's what we call it. {...} If [it isn't wide enough for] a barge to turn around, it's not a canal. So now, just the 450 m at the Kitahama [end] remains at the width of forty meters. That's a canal. You see? And the remainder [of the canal], that's now half that width? That's not a canal." Interview with Mineyama Fumi in Otaru, September 1, 1997. Information in parentheses added by the author, information in brackets by the translator.

an elongated axis of time, which extends in both directions toward past and future generations of the region.

Mineyama Fumi's repudiation of change and her strong desire to save the canal in its entirety characterizes her as CNG -. Her preference for avoiding all-out conflict in favor of dialog makes her POA +. At the same time, over the course of the protracted battle to save the canal, and in her own written accounts of her activism, one can detect Mineyama's growing tolerance of change and willingness to engage in direct political confrontation.

*5.4.3 The Community Development (*Machizukuri*) Activists*

A third camp within the preservation movement pursued the goal of *machizukuri*, or community development, and proposed using the canal and surrounding areas to develop the local tourism industry. While they were often aligned with Mineyama Fumi and the other "pure preservationists," they offered a different rationale for the canal's preservation. While Mineyama and her associates saw the canal's preservation as an end in itself, the *machizukuri* contingent viewed it as a means of revitalizing the regional community. Their sentimental desire to save the canal in its entirety did not prevent them from making a more cool-headed assessment: a successful, broadly-supported bid to save the canal in its entirety would require an economically-oriented proposal for "preservation-based redevelopment," or "tourism development." Indeed, they recognized that the survival of not just the canal but the preservation movement itself depended on the development of such an approach.

Members of the *machizukuri* group played key roles in the latter half of the preservation movement, as they stressed the importance of "urban revival," "tourism development," and "community development." While members originally sought to save the canal in its entirety, they developed a logic and argument that went beyond this original goal. Rather than focusing exclusively on the "irreplaceable" nature of the canal as a physical place, they began to discuss the "economics of preservation," and "developing tourism as a vehicle for urban revitalization."[54]The *machizukuri* group included the young urban planners collectively known as the "Hokudai (Hokkaido University) trio," and also the "Otaru three:" Yamaguchi Tamotsu, Ogawara Tadashi, and Sasaki Kyōjirō. As we have seen, their rationale for preservation successfully mobilized local business circles and more conservative elements of Otaru society, giving the preservation movement a significant boost in support. Ironically, in order to save the canal as a distinctive feature of the Otaru townscape, the *machizukuri* group framed their appeal not as a matter of distinctiveness, but rather in universal

[54]Not only was the incorporation of economics and development into the discussion met with resistance from many of the movement's original members, who had stressed the importance of the Otaru canal as a symbolic representation of their hometown, but even many of the central figures in the preservation movement's later phase admit they were initially hesitant to frame the issue in this way. Interview with Yamaguchi Tamotsu in Otaru, September 13, 1988.

terms designed to garner broad support. Doing so brought them to a position of dominance within the movement.

Notwithstanding the universal orientation of the *machizukuri* group, they did not immediately converge upon the generalized language of postwar democracy and ahistorical "citizens." Significantly, they remained attached to the specific landscape of the canal itself. This finding represents a decisive difference from the conclusions drawn by Nitagai Kamon in his study of neighborhood movements in postwar Japan and Okuda Michihiro's community theory regarding the potential and dynamics of community in postwar Japan. Both Nitagai and Okuda posit an idealized image of the citizen in postwar democracies, divorced from specific environmental or historical contexts (Horikawa 1998). Yet the distinctive environment of the Otaru canal and the specific historical context of their preservation campaign were of critical importance to community development activists like Yamaguchi Tamotsu and Ogawara Tadashi. Specifically, their earlier experiences of anti-*Anpo* protests and other student demonstrations in the 1960s, and their desire to avoid repeating earlier mistakes, prompted them to develop new strategies for saving the canal. This enabled the preservation movement to evolve in new ways. The past experiences of participants are thus not mere prehistory, but rather a vital strand woven into the fabric of their activism and its evolution.

From Preservation Activist to City Council Member: A Life History of Yamaguchi Tamotsu

Yamaguchi Tamotsu was born in 1947 to a family in the lumber business in the former village of Kanayama-chō, in what is now the city of Gero, in Gifu prefecture. Yamaguchi first arrived in Otaru on the snowy evening of December 14, 1975. Yamaguchi was twenty-eight when he stepped off the train at Otaru station and has lived in Otaru ever since. Yamaguchi first ran a coffee shop and then opened his own woodworking business. He also married and raised a family, while also participating in the movement to preserve the canal. Yamaguchi later served three terms on the city council before returning to his woodworking business.

The Japanese student movement of the 1960s and Yamaguchi's early encounters with European civil society both influenced his involvement with the Otaru canal. Yamaguchi dedicated himself to the student movement from his time at Gizan High School in Gifu and continued until he dropped out of the law department of Ritsumeikan University in Kyoto. He reflects: "I was part of the student movement from high school onward. My brother was part of the *Anpo* protests, and in some sense his influence prompted me to read certain books and so on" (Author interview with Yamaguchi Tamotsu in Otaru, September 4, 1998).

As a high school student, Yamaguchi mounted a campaign against his high school's prohibition against long hair, and after a tenacious two-year campaign succeeded in overturning the school's regulations. This early experience in activism represented a starting point for Yamaguchi; he realized that by acting independently and voicing dissent, one can change the status quo:

(By the time they reach high school,) people develop a sense of justice regarding what you could call the inconsistencies [or contradictions] in society. I think that's the starting point. You learn things, and discover things that seem strange to you. If you turn a blind eye to what you find strange, somewhere along the line you'll have to suppress your own feelings. You sense that something is wrong, and those feelings build up. Resentment builds up. For example, there was what happened in Kobe—that school gate accident. A student arrived just after the starting time, and was killed when [a teacher slammed] the gate shut on her head. I expected students to revolt when that happened. But they didn't, and I think student records were the reason—[students'] concern that it might affect their ability to move up [to the next school]. This is how people are controlled—we've all been controlled, ever since our generation. Ultimately those students in Kobe weren't able to say anything about what had happened, although I'm sure there was a lot of emotion. The student [who died] must have had friends, people with whom she was close. I thought one of them would step up and speak for the group, and express what they were all feeling, and that they would band together for change. That didn't happen. *I can't help thinking society should be more straightforward: people should be able to say that what's right is right—and if something is wrong, try to fix it.* But I'm afraid that's not how things work, even though it seems so simple. {...}

I still believe that it is by speaking up when you feel something close to you is inconsistent that we can change society from close to home.

Interview with Yamaguchi Tamotsu in Otaru, September 4, 1998. Emphasis and information in parentheses has been added by the author, information in brackets by the translator.

Yamaguchi's logic is simple: "it is right to try and fix what is wrong" in society, and it is important to say "what's right is right." And "it is by speaking up when you feel something close to you is inconsistent that we can change society from close to home."

Yet while Yamaguchi successfully changed the rules at his high school, he was unable to change larger contradictions in society in the way he had hoped. At university, Yamaguchi grew disillusioned with the sectarian strife of the student movement. Not only did his longed-for social transformation fail to materialize, but as a so-called "non-sectarian radical" he found his own values destroyed as his attempts at activism, his "starting point" to achieve social change, were hindered by sectarianism and dogmatism. In 1970, in his fourth year, Yamaguchi withdrew from Ritsumeikan with the plan of traveling to France in an attempt to rebuild his own value system. Yamaguchi did not undertake this journey lightly: he studied French at the Institut Francais in Kyoto before his departure. Clearly, this was not to be a student's cheap European tour, but rather an intentional voyage of self-discovery.

Yamaguchi hitchhiked across Europe, supporting himself with odd jobs at each destination. On his way to Stockholm, Yamaguchi was dropped off near a park, where he happened to encounter the company of "Tivoli," a traveling circus. He was subsequently hired on as a painter. Thereafter, he used a rented apartment in Sweden as his base while traveling with the circus company through France, Belgium, and other parts of Europe.

Along the way, Yamaguchi had the opportunity to observe European civil society, which he saw as rooted in communal spaces:

Over there (in Europe) I belonged to all sorts of groups. I was once part of an anarchist group, for example, and also in the Fourth International. The Fourth and other groups with links to the Japanese student movement were just like Japan. Extremely dogmatic. Leninist. They were basically carrying on these ideological disputes. But the anarchists were different. They were an interesting bunch. Put simply, they had what you could describe as their own way—a policy—for living. And if they didn't feel their taxes were being used in rational manner, for example, they refused to pay that portion. That's the kind of thing they would do. And they didn't mind if their electricity was shut off or what have-you, because they believed they were in the right.

Interview with Yamaguchi Tamotsu, September 4, 1998. Information in parentheses added by the author.

Independent individuals with a "policy" for "their own way of living," made a deep impression on Yamaguchi, who became increasingly aware of the different structures of Japanese and European society.[55] This included differences in the organization of communal spaces:

Public places, libraries—these were really fun places. Elderly people would go there, not [just] students preparing for exams (as in Japan). You were allowed to drink tea and it remained open until 9:00 (p.m.); you could buy a cup of coffee for one hundred yen. Old men would sit there basking in the sun. They sold bread; you could eat a sandwich. High school students weren't interested in reading books, but right by the entrance there was an area where they could put on headphones and listen to—well, I suppose nowadays it would be CDs, but back then it was records. The library was a place you went to enjoy yourself. Old people treated it like a coffee shop, while researchers had their own area in the back to work, and young people would come to listen to music. This allowed people to mingle. And the library was *right* where it should be—*smack dab* in the middle of town, in the easiest place for people to reach.

Interview with Yamaguchi Tamotsu, September 4, 1998. Emphasis and information in parentheses has been added by the author, information in brackets by the translator.

It was not simply the existence of a public space where individuals could meet and interact that impressed Yamaguchi, but the fact that this space had intentionally been provided "right in the middle of town, where it should be," in "the easiest place" for people to reach. Yamaguchi's words reflect his faith that the right public facilities would naturally be located in the best place.[56] This experience would influence Yamaguchi's later involvement in the Otaru preservation movement.

Yamaguchi returned to Japan in 1975, and his newly acquired language skills landed him a job at the Swedish Center, affiliated with the Swedish Embassy in Tokyo. However, his stint as a salaryman proved short-lived:

In Japanese cities, light is always reflected in the same direction. From glass, the surface of a wall… It's different with stone: natural stone refracts light differently. Natural materials are

[55] Yamaguchi's understanding of European individualism is not nearly as simple or idealized as it may sound here. What is significant is his reaction to it at the time, as a young man in his twenties.

[56] This refers to the concept of "amenities," which encompasses the comfort, harmony, and relevance of an environment. Yet even in England, where the concept originates, it is "difficult to define, even when it is felt." Originally described as "the right thing in the right place," in Japan the term has come to be interpreted as a synonym for "comfort," but we must recall that the original meaning includes the ability to criticize the status quo and press for the provision of essential facilities or services. See Horikawa (2012).

> all different: trees, grass, boards—wooden planks. But houses made with aluminum sash, glass… the refraction of light—these materials don't fully absorb light and what's more, they reflect light in the same direction. It's very…well, frightening isn't the right word, but it feels oppressive. This was the first decisive difference I felt (from Europe).
>
> Interview with Yamaguchi Tamotsu in Otaru, September 8, 1997. Information in parentheses has been added by the author.

Yamaguchi felt "oppressed" by Japanese cities, which he found to be "decisively different" from the cities of Europe. Japan's cities were "dirty" and "no good, frankly." (Interview with Yamaguchi Tamotsu in Otaru, September 4, 1998). He decided to immigrate to Canada and briefly worked as a welder at a dam construction site in Gifu in order to gain some vocational skills before undertaking the Canadian immigration screening process (*Asahi Shinbun*, Hokkaido edition, June 7, 1984). The final step in his application was scheduled for February 1976.

Before taking this final step, Yamaguchi embarked on what he thought would be a farewell trip across Japan. He began his journey in Hokkaido and stopped off in Otaru to visit an acquaintance. The date was December 14, 1975. "It was the day of the Chūshingura raid [a famous event in the well-known story of the Forty-seven Rōnin], so I remember the date well. The sea was choppy, a flurry of whitecaps." (*Asahi Shinbun*, Hokkaido edition, June 6, 1984). This is how Yamaguchi recalls his first impression of Otaru:

> At the old train station, everyone was crowded together around the heater (in the waiting room), holding out their hands toward the heater to warm themselves, like this. There were old men, high school students… The sight of them made me happy. It made me realize that there were still some good sights in Japan, after all. That was my first impression. The station was on a hillside (that sloped all the way down to the port), so when I looked out I could see the sea. I could see the whitecaps. There were boats out on the water. This ordinary scene, in a city of around two hundred thousand—at that time there were about two hundred thousand residents—made me think that this must be a good place. I had already given up on Japan and completed my immigration procedures for Canada when I came to Hokkaido. I had never been to Hokkaido before, so I had planned a six-month trip from Hokkaido to Okinawa. After that I would go to Tokyo and complete the final round of paperwork. I was at that stage. I had finished my final interview when Otaru caught hold of me. I was charmed, and I stayed on. I rented a room. And as I went through the town *I was intrigued by the alleyways—there were signs that real, everyday lives were being lived there.*
>
> Interview with Yamaguchi Tamotsu in Otaru, September 4, 1998. Emphasis and information in parentheses added by the author. Information in brackets has been added by the translator.

Comparing Yamaguchi's recollection of his arrival in Otaru with his earlier comments about Japanese cities, one discerns the contrast he drew between the "suffocating," standardized, and impersonal life in Tokyo and the distinctive landscape of Otaru, with its "intriguing" streets and signs of "real, everyday" life.

In 1976, Yamaguchi opened a coffee shop called Merry-Go-Round[57] in the Temiya district of Otaru and gradually grew more familiar with the city—and the

---

[57] The name of Yamaguchi's coffee shop was inspired by Tivoli, the traveling circus group that he had joined as a painter during his time in Europe. See also Edagawa (1985: 12–13).

canal. Writing two years after his arrival in Otaru, under the pen name "Matakichi," Yamaguchi reflected:

> I was delighted because (the canal) *was the last vacant area in the city*, left here by chance. It is *the last park* to forget to charge admission or erect a fence.
>
> Machikado kara no Tayori: Merry-Go-Round Tenshu/Matakichi, *Zappu* (ZAP) #5, 1978: 2. Emphasis and information in parentheses added by the author.

Yamaguchi was "delighted" by the survival of this "last park" that had "forgotten to charge admission." His understanding of the canal remained unchanged, although he would later express himself even more clearly on the subject.

> In Otaru, it's a common sight to see old men put their grandchildren and a fishing rod on the back of their bicycles and head to the port to fish. The canal is just a five-minute walk from the station, and it was such a wide expanse of open land—overgrown with weeds, of course. Somehow it was logical that this was the case. At that point [Japan was] in a period of [high] economic growth—ten percent growth—so it didn't seem possible that there could be any vacant land left. Old alleyways were disappearing, and vacant land was disappearing from the cities. It was a time when places whose ownership wasn't clear, places that, amazingly enough, anyone could still use were disappearing quickly. I felt keenly how *right* this place was.
>
> Interview with Yamaguchi Tamotsu in Otaru, September 4, 1998. Emphasis added.

Because the canal functioned as a city park and a space where all Otaru citizens could freely access the water, Yamaguchi saw its existence as entirely "right" and rational from the perspective of civil society. Yamaguchi's elaboration of the concept from the vantage point of civil society challenges the common assumption that rationality, more commonly associated with bureaucratic institutions, has a single definition.

Yamaguchi continued to observe the canal, and appears to have intuited its necessity to Otaru:

> The port is what originally brought this town into being. In that sense, the port is a kind of cornerstone of this town. (For example,) The grove of a village shrine is located right in the center, and if that grove is destroyed, the (village) community collapses entirely. In the same way, I felt the port was immensely important to Otaru. It became clear that the port would be cut off from the town by the six-lane road (in the city's plan). I sensed there would be some form of divine retribution if that happened. The topography of the town was so good—on the palm of my hand, here (*pointing to his palm*) is the port, and here (*pointing to the fingers extending from the palm*) were the different communities. On the north side of Otaru there's what is called the Echigo group—this is where a lot of herring fishermen from Niigata settled. {…} And people living in the town (center) are (originally from) Hokuriku—they're merchants. {…} That was where the merchants lived. If you look at it this way, there are differences (in the people and personalities of each district). They've *very cleverly segregated themselves* into different areas. Of course (in reality) there are finer distinctions as well. *Yet all of these areas were connected to the water*, and would be *cut off* by (the Rinkōsen road). That's an unhappy fate for the town, and *extremely shameful from a design perspective*, you know.
>
> Interview with Yamaguchi Tamotsu in Otaru, September 4, 1998. Emphasis and information in parentheses has been added by the author.

The thrust of Yamaguchi's argument is clear. First, much like the grove of a village shrine, the canal was a central presence for the community and a place where people could freely assemble. The canal area was "the last park to forget to charge admission or erect a fence," a place where older Otaru residents could spend time with their grandchildren as they pleased. Yamaguchi recognized that this was how local residents understood and experienced the Otaru canal.

Yamaguchi makes another important point: the physical space of the canal functioned to regulate between diverse communities. As he points out, settlers from other parts of Japan formed smaller communities comprised of people from the same region. These regional divisions likely overlapped with class and occupation-based distinctions. The canal governed the relationships between the various communities. While the fishermen in their seaside neighborhood and the merchants living in the heart of the city may have "cleverly segregated themselves," every neighborhood remained "connected to the water." The canal can be understood as an interface that skillfully linked the "port system" along the coast to the "urban residential system" closer to the mountains. This reflects a form of city design, or put differently, a system of spatial control, developed over the long history of the lived city. Yamaguchi saw the destruction of this city design as "shameful," and believed it threatened the community with "total collapse." This profound realization led him to dedicate himself to the cause of preservation.

This is when Yamaguchi encountered the "Hokudai trio." Yamaguchi had started a folk singing group. Three engineering students in the city planning program at Hokkaido University learned of Yamaguchi's budding reputation as a talented live performer and visited him at his boarding house, where they invited him to join their efforts to save the canal. Yamaguchi recalls this period, in the early October of 1976, as follows:

> The idea that building this new road would improve the economy was so ridiculous. That's why there were people saying as much all around me. The Hokudai trio weren't even Otaru natives, although they were studying city planning, and they wanted to put an end to the Rinkōsen plan. They weren't even from Otaru, and yet they were the ones who had identified the problem [with the road plan]. And it just so happened that I had also come to Otaru and involved myself in all sorts of things, so they heard about me and sought me out. And as we talked I decided that I would also help, and joined (the APOC).
>
> Interview with Yamaguchi Tamotsu, September 4, 1998. Information in parentheses has been added by the author.[58]

Having met companions expressing similar feelings about the issue, Yamaguchi joined the APOC. Yamaguchi had already perceived the reasons why the canal should

58 What did they actually discuss on that long-ago day? According to an article that appeared in the *Asahi Shinbun*, "Ishizuka and his friends {…} explained their movement to save the canal. Yamaguchi says: "I was impressed by how they stressed [the importance of] urban amenities" *Asahi Shinbun*, Hokkaido edition, "Unga Mondai Jyūnen no Ayumi," Part 2, No. 32, June 21, 1984. Yamuguchi's repeated references to the Hokudai trio's outsider status ("they weren't even from Otaru," "they weren't even Otaru natives and yet they were the ones who had identified the problem") reflects the strong tendency of Otaru society at the time to exclude outsiders.

be saved; he was now able to channel these into a clearly defined movement for preservation. Little did he know that he would eventually come to lead the movement.

Yamaguchi soon became acquainted with Fujimori Shigeo, and learned how to organize a movement and devise a strategy for saving the canal:

> You see, for me, *it wasn't the fact that the canal was unused that was irrational.* How can I put it? *I thought the existence of an unutilized space was a good thing.* There weren't other places like that. During Japan's high economic growth period, only Otaru had that. That's why I fell in love with the place. And then they proposed turning [the canal into a] road. I thought that was wrong. {...} There were many people who loved that desolate landscape...that original, indelible image of the landscape. All the artists felt that way. Because it was so desolate, it was filled with wild flowers—flowers that no one could name would bloom there. Rotted-out boats floated on the canal, and there were the warehouses with the *paint peeling off the walls...* {...} *They said this made the scene wonderfully nostalgic.* You know, people would come from the cities. People would come to take photographs. *But I couldn't take this view. If you asked me if [the canal] could survive in that form, well, I would say it definitely could not. For a movement to succeed, a majority of people need to believe in what you are doing.* {...} I knew that very well. I had been part of the student movement, and before that I'd protested the [high school prohibition on long hair]. So next, within the movement, I started thinking about what we needed to do for the canal to remain. And I realized that the canal had originally been a pristine, beautiful place. Waterfront areas aren't just for places for humans—there are fish, and back then it was completely natural for there to have been fish, because it was [connected to] the sea. When the buildings were first constructed, their stone walls had been beautiful as well. Properly speaking, *that* was the original, indelible image [of the canal]. I realized what we needed to do—if we [remade the canal according to this earlier image] people would definitely gather there. {...}
>
> Overseas there was a waterfront development project in which an old chocolate factory was turned into a shopping center (Yamaguchi is referring to the redevelopment of San Francisco's Pier 21), and we said we could do that in Otaru. So we made a plan. If they said it was a problem that ships no longer came [to the port], well, we could say Otaru could be turned into a yacht harbor. The big city of Sapporo is right there, so we devised that sort of plan.
>
> *But to tell you the truth, I was reluctant [to emphasize tourism development]*. In Japan, turning a place into a tourist destination usually means making it into a cheap and vulgar hot springs resort, and I wondered if we would actually be able to create something beautiful. Even though I worried about what kind of strange businesses would rush in [if we succeeded], I also knew that we had to paint this kind of picture [of redevelopment] in order to win the support of the majority of people. Before the war, Otaru had a bigger population than Sapporo, and was truly an economic center. {...} In terms of bank account savings, Otaru led all the cities in Hokkaido right up until 1950. {...} So locally there was a strong sense of resistance to Sapporo, and there really was a general feeling that Otaru was fading—that it had stalled, and that something needed to be done about it. Well, we saw this as a chance to revive Otaru with [public] support, for the success of our movement. That meant we had to convince people that [our plan for redevelopment] would make more money than the new highway. {...} Because everyone (warehouse owners and landowners) had been told that the value of their property would go up if the road was built... So we had to refute that idea, and tell them that bringing people to the area would make them a lot more money than building a road.
>
> During the high economic growth period they destroyed all the (old townscapes), and only Otaru's remained. Otaru's townscape remained simply because we were behind the curve [of Japan's economic growth] and because of what Otaru had been in the past [when it was thriving]. That [scenery] was what defined Otaru, and so we told people that [the old

> townscape] was what we needed to make use of in our community development—that it was this townscape that would make us money.
>
> *But this is just what we had to say from the perspective of our campaign—it wasn't completely heartfelt as a social movement, and I felt somewhat guilty about this. But we felt we had to take this tactical approach.*
>
> Interview with Yamaguchi Tamotsu in Otaru, September 4, 1998. Emphasis and information in parentheses has been added by the author, information in brackets by the translator.

Although artists may have found the "rotted-out boats floating on the canal," and "the warehouses with the paint peeling off the walls" to be "wonderfully nostalgic," Yamaguchi himself "couldn't take this view," because he knew this sensibility could not possibly save the canal from destruction. And while Yamaguchi personally didn't feel the fact "that the canal was unused...was irrational," but rather than the "existence of an unutilized space was a good thing,"[59] he nevertheless made the strategic choice to emphasize tourism development. Thus, tourism became the focus "for the success of the movement," although this was not entirely "heartfelt," and Yamaguchi felt "reluctant" to frame the issue in this way.

In the above passage, Yamaguchi makes three main points. First, he did not share the aesthetic school's rationale for saving the canal. Second, he wanted to restore the original landscape of Otaru ("the canal had originally been a pristine, beautiful place...that was the original, indelible image"). Third, he stressed the strategic utility of framing the preservation issue as one of tourism development ("so next, within the movement, I started thinking about what we needed to do for the canal to remain").

As we have seen, however, the preservation movement was consumed by the fierce struggle over the canal, and ultimately fractured due to internal discord. What was the fate of Yamaguchi's proposed "tourism development strategy?" What adversaries emerged to oppose his tactics? Two useful keywords emerge from Yamaguchi's account: the word "commie" (*aka*) and the phrase "essence of the administration."

> In the end, the citizens' movement was obstructed (by the mass media, academics and apathetic citizens). They called us 'commies'—Communist Party members. I denied having any connection to them. You (*speaking to a young graduate student present*) are too young to understand this, but I was branded as a communist for (becoming involved in this kind of movement). I hate the Communist Party and the Communist Party hated me. {...} The LDP lot really had no idea—they thought I was a communist, and started spreading rumors to discredit us. But in fact the JCP was the only (party) to oppose filling in the canal. We (the APOC) would circulate a petition (to save the canal), and then the JCP would do the same, and this made us communists (in the eyes of the public). *This forced us to say things that the communists would never say.* {...} The JCP would say something like, 'the canal is a historic structure, so we should turn it into a public park.' They (the JCP) were happy to talk about 'culture.' Well, would (the canal) survive in that case? Citizens didn't think so. They (the JCP) might have been right, but we had to think about (what would actually) save the canal, you see? And so we had no choice but to say 'you'll make a profit this way.' *We had to talk about the JCP's biggest weak point.* You know the term 'redevelopment?' The

[59]Here Yamaguchi is referring to the use of space according to what *civil society* finds rational. To elaborate further, the survival of spaces that have not been fully rationalized but rather used and mastered at the discretion of the individual is precisely his definition of a civil society-based rationality.

JCP argued that 'redevelopment is a capitalist argument.' We refuted that idea. We said that 'redevelopment had to make sense to citizens.'

Interview with Yamaguchi Tamotsu, September 4, 1998. Emphasis and information in parentheses has been added by the author.

What Yamaguchi describes here with such detail and candor is how he sought to counter the label of "communist" and make an effective appeal to civil society. In order to clearly distinguish themselves from the JCP, members of the preservation movement had "to say things the communists would never say." They appealed to civil society by focusing on the profits of tourism development—in other words, "the JCP's biggest weak point." This elicited an unanticipated level of support from Otaru citizens.

However, another daunting hurdle awaited Yamaguchi and his colleagues:

(Mayor) Shimura's statement that 'this (issue) concerned the essence of the administration' sums it all up. (Shimura) has the mentality of a low-level clerk, and that's why he couldn't tolerate the idea of returning funds to the national government or Hokkaido (prefectural government). He had been careful to avoid a deficit, and *skillfully* extracted those subsidies—as an administrative official, he was talented. I suppose he was proud of himself for that. So, in his world, the idea of relinquishing all that [funding] halfway through really came up against the fundamentals of (bureaucratic logic). In this sense, the national government has really ruined the regions.

Interview with Sasaki Kyōjirō and Yamaguchi Tamotsu in Otaru, June 5, 1990. Information in parentheses has been added by the author, information in brackets by the translator.

The mayor stubbornly stuck to his argument regarding 'the essence of the administration.' What he meant by this was that his administration would be discredited if it abandoned a plan that had already been subsidized. He told us: 'If you (the preservation activists) cancel this (public works project) halfway through then we'll have to return the subsidies and bankrupt the city's finances. What do you have to say to that?' I thought the future was more important than his concern. He felt threatened (by the petition signed by 98,000 citizens), and countered with his argument that (a change in plan would) 'affect the very essence of the administration.' Their [bureaucratic] logic never involves "the basis of the citizens' future" though, does it? He held out until the end, sticking to that one-line reply that the 'administrative system doesn't work that way.' He never changed his position. That's how much distance there is between the feelings of citizens and the administration.

Interview with Yamaguchi Tamotsu in Otaru, September 4, 1996. Parenthetical information has been added by the author, information in brackets by the translator.

While preservation activists had somehow managed to survive the charge of being "commies," they were soon obstructed by Shimura's commitment to "the essence of the administration," as we have seen in Chapter 4. Had the city government suspended the Rinkōsen project, it would have been forced to return the subsidies, and officials feared this would jeopardize the city's eligibility for future subsidies as well.[60] As detailed in preceding sections of this chapter, the preservation campaign's momentum was stalled by the solid resistance of the city government, and ultimately destroyed

[60]The national subsidy system wields a decisive level of influence over the local sphere. The fact that evaluation criteria are aligned with national level (policies) undermines the autonomy of the local government. See Chapter 3 and Horikawa (2000).

by internal power struggles. Then the canal issue went quiet. Yamaguchi recalls this period of silence:

> Yamaguchi Tamotsu: "But you know, Kyojirō, even after the (canal) fight ended, it was good to have people that understood—who you didn't have to explain to. A lot of things were said about me, and the reason I never fought back was because I didn't want to be taken for one of that lot and because there were two people (referring to Sasaki Kyōjirō and Ogawara Tadashi) who understood me.
>
> Sasaki Kyojirō: "Damn right."
>
> Interview with Yamaguchi Tamotsu and Sasaki Kyōjirō in Otaru, June 5, 1990. Information in parentheses has been added by the author

Labeled by some as an "extremist," and by others[61] as "the man who sold the canal," Yamaguchi "absolutely refused to refute or explain" the rumors about him that circulated through Otaru. After a long period of inaction, he gradually became involved in community development activities: he [helped] establish the Otaru summer festival in 1988, joined a 1997 conference on tourism promotion, and in 1999 headed the executive committee of the first annual Otaru Snow Light Path Festival. Local tourism grew rapidly in the years after the preservation movement disbanded, and Yamaguchi watched as tourism development, a concept that he had "been forced to adopt as a strategic approach," became a reality in Otaru. His renewed involvement with community development was based upon his firsthand observations of the potential—and problematic aspects—of tourism.

The potential for tourism development was clear, given the growing number of visitors to Otaru. What were the problems? The volume of freight handled at the port of Otaru was in a state of steady decline. Tourism, meanwhile, continued to grow. Yet the city government lacked a department dedicated to tourism and consequently, a tourism policy. Although Otaru was transforming from a commercial port city to a tourist destination, tourism was still not understood as a key industry by its leaders.[62] Yamaguchi identified this problem, and in 1997 became assistant head of the newly reorganized Otaru Tourism Promotion Council's research committee:

---

[61]This accusation came from preservation activists who had supported the effort to recall Mayor Shimura.

[62]Yamaguchi says: "…I myself complain about this or that with regard to the townscape, but the fact remains that (the Otaru townscape) was an asset, and tourism has become an important industry" (Interview with Yamaguchi Tamotsu in Otaru, September 8, 2007). Information in parentheses has been added by the author. This is supported by the words of Akama Hajime, executive director of the Otaru Tourism Association: "This was a port city, and has developed as a port city—starting with the *kitamaebune* [Edo and Meiji period ships that traveled from Osaka northward to ports in the Hokuriku and Hokkaido] and all the way up through the time when Otaru was a coal-shipping port. The national government knew that and (historically has) invested a great deal (in the port). Because of that, warehouse owners and forwarding agents were the central players [in Otaru]. And even now they still think they are the main players—people in the warehousing business and industries like that. That's why at city hall, it's the port bureau that has political power now. But it's actually tourism—viewed from the outside, (Otaru) is a tourist town. (But the people of Otaru) don't think of it that way. {…} I was once asked if people could actually expect to make a living on something as tawdry as tourism" (September 12, 2007 Interview with Akama Hajime in Otaru).

> Otaru tourism had been ignored for ten years. The annual number of visitors increased from one million to six million last year, before we could even get our bearings. We became a big tourist city. In Sakaimachi, in the area around Kitaichi Glass, all these vulgar shops selling anything—crabs, melons—popped up to try and take advantage of their proximity to Kitaichi. So, to start, we made restrictions (through an ordinance) regarding the appearance of buildings, so no weird buildings could be built. But there were no regulations on the type of commercial activity that could be conducted—(merchants) were free to run any type of business they liked. Well, I didn't agree with that, and (my concerns) were around this point; I didn't care for the Dejima [model of] tourism.[63] I want to make Otaru into a more extensive and atmospheric tourist city. But my thoughts alone wouldn't change anything... So we formed the Tourism Promotion Council, but at first they just wanted to run a tourism campaign—to invite people to come visit Otaru. I said that sort of thing was pointless. There's nothing wrong with (a tourism campaign), but we also need to take the time to really think about tourism. I gave my opinion that we needed to draw up a plan for tourism. A separate survey and research division was set up, and I pushed for us to closely analyze Otaru tourism—so we all debated and put together a report last year (in fact, the report came out in the same year as this interview: 1998). The report includes a lot of [new] measures, things that need to be done. So from last year we've been putting together a business plan (for implementation). I'm currently the assistant director of this group. {...} And surprisingly, even though I fought with the administration over and over for a decade, they've finally joined our circle on good terms, and listen to what I have to say.
>
> Interview with Yamaguchi Tamotsu in Otaru, September 4, (1998). Information in parentheses has been added by the author, information in brackets by the translator.

Yamaguchi joined the Tourism Promotion Council as a venue for addressing the fundamental issues confronting the long-neglected tourism industry. There he tried to demonstrate through data the fact that Otaru had already become a tourist city, and develop a basic direction for tourism policies. Naturally, this was an outgrowth of his previous efforts to save the canal. As Yamaguchi writes in "Looking Back on Tourism in Otaru: Ascertaining Our Origins and Issues," in the 1998 report produced by his research group, *A Plan for Otaru Tourism*:

> Otaru's prosperity during the Meiji, Taisho, and early Showa eras allowed it to acquire, and it has since managed to retain, an incomparable number of modern Japanese urban legacies to an extent not seen elsewhere in Japan. Therein lies the origin of Otaru tourism. {...}
>
> In the current age of low economic growth, tourism alone continues to grow and attract investment. There is no doubt that this has had an immeasurable impact on the stagnant Otaru economy.
>
> The question of whether we can develop this into a citywide strategy depends upon the next decade of Otaru city planning. Looking back, the past decade of tourism was propelled by sheer energy alone. However, the success of tourism over the next ten years depends on [the establishment of] a long-term vision.
>
> Otaru Kankō Yūchi Sokushin Kyōgikai Chōsa Kenkyūbu ed, (1998: 13).

The fact that Otaru "managed to retain" an "incomparable number of modern Japanese urban legacies" represented the "origin" of local tourism. Yet "sheer energy alone" was not enough. The city needed a strategy and long-term vision for tourism.

---

[63] See Footnote 30 for an explanation of the term "Dejima tourism."

Yamaguchi found a path to put his proposals into action, made possible by a change in his relationship with the administration. Yamaguchi's central role in planning the Otaru Snow Light Path Festival (Otaru Yuki Akari no Michi) was part of this changing relationship with city officials.[64] Local tourism was largely confined to the summer months, and Yamaguchi recalls how the city would fall silent in the fall and winter. The Snow Light Path Festival was both an attempt to boost tourism during the quiet winter season and also a strategy for revitalizing the local community.[65]

Over time, however, Yamaguchi confronted the limits of activism and the vague nature of third sector initiatives. He realized the limitations of addressing a single issue—the pros and cons of filling in the canal, for example. Regardless of the demands made by an opposition movement, final decision-making power rests with the city government, which also controls the disclosure of information. The sheer amount of energy required to press the city government to disclose the contents of a single document underlines the limitations of activism. Under "third-sector" partnerships between public officials and private citizens, such as the Tourism Promotion Council and the Otaru Snow Light Path Festival, questions of responsibility and jurisdiction remained ambiguous—and largely subject to the wishes of the administration. The fate of events organized because the administration "finally joined our circle on good terms" would likely have been more precarious had Yamaguchi's relationship with municipal officials remained on more difficult footing.

In April 2003, Yamaguchi made a successful bid to join the Otaru City Council. Upon entering the city council race, Yamaguchi shared his aspirations:

---

[64]This is a winter event first held in 1999, named after the collection of poems *Yuki Akari no Michi* by Itō Sei, a poet with a deep connection to Otaru. In an interview with a reporter from the *Hokkaido Shinbun* during the lead-up to the first event, Yamaguchi remarked: "during the time of our opposition to the land reclamation of the Otaru canal, we pushed back against the city administration. Many (of those preservation activists) are now on the organizing committee for the Otaru Snow Light Path Festival. At that time, we believed that preserving the canal was a better way of attracting people and profiting economically than paving it over to construct a new road. That's exactly the case today, and tourism has become the engine of the Otaru economy. The city government and I share the desire to bring new energy and cheer to the city of Otaru. Ten years have passed since [the end of] the canal preservation movement, and we have now met each other halfway, and come together to produce this event. It's a good relationship, in which the government can support ideas coming out of the private sector. "Nichiyō Intabyū: Yamaguchi Tamotsu-san Fuyu no Ibento Shimin Sanka de Machi ni Kakki o." *Hokkaido Shinbun*, December 6, 1998.

[65]The proposal for this event explains the purpose of the festival as follows: "This [event] was motivated by the desire to do something in winter, when liveliness (derived from tourism and other sources) dissipates, and at night, when [the town] is dark and deserted... {...} Many people have been attracted by the "Otaru spirit," typically conceived of in terms of 'warmth,' 'nostalgia,' and a 'human atmosphere.' Yet one could equally argue that it is winter, when the town is blanketed in snow and the nights are shrouded in black, that brings out the best of Otaru. This is because we can create these unique and coordinated lights. {...} 'Lights that bring history to life,' 'lights filled with warm feelings,' 'lights [that reflect human] relationships,' 'lights made by everyone.' This light environment itself is a visualization of 'the Otaru spirit.' Consequently, in direct contrast to 'flashy' and 'gaudy' festivals, we seek to create a winter festival that offers 'depth,' 'sparkle,' and 'inspiration,' and have thus named it the Otaru Snow Light Path Festival (Otaru Yuki Akari no Michi Jikkōiinkai 1998: 5).

Yamaguchi Tamotsu's Hopes

{...} And then I learned about the movement [to save] the canal. The landscape of the canal and old warehouses was the pride of Otaru. The youth [of Otaru] continued [with this effort]. A barge doubling as a beer hall appeared on the dirty and foul-smelling canal that people wanted to cover up. They showed nostalgic old movies in one of the antiquated warehouses that had been described as the very symbol of a dying town. They cleaned up the overgrown and garbage-filled banks of the canal and set up over two hundred stalls. And countless people turned up. I believe that the rebirth of Otaru began with the Port Festival. When ordinary citizens speak their mind with conviction and take responsible action, they can absolutely change their town...and our town has changed. Our city of Otaru has been transformed into a major tourist destination known by everyone in Japan. But have our efforts at community development really been fulfilled? Or have we just now arrived at the starting line? I believe our real endpoint will be the establishment of a true civil society, in which citizens and the administration come together to discuss the future of our town with an equal sense of responsibility. We are now finally establishing a citizen-based tourist organization, with the municipal government providing behind-the-scenes assistance, and a number of new enterprises have emerged. We are thus in the process of creating a 'participatory form of community development.'

However, there are still many issues that need to be addressed. Other fields, such as education, welfare, and city and transportation planning, are all the more in need of citizen participation. Let us establish a clear structure or system—something that could even be described as the 'Otaru way.' We are the ones that can act as the producer of this type of change. We are now the generation bearing the responsibility for our society. I believe that if we continue to spread our message from this small provincial city, and take responsible action, we can absolutely change our country.

Finally, I'd like to explain my acceptance of the DPJ's endorsement. To be honest, I have never belonged to the LDP—but this does not mean that I intend to offer my unconditional support to the DPJ or Rengō [Japanese Trade Union Confederation]. I have come to believe that by remaining so inward-looking, Rengō and other unions have come to ruin our country and regions. It is essential that the unions join civil society and start to make clear, tangible contributions to our society.

There are all sorts of things that cannot be achieved by individual citizens, but only by organizations. If the DPJ and Rengō can begin to change in this way, we may be able to change the LDP as well. This is where I would like to dedicate my efforts.

Blog post by Yamaguchi Tamotsu, February 22, 2003. https://blog.livedoor.jp/yamaguchi_tamotsu_2005/archives/50036123.html Last accessed July 27, 2017.)

In this post, Yamaguchi first assesses the fruits of the preservation movement and the Port Festival, but concludes that "we have only now arrived at the starting line" when it comes to community development, because the type of civil society in which "citizens and the administration come together to discuss the future of our town with an equal sense of responsibility" has not yet been achieved. However, "the process of creating a 'participatory form of community development'" was underway, thanks to the establishment of "citizen-based organizations" supported by the municipal government. As part of "the generation that now bears responsibility for our society," Yamaguchi sought to further this process as a member of the city council. This summarizes Yamaguchi's political ambitions, but equally reflects the opinions he first developed in the early days of the canal war. When asked to summarize the canal preservation movement some years earlier, Yamaguchi responded:

> I don't think that we lost to the administration, because we were so forceful on the subject of community development. In terms of substance, we didn't lose. But we did lose out to the established political parties. Those guys told us that 'we're the professionals when it comes to this kind of movement, so leave things to us.' But they didn't listen to what we had to say, and ended up crushing us. And we had been so careful to stay out of their way. You know, we strategically stressed "the economy" and "tourism," but in our hearts all we were really doing was filling a gap left by the existing political parties by talking about the basic infrastructure of everyday life for ordinary citizens. Wouldn't you agree? Developing social capital isn't something that national or municipal administrations automatically take on. Politics defeated us in the end. The citizens [of Otaru] lost out.
>
> Interview with Yamaguchi Tamotsu in Otaru, November 29, 1989. Information in brackets has been added by the translator.

In this passage, Yamaguchi expresses the pride he took in his steadfast advocacy of community development, but also his chagrin that an issue he believed to be completely unrelated to the traditional conservative-progressive political divide could ultimately only be interpreted and resolved within that same political framework. When Yamaguchi concludes that "politics defeated us in the end," he means that the preservation movement finally had no choice but to bring the issue into the political arena for resolution. His reference to "the basic infrastructure of everyday life for ordinary citizens" explains his decision to participate in the Revitalization Committee following the realization that "if the canal cannot be saved, we have no choice but to limit our sights to the port development plan and pursue a form of redevelopment that focuses on community development" (Interview with Yamaguchi Tamotsu, Ogawara Tadashi and Sasaki Kyōjirō following the final meeting of the Council of One Hundred, September 1, 1984).[66] It also anticipates the formulation of the Otaru Port Plan discussed further below.

Yamaguchi made a strategic decision to run as an independent in his first bid for city council, but also felt he needed to affiliate with a faction in order to secure the right to issue questions at the assembly. Believing that armed with the authority of a city councilman with exceptional access to information he would be uniquely positioned to pursue community development, Yamaguchi accepted the endorsement of the DPJ.

As a city councilor, Yamaguchi endorsed and implemented policies to improve the appearance of the station area by removing a pedestrian bridge and extending municipal facilities to the port area. Yamaguchi believed that bringing commercial establishments and residences to the economically valuable port area would contribute to both the redevelopment of the Otaru port area and to the community development of the waterfront area. During Yamaguchi's tenure on the city council, a new panel charged with developing a future vision for the port was established, and an official decision was made to extend city facilities to the port's Wharf No. Three.

---

[66] In the final phase of the fight to save the canal, the activists who stubbornly insisted on preserving the canal in its entirety were labeled the *unga gyokusaiha*—in other words, those unwilling to make any compromises on the terms of the canal's survival, even if they went down in flames. Here Yamaguchi's words underline the difference between his position and that of those who steadfastly refused to consider any compromise on the issue.

Yamaguchi won reelection to the city council in 2007, and again in 2011. He retired in April 2015, having served three full terms on the city council. Having first pursued the goal of community development in opposition to the city government, Yamaguchi ultimately pursued the same goal from within the city council.

From the Student Movement to the Family Soba Shop: A Life History of Ogawara Tadashi

In October 1948, Ogawara Tadashi was born to Akira, the proprietor of the Otaru soba noodle restaurant Yabuhan, and his wife, Toyoko. Their long-awaited first son, Tadashi, is the subject of this fourth life history.

Several words are key to understanding Ogawara Tadashi's life history: the "cultivation" of Yabuhan, the restaurant he would inherit from his parents, and the Japanese "student movement" of which he was a part during his university days. The intersection of Ogawara's efforts to "cultivate" Yabuhan into something more than it had been in Akira's day and his ardent support of the "student movement" dedicated to social change inspired his later community development activism in Otaru.

A survivor of the New Guinea front in the Pacific War, Ogawara Akira later quit his office job with the Otaru labor standards office to start up his own soba shop in 1954.[67] Having first apprenticed at the Yabu soba restaurant in the Kanda district of Tokyo, Akira reportedly named his restaurant "Yabuhan" to reflect his desire to make his restaurant "at least half (*han'bun*)" as great as the one at which he had learned his trade. Yabuhan prospered, but while the young Tadashi, who was viewed as the successor to the family business, sometimes helped out, he never believed the restaurant would be part of his future. Meanwhile, in contrast to his future friend and ally Yamaguchi Tamotsu, the rugby-obsessed Tadashi remained completely uninvolved in the student movement throughout his time in high school. Indeed, given that Akira once managed the campaign of an LDP-affiliated candidate for city council, the Ogawara household was likely a conservative one.

His impending graduation from high school forced Tadashi to think concretely about his future. Unwilling to carry on with the family business, he searched for a way out of Otaru. Citing the influence of an uncle, he settled upon the field of architecture and, after one year of additional study, enrolled in the architecture and engineering department of the Shibaura Institute of Technology in April 1968. It would seem that Ogawara Tadashi had successfully escaped Otaru.

At the time of Ogawara's matriculation at the Shibaura Institute of Technology, the university was in the throes of the student movement. As a high school student, Tadashi had been consumed with rugby and a rebellious determination to avoid taking over his father's business. But once at university, he quickly underwent an enormous transformation: the unsophisticated, provincial teenager became a passionate activist, newly alive to social issues. As Ogawara himself recalls:

[67] Details from Ogawara Tadashi's early life history up until the time of his expulsion from university are based on Tadashi's own detailed account of this period on the Yabuhan restaurant website, entitled "Ogawara Tadashi: The Four Decade Story of the Otaru Soba Shop Yabuhan," https://www.yabuhan.co.jp/yabuhan_2011/wp-content/031_01_yab_story_2010.html (Last accessed on July 27, 2017).

> The (Shibaura Institute of Technology) entrance examination stated they had a quota of sixty students. But when I arrived, I discovered they had accepted one hundred and twenty—leaving me to wonder what number I had really been [on the list]? That was the state of the place—of course, it was the same at any university, because our generation was the largest, in terms of population. I went off to university at a chaotic time. I had to duck under the barricades to take the entrance exam. And when I started, there were barricades surrounding the school entrance ceremony. And I was completely naïve—this country bumpkin, this pretty boy (*laughing*) who got caught up in it all, right off the bat. I demonstrated at Sanrizuka [in opposition to the construction of Narita Airport] and oh! (now that you mention it) the U.S. army had a field hospital in Ōji. You know the Ōji station on the Yamanote line (in fact, the Keihin Tohoku line)? There was a U.S. military field hospital there, and that's where American soldiers injured in Vietnam were taken. The hospital was going to be expanded, and every day I would go to [anti-war and anti-U.S. military] demonstrations. You had the campus for third and fourth year students in Tamachi, and then just past Ōmiya, in Saitama prefecture, there was the general education (campus for underclassmen). We were constantly traveling back and forth between the Tamachi and the Ōmiya campuses on the Keihin Tohoku line. Everyone would get off midway [between the two], at Ōji. We'd get on the train having been told that today there would be a debate with the third years, with our superiors, [in Tamachi], but then we'd be told to "get off at Ōji!" When we arrived, the area around the station was in total uproar. Someone told us to 'go for it, everyone!' and we'd run out (to confront the riot police.) {...}
>
> There was the Sanrizuka battle, the conflict surrounding Okinawa's reversion, the *buraku* liberation struggle {...} and then there was the opposition to the deterioration of the so-called immigration measures [addressing the status of resident Koreans] ... you could say I was involved in every possible form of activism during the seven years and eight months that I was at university. Finally, we created the university cooperative association, and just as we got that (on track) my health completely gave out. I'd also just been dumped by my girlfriend, and, well, I was exhausted {...}
>
> Interview with Ogawara Tadashi in Otaru, August 31, 1999. Information in parentheses has been added by the author. Information in [brackets] has been added by the translator.

Regardless of whether or not Ogawara was the "pretty boy", he claims to have been, the story of how a "naïve" young "country bumpkin" starting university in Tokyo, where he knew no one, immediately became "caught up" in the student movement is easy enough to understand.

Ogawara grew increasingly absorbed by activism, sacrificing both food and sleep to read indiscriminately and engage in fierce debates, all in an effort to remake himself. In a book on the student movement at Shibaura, *Mōhitotsu no Zenkyōtō*, Ogawara recalls this period as follows: "in order to win the debates raging at Shibaura, we'd use any tool available. There was nothing wrong with that" (Shiba Kōdai Tōsōshi o Katarukai 2010: 199–201). Ogawara's recollections suggest that even as an underclassman in Ōmiya, he had already acquired the ability to out-argue the upper-classmen in Tamachi and that he had adopted a pragmatic approach to achieving his goals—and "there was nothing wrong with that." Ogawara's mastery of basic accounting and finance, gleaned from his years helping out in the family restaurant, led to his appointment as treasurer of Shibaura's student executive council in May 1970.

Significantly, Ogawara possessed practical skills in addition to ideals and ideology. The following episode demonstrates his stubborn bent:

> At around four in the morning on May 5, (19)71, a police detective showed up at my boarding house. He was there to arrest me for an 'incident' on April 13, when I had fought with a guard during the entrance ceremony guidance [session]. I had no means of escape. My landlord was in a complete panic. My first concern was for the student council account passbook and seal, which I had in my possession. I told [the detective] that I would personally witness his inspection of my rooms, and as I got dressed I slipped the passbook and seal into the pocket of my blazer and then told him I needed to use the toilet. I hid the passbook and seal there. Since the bathroom was shared [with other tenants], it wasn't included in the police search.
>
> Later, when I spoke to my lawyer, I was able to tell him where I had hidden the passbook and seal, and was thus able to protect the student council accounts. I was always conscious of my duty to protect the student council finances.
>
> Shiba Kōdai no Tōsōshi o Katarukai (2010: 240–241). Information in parentheses has been added by the author, information in brackets by the translator.

Ogawara's quick-wittedness and a strong sense of responsibility are both apparent in this episode, which reads like a scene out of a spy movie.

A year before Ogawara's arrest, in June 1970, Yabuhan was destroyed by a fire that started in a neighboring house. Upon hearing that his family home had been destroyed, Ogawara raced back to Otaru. In the short time he spent at home, he observed how Akira busied himself with the reconstruction of the shop:

> Akira had managed to survive the war despite enduring food shortages and malaria on the New Guinea front. Far from being crushed by the total destruction [of his home and livelihood], he encouraged his family and staff, and started down the path toward reconstruction. In 1954, he had started his noodle business with nothing but the clothes on his back, and [later] overcame his anguish at losing everything, just as he was set to repay his bank loan for the purchase of the land and building, in order to re-open his noodle shop.
>
> His son, Tadashi, returned to Tokyo having witnessed firsthand the formidable nature of [his father Akira's] unflinching efforts to overcome the destruction of the spreading fire.
>
> Ogawara Tadashi, "Otaru Sobaya Yabuhan no Kenchiku Tenpo Yondai Monogatari," https://www.yabuhan.co.jp/yabuhan_2011/wp-content/031_01_yab_story_2010.html; Last accessed July 27, 2017.

Akira had just set his sights on rebuilding Yabuhan when Tadashi returned to Tokyo and reimmersed himself in the student movement. Yet in a foreshadowing of his own future, Tadashi took to heart Akira's "formidable" management abilities and approach to organizing a variety of tasks.

As the student movement reached its climax in June 1971, Ogawara addressed a packed audience at a student rally and received a standing ovation:

> A special rally was held on the day after my release. Not even a month had passed [since my arrest] but the mood on campus had changed completely—it felt like everyone was ready to fight. The transformation was so great that it made me wonder what exactly had happened while I was under arrest. There was no time to report back on my detention, or listen to the details of what had happened during my absence. One chaotic episode followed another, as my mind worked furiously to catch up with these extraordinary changes. At the rally as well, without any idea of what to talk about I lost myself in working up the crowd. All the same, a huge round of applause broke out and I was overjoyed to realize I had so much popular support.
>
> Shiba Kōdai Tōsōshi o Kataru Kai (2010: 222–224).

Clearly, the "naïve" young "country bumpkin" from Otaru was already able to captivate an audience.

But Ogawara's transformation left him with little aspiration—or ability—to secure mainstream employment. In April 1973, Ogawara joined the public relations department of the Shibaura student cooperative association in an effort to continue his political activism while working on campus. But the student movement, which had originally spread like wildfire across the campuses of Japanese universities, now began to wane, leaving Ogawara with little in the way of future prospects.

Although he still had no intention of taking over the family business, the effects of Ogawara's complete dedication to the student struggle, poor eating habits, and unhealthy lifestyle began to catch up with him. He was diagnosed with a duodenal ulcer and after seven years and eight months of enrollment, he was expelled from the university. In April 1974, he finally returned to Otaru—and the family restaurant. Worn out, Ogawara returned to Otaru still burning with the desire for social transformation, but increasingly disillusioned with the student movement itself. He returned home with no clear accomplishments, and with nothing waiting for him.

While recuperating at home, Ogawara assisted with Yabuhan's deliveries, but his lack of physical stamina left him immediately exhausted by this task. He recalls these early days following his return to Otaru:

> {…} You know, I spent my prime at the student cooperative. When I was twenty-six, twenty-seven. That was when I was at my boldest {…} I never graduated from university, so—well, in terms of social background, I was [just] a high school graduate. Because I was expelled after attending [Shibaura for] seven years and eight months, you see? And then I come back here to that 50,000 yen salary I mentioned? In the restaurant world—the soba restaurant world—the president's son doesn't automatically become a top executive. You're told to start off by making the deliveries. {…} That's how you earn your credentials as an artisan. In this world, if you announced you wanted to start off by making the noodles, you'd be told off immediately. I had this internal conflict as to how I could prove myself in this line of work. And then there was the idea that the owner's son had come home to take over the family business. Outsiders viewed me as "the next owner," "the next guy to run the soba restaurant." My standing was higher with outsiders than in the kitchen. In the kitchen, I was just the delivery boy. I could understand this gap because I had just barely managed to survive in that other world during my student activist days, and that's why I was able to stick it out. And socially speaking, I wasn't interested in being seen as the next [soba] shopkeeper, but [thought about] what I could do in the future. What kind of friends should I seek out? Or rather, what kind of relationships could I establish if I were to run a soba shop in this town? [These questions] really weighed on me. That's why I looked around for other guys like myself. And it was through the movement to save the canal that I met them.
>
> Interview with Ogawara Tadashi in Otaru, September 12, 2011. Information in parentheses has been added by the author, information in brackets has been added by the translator.

The keyword in this portion of Ogawara's narrative is "gap." Inside his father's soba shop, Ogawara had to earn his "credentials," because the "president's son doesn't automatically become a top executive" but rather "starts off by making deliveries." Outside this world, however, he was fawned over as the Yabuhan's "next owner." In other words, his standing outside the restaurant (as its "next owner") was much higher than inside (where he was merely the delivery boy and dishwasher). Ogawara says that what allowed him to perceive this gap was his experience of "barely surviving" as a radical student activist.

During these gloomy days following his return to Otaru, Ogawara was "intensely aware" of "what kind of friends" he should seek out if he were to "run a soba shop in this town," and "looked around for guys like himself." When Ogawara started thinking about how to further improve Yabuhan, he had to first understand his role as the owner of a shop located within the local community, as the following excerpt from the Yabuhan website makes clear:

> There is a limit to what the efforts of a single soba restaurant—of what any individual establishment—can produce. Without re-energizing the town of Otaru itself, there could be no prospects for business, and the second-generation owner [Ogawara Tadashi] participated in "community development based on the Otaru canal."
>
> https://www.yabuhan.co.jp/yabuhan_2011/wp-content/031_01_yab_story_2010.html; Last accessed on July 27, 2017. Information in brackets has been added by the translator.

In one sense, thinking about one's business in relation to the local community is entirely natural, but we should take heed of the other circumstances that drew Ogawara to this conclusion:

> It was astounding how deserted the shop was. When Tadashi returned and passed through the *noren* curtain hanging over the door [of the shop] he found the wait staff playing ping-pong on one of the tables. After opening the shop a third time, Akira again busied himself with various industry associations, leaving the restaurant in the care of a single proprietress (Toyoko). Without a unified "command tower" controlling both the kitchen and the dining area of the restaurant, however, staff morale was suffering. The new [renovations] had not infused any pride into the restaurant staff. This was the state to which the Yabuhan had been reduced. {...} Employees rebuffed Tadashi's attempts to convene a staff meeting, either stipulating that meetings be held during working hours or accompanied by overtime pay, or arguing that part-time workers had no obligation to attend. Tadashi realized that even Yabuhan was not immune to the apathy and indifference of a town in fatal decline.
>
> Ogawara Tadashi, *Otaru Sobaya Yabuhan no Kenchiku Tenpo Yondai Monogatari*, https://www.yabuhan.co.jp/yabuhan_2011/wp-content/031_01_yab_story_2010.html; Last accessed on July 27, 2017

That the employees who had been "playing ping pong" on the Yabuhan tables asked for "overtime pay" for attending a meeting aimed at improving the restaurant made Ogawara realize that "even" Yabuhan was "not immune" to the "apathy and indifference of a town in fatal decline." All the more reason for Ogawara to press forward with his own ideas for Yabuhan's development, which included creating a new manual for welcoming customers, and promoting a change of mindset among Yabuhan employees. In such an economically stagnant region, however, those crucial customers failed to materialize. Inevitably, Ogawara's realization that "without reenergizing the town of Otaru itself, there could be no prospects for business" led him to the task of bringing new dynamism to the local community.

Ogawara was drawn to the campaign to save the canal by his strong desire to improve Yabuhan and his quest to find "guys like myself:" other veterans of the student movement. First, during his search for "guys like myself," Ogawara became aware of the [broader] movement to preserve traditional townscapes. After all, 1974 was not only the year in which he returned to Otaru, but also the year that the canal issue began to gain prominence, as members of the newly established APOC began mobilizing to protect the canal:

> I suppose the 'towncape' [*machinami*] preservation movement is about city dwellers identifying their own neighborhood, their environment, as an issue to be addressed {...} If that is what 'townscape preservation' was really about, well, it marked the first time that city dwellers spoke out against Japan's urban policy. Now, at that time I still liked to refer to [everything as] a "revolt." But as a veteran of the All-Campus Joint Struggle League [I] did see this as a potential 'sign of revolt.'
>
> Interview with Ogawara Tadashi in Otaru, August 31, 1999. Information in parentheses has been added by the author, information in brackets has been added by the translator.

If we follow Ogawara's definition, the townscape preservation movement represented the first "sign of revolt" against Japan's urban policy. Both Ogawara's choice of words ("revolt") and his determination to make sense of the movement within its historical context clearly emulate the approach of the Japanese student movement of which he had once been such a passionate participant.

However, this was not an unreservedly positive description by Ogawara, who had been subject to multiple arrests in the course of the "battles" of his university days, but rather a more skeptical assessment:

> There was the Sanrizuka residents' movement opposing the new [international] airport, for example, or the old ladies' protests against the U.S. army's live-fire ranges at [USMC] Kitafuji, the antinuclear protests at Kashiwazaki in Niigata—I was constantly going to this type of residents' movement. But the historic townscape-type of citizens' protest—well, my first experience with that kind of thing was when I returned [to Otaru]. And because I had participated in those kinds of protests, the APOC's townscape preservation movement—at that time, that was really the only term for it—well, it made me think that while I had come back to Otaru, and kept on living, training at the soba restaurant with some determination, that this alone was pretty trivial, and that I should focus on the town {...} When I decided I wanted to live in that town in a more energetic fashion, I discovered a movement that I felt could be interesting to get involved in. It was my experience in the (student) movement that prompted me to look into it. {...} I would talk with people in the bars along Hanazonochō in Otaru and wonder what this 'townscape preservation movement' was all about. And I wondered whether it stemmed from neighborhood anti-pollution movements—I wondered whether that had been the origin. To be sure, townscape preservation wasn't a clear-cut form of anti-authoritarian protest—that association from back then, the *Machinami Zemi* {...}—well, that was a movement started by a bunch of retired old men. After all, campaigning to save historic old buildings isn't the kind of movement that would really clash with the political powers to a great extent—although this didn't end up being quite the case with the Otaru canal. So I had my doubts as to whether this townscape preservation thing was really something I could get excited about.
>
> Interview with Ogawara Tadashi in Otaru, August 31, 1999. Information in parentheses has been added by the author, information in brackets has been added by the translator.

Ogawara's "experience in the student movement prompted" him to "look into" the APOC, but his initial assessment of the preservation movement was complicated. On one hand, Ogawara wondered whether the movement stemmed from anti-pollution movements elsewhere in Japan, and might thus represent a "first sign of revolt." Yet he also suspected that preservation was an old man's cause, unlikely to provoke any real clash with the authorities. Needless to say, the youthful Ogawara had little appreciation for any movement that failed to "really clash with the political powers."

Still dubious, Ogawara nevertheless ventured into the APOC's sphere:

> I felt this was interesting—this could become interesting, depending on how I got involved. But I didn't feel there was any way I could involve myself with the APOC at this time—the 1974 APOC. I attended one meeting, and even though there were only fifteen people present, there were still different factions of sorts. Ms. Mineyama was more detached from it all, but there were two female vice-presidents of the association: Ms. Morimoto and Ms. Kitamura—old ladies who tried to win you over to their side. They would try to make me into an ally, asking me if I was single and if so, if they could set me up. I'd be summoned in secret to a coffee shop and asked: 'Don't you want to get married?' I'd reply with, 'yes, but…' So that was the level (of conversation) we'd start out at… {…} And I thought to myself how vulgar it all was.
>
> Interview with Ogawara Tadashi in Otaru, August 31, 1999. Information in parentheses has been added by the author, information in brackets has been added by the translator.

Here once again one can discern a "gap." Let us recall the extent of Ogawara's experience with the student movement:

> I returned to Otaru in 1974. I was a veteran of the (student) movement, so these so-called citizens' movements like the APOC felt extremely…what can I say? Unsatisfactory. After all, I had been fighting political authority, and clashing with the riot police... {…}
>
> Interview with Ogawara Tadashi in Otaru, August 31, 1999. Information in parentheses has been added by the author, information in brackets has been added by the translator.

Given his previous experiences, the APOC's activities were "unsatisfactory" to Ogawara. He continues:

> "But then I had been involved with extremely radical—well, what I thought of as radical—protests against authority. I lived in a world where this was a matter of life or death. So I looked at this campaign to save the canal, and I figured getting involved with such a sad little movement wouldn't get us taken seriously by those promoting (the construction of) the new road, and that I'd be branded a communist—even though I wasn't—and become even more isolated."
>
> Interview with Ogawara Tadashi in Otaru, September 12, 2011. Information in parentheses has been added by the author.

Given the "gap" between the type of activism Ogawara had experienced in Tokyo and the APOC ("I thought to myself how vulgar it all was"), Ogawara dismissed the preservation effort as a "sad little movement" that wouldn't be "taken seriously" and would leave him "all the more isolated." In other words, if the first "gap" refers to disparate assessments of Ogawara himself (outside > inside the soba shop), the second gap refers to his differing assessments of activism (the student movement > the canal preservation movement). Given Ogawara's attitude toward these gaps, his POA orientation is negative.

Having turned his back on the APOC, Ogawara cast about for something else to do. As he contemplated starting his own organization, he encountered another young man named Sasaki Kyōjirō:

> At one point I thought about setting up a separate organization from the APOC {…} In 1975, in the spring of the year I returned to Otaru, this guy Sasaki Kyōjirō opened his café, Kyōjirō. It was a spaghetti restaurant. There was something about him that I liked. How can I put it? He had this gloominess to him. We didn't talk about anything significant, but {…} when I invited him to (my) wedding, he showed up…in a bow tie and tuxedo jacket. There was no one else like him [in Otaru] at that time. That made me realize that he must be interested in [getting to know] me. And as we kept talking, [I realized that] Kyōjirō and I had the same hopes. Kyōjirō had once drifted through Europe, and he told me that wherever he went—the French countryside, a medieval city—he had this nagging sense of *déjà vu.* He came back to Japan still troubled by the fact that he couldn't figure out where he had seen that kind of scenery before, and then suddenly it came to him: he had been reminded of (the landscape) of his own hometown. So he trained in Sapporo and worked with the dream of running his own café in Otaru, and opened up his current café, Kyōjirō. He had the same hopes that I did {…} At that point I was running deliveries every day. Deliveries alone took up half the day, and I felt that this was a pretty tedious way to live. And I felt that in order to make Otaru a more interesting place to live in we had to learn more about Otaru, and so I started the 'Otaru Research Society.' At that time the owner of Kitaichi Glass still ran a small shop along the national highway (National Highway 5). We brought in Prof. Shinozaki from the (Otaru) University of Commerce as a kind of academic member. And then there was Tsuchiya—Tsuchiya Shūzō, from the Otaru Board of Education. He's still around. And someone who had been a museum curator, and four or five others. We took over—we borrowed—that meeting room on the second floor of the Chamber of Commerce and Industry, and hung a big map of Otaru on the wall. Once a week we'd [go over the map] and look at what kind of building each had been. That's all we did. It was nothing more than a campaign to raise awareness. But that didn't turn out to be so interesting after all, and Kyōjirō brought up the APOC—he consulted me because he wanted to go to the APOC.
>
> Interview with Ogawara Tadashi in Otaru, August 31, 1999. Information in parentheses has been added by the author, information in brackets by the translator.

Together Ogawara and Sasaki, kindred spirits, launched the Otaru Research Society. Reviewing the various buildings of Otaru, however, was "nothing more than a campaign to raise awareness," and in the end, not "so interesting." This harsh assessment is understandable, given Ogawara's experiences in the student movement, where battles could actually be fought and won.

The stalemate at the research society prompted Sasaki to "go to the APOC," but Ogawara opposed this plan, believing the APOC's approach could achieve nothing:

> The administration's plan [to pave over the canal] seemed like it would simply make an already shabby town even shabbier. {…} The efforts of each individual soba shop or other business couldn't do much, and [there wasn't much hope for our kind of business] if the town as a whole didn't improve. All of these different feelings sort of came together [in me]. I came back (to Otaru) at (the age of) twenty-seven, got married at twenty-nine, and I was churning with these kinds of feelings. The one thing I did know for certain was that the APOC—that structure of activism—wouldn't be able to achieve any sort of breakthrough.
>
> Interview with Ogawara Tadashi in Otaru, September 12, 2011; Information in parentheses added by the author, information in brackets by the translator.

Despite his various misgivings, the one thing Ogawara "knew for sure" was that the "structure" of the isolated APOC could not achieve any sort of "breakthrough." He continues:

> Those guys [in the APOC] had no chance of winning. By starting up afresh with a brand new movement that could make a favorable impression (on city residents), we could also lift up this old movement whose numbers had plummeted. I mean, the APOC was so isolated: at that point there were around 2,800 people working at City Hall, and the APOC had just fifteen or so members. {…} Without getting the public involved, we had no hope of victory. And given the old image of the APOC—well, its established image—they couldn't make any breakthrough with the public.
>
> Interview with Ogawara Tadashi in Otaru, September 12, 2011. Information in parentheses added by the author, information in brackets by the translator.

Ogawara had begun to think strategically: because the APOC's public image was now "established," the group had "no chance of winning." He realized he needed to start a new type of movement—one that could involve the citizens of Otaru.

Ogawara's next move was inspired by the success of Kurashiki Ivy Square—a project that combined "preservation with redevelopment" in the city of Kurashiki in Okayama prefecture.[68] The second source of new inspiration was his encounter with another young resident of Otaru, Yamaguchi Tamotsu.

> As I studied more, I realized that townscape preservation in Japan usually meant the type of thing you see at Meijimura in Inuyama—what you might call 'cryopreservation.' The Historic Village of Hokkaido [Hokkaido Kaitaku no Mura] would be another example. Buildings that seemed destined to be torn down and couldn't be preserved in their original location would be dismantled and reconstructed in one big village somewhere else. There are many similar cases—for example, when they took just the lobby and floors of [the old] Imperial Hotel and [used them in the new hotel]. That type of preservation in Otaru didn't excite me. If that canal, the whole space of that area became a set piece, or a kind of [outdoor] museum that merely showed tourists 'these were the type of buildings we had here in the past'—well, that didn't sound very promising. And it just so happened that when I was organizing within the student movement I had made my through different universities across the country—and [so] I had seen Kurashiki Ivy Square and places like that. That felt like a more interesting way to do things. {…} You know, they turned an old brick warehouse into a [chic] hotel for young people. That felt like an interesting idea to bring to Otaru. I had caught a whiff of this sort of thing, and then the editor of a regional newspaper—there's this newspaper called the *Hokkaido Dokusho Shinbun*—{..}, he told me: 'I'll give you the whole front section, so write something. Well, it wasn't the whole front section, just the last page. So I thought to myself, 'alright, I'll give it a try,' and I wrote a piece called 'Farewell to Cryopreservation.' Anyway, [my point was that] Otaru {…} was this treasure trove of [historic] buildings and we could do a far better job of utilizing them. Instead of treating them as mere museum pieces, we could turn some into lodging, others into ateliers, and thus could come up with a much more vibrant and exciting way of using them. There's that biblical expression, "new wine in old wineskins." I remembered that [passage] from the Bible I read while I was locked up [by the police, during the student protests] and I used that phrase as a kind of analogy, [saying,] let's plant and nurture things that can bring "new energy to an

[68] Kurashiki Ivy Square was a hotel built using a group of brick warehouses that were once part of a cotton mill in Kurashiki's Honmachi district into a hotel. Opened in May 1974, this was a pioneering case of "adaptive reuse" (Kihara Keikichi 1982).

old town." If we did that, perhaps Otaru could overcome its status as a dying town. These are the kinds of thoughts I dashed off impulsively [in the article].

Author interview with Ogawara Tadashi in Otaru, August 31, 1999. Information in parentheses added by the author, information in brackets by the translator.

Ogawara's observation of Kurashiki Ivy Square left him with the sense that there was the potential for something other than the "cryopreservation" of reconstructed buildings, like that on display at "Meijimura." Kurashiki Ivy Square represented a style of preservation more in keeping with the times, which was capable of drawing crowds and creating a lively, bustling atmosphere.[69] In his article for the *Hokkaido Dokusho Shinbun*, Ogawara laid out the specifics of his vision. It is a startlingly accurate review of the development of historic preservation efforts in postwar Japan, from "static preservation" to "adaptive reuse"—in other words, the act of bringing "new energy" to "an old town."

The article brought Ogawara into contact with yet another key figure in our story: Yamaguchi Tamotsu. The two were introduced by Sasaki Kyōjirō:

And that (article) brought me back into contact with the APOC, and (Sasaki Kyōjirō) told me 'you know, there's one guy in the APOC who talks like you.' At any rate, Sasaki had told me that I needed to meet this Yamaguchi Tamotsu, and we immediately hit it off. How can I put it? It was this chance encounter, a fortuitous meeting—there was a newspaper reporter who expressed it like this: 'Through individual encounters, these citizens' movements form human relationships that, like the strands of a rope, come together into a single entity for activism.' Well, that's exactly what it was like. {…} [Yamaguchi and I] found each other just as we were both troubled by the question of how to expand the scope of the APOC's activism, because the APOC had become a fringe group. There weren't many other people who had learned through the student movement how to set up a campaign, so that put us on good terms. I'm also the type who gets fired up when I have comrades around me, so we decided to [run with our ideas]. And that's how we became what they now refer to as the *reisan*—that's an abbreviation of *rei no sanningumi* [the usual three]: 'Ogawara, Yamaguchi, and Sasaki.' But I shouldn't say my own name first! We became 'Yamaguchi, Sasaki, and Ogawara.' {…}

Interview with Ogawara Tadashi in Otaru, August 31, 1999.

Ogawara "immediately hit it off" with Yamaguchi, a kindred spirit and another veteran of the student movement, who "knew how to set up a campaign." Together with Sasaki Kyōjirō they became "the usual three" and formed a strong bond.[70]

---

[69]Ogawara also married during this period (in November 1977). His bride had hoped to spend their honeymoon on a beach in Okinawa, but Ogawara privately decided they would visit Kurashiki instead. The reason for his decision was to revisit Kurashiki Ivy Square, which he had first seen during his student canvassing days. That Ogawara turned his own honeymoon into an observation visit may recall the pilgrimage across Japan by the "reckless" Fujimori Shigeo. It is worth remembering, however, that it was in Kurashiki that Ogawara first developed an image of what the district surrounding the preserved Otaru canal could become.

[70]The newspaper article Ogawara refers to in the preceding excerpt is from "Unga Mondai no Jyūnen" [The Canal Issue Ten Years On], which appeared on June 9, 1984 in the Otaru city edition of the *Asahi Shinbun*. The article was authored by Shin'chi Mitsuo. The actual expression used by the reporter is: "Relationships are woven together like the strands of a rope."

Woven together like the "strands of a rope," the three young men pushed forward with their plans. Using Kurashiki Ivy Square as a reference point, they began to envision a new type of campaign that would bring life and energy to Otaru's waterfront. Together with their friends, they began to put this plan into action:

> That's why we started the Port Festival. I mean, this was a youth campaign. Somehow, the mass media will root for the underdog when it comes to youth movements—even if they don't support you, they'll pay attention to you. The other thing about youth movements is that young people will be even more enthusiastic when they're in the spotlight. Right? (We set out with the feeling that) we young people would start something different.
>
> Interview with Ogawara Tadashi in Otaru, September 12, 2011. Information in parentheses added by the author.

As we have seen, the Port Festival came about in large part through the efforts of "the usual three," with Sasaki playing a central role. But how did the three men view the festival? In the words of Ogawara:

> So this is how I got involved in planning things like the Port Festival, and served as the second chairman of the executive committee in its second year. {…} This wasn't merely an event where young people could drink, make merry, and get rowdy. Through this event, we could express our efforts at community development, a new kind of community development in Otaru—one that began with the canal, with the campaign to save the canal, with the preservation of the canal. So those four hundred booths at the festival were a small experiment of sorts, to demonstrate what kind of tenants could be positioned along the canal. [We were trying to show] how many people would come to the area if we were to line the canal with shops. We did it to prove this point. And we set up a beer hall on a barge, and rented out the stone warehouses to use for seminars and entertainment halls. That is, we were suggesting that these stone warehouses could be used as an assembly hall, or as a community center, or as a theater. And those four hundred booths, made with just a plank of plywood each, were a kind of experiment that proved that the canal wasn't just a historic environment, but a place that could bustle with people, and flourish, and that this could send economic ripples [through Otaru].
>
> Interview with Ogawara Tadashi in Otaru, August 31, 1999. Information in brackets has been added by the translator.

Ogawara's message is even clearer if we take these comments together with the following passage from the Yabuhan website:

> [The Port Festival] became an opportunity for the youth of Otaru to get involved with the campaign to save the Otaru canal. [The idea was that] not only young people, but the majority of Otaru citizens, [think about the canal], although they might not give voice to this. 'Identity' was very far from being a common term at the time, but not even knowing that there was such a word, the young people behind the Port Festival proclaimed: 'the Otaru canal was proof of what an Otaru person should be, and a passport [an identity].' They decided to continue the event. {…} The festival was held annually for seventeen years, through 1994. For two days each year the festival welcomed three hundred thousand visitors.
>
> The Port Festival. This event proved that using tourism as an opening for the [adaptive reuse of] the Otaru canal and its surroundings could dramatically improve the economic potential of the city. The event demonstrated the enormous historic value that lay hidden within the space of the canal and the surrounding area. This wasn't a PowerPoint slide of a hypothetical scenario. The event proved what could actually be achieved.

> Ogawara Tadashi, "Otaru, Sobaya Yabuhan no Kenchiku Tenwa Yondai Monogatari." https://www.yabuhan.co.jp/yabuhan_2011/wp-content/031_01_yab_story_2010.html. Last accessed July 27, 2017.

Ogawara saw the Port Festival as an "event that proved" that "[adaptive reuse] was not a mere "hypothetical scenario," and could dramatically improve "the economic potential of the city." Descriptions of the Otaru Port Festival as a kind of "flea market" are thus way wildly off the mark.

The young organizers of the Port Festival were astounded by the tremendous turnout. Could they really hold an event that elicited such a resounding response only once? Ogawara, Sasaki, and their friends formed the Otaru Yume no Machizukuri Executive Committee[71] to transform the Port Festival into an annual event:

> ...We needed an executive organization to ensure that [the Port Festival] would not be a one-time event—that it could be continued. What we came up with was the Otaru Yume no Machizukuri Executive Committee, which we called the Yume-machi committee for short. The idea behind the name was our desire to create a place in which every resident, each person, could have their own vision [*yume*] for the future of their town [*machi*]. It was about self-determination and civil autonomy for one's own town, and we (started from a critical position, asking) whether these had really existed (up until now). This wasn't about having the town's most prominent residents convene a deliberative committee and come to a decision. We were saying that if each and every person would think about how to improve their town or their village, these places would become amazingly strong. That's how we decided on the name.
>
> Interview with Ogawara Tadashi in Otaru, September 12, 2011. Information in parentheses has been added by the author.

The guiding concepts of the Yume-machi Committee were "self-determination" and "citizen autonomy." The kernels of these ideas, which Ogawara and his colleagues would later develop into a broader philosophy, were already evident in the process leading up to the very first Port Festival. Let us first learn how Ogawara and his colleagues discovered the term "community development," and then examine the ideas that informed their organization of the Port Festival.

> The term 'community development' didn't exist, although there was [the term] 'historic preservation.' {...} Using the one word to sum up something like 'identity' came quite a bit later. It was the same with [the term] 'community development.' The only term we had [before that] was 'urban planning.' But 'urban planning' sounded so bureaucratic, and everyone had a negative response to [the term]. And this is when *Jurist* came out with a special edition on court cases involving residents' movements across Japan. This was in *Jurist*—in a special edition of *Jurist*. And it included a summary edition of various residents' movements nationwide, and when I looked through this at Kinokuniya (bookstore), I came across the term *machizukuri* [community development]. The word shone up at me from the page like a beam of light! This was it!
>
> Interview with Ogawara Tadashi in Otaru, September 12, 2011. Information in parentheses has been added by the author, information in brackets by the translator.

[71] See Footnote 17.

Here Ogawara is probably referring to a 1977 edition put out by the editorial department of *Jurist*. The term 'community development' has long since made its way into everyday conversation, but for Ogawara (and likely many others), encountering the term for the first time felt like seeing "like a beam of light." Ogawara and his friends introduced this new term to their organization. First, however, the young organizers had to apply their new vocabulary to the success of the Port Festival, and then link it to their future goals for community development. Ogawara's recollections illuminate the process by which this occurred. As it turns out, the story of the Port Festival has unlikely origins on the island of Izu Ōshima, far removed from Otaru.

> In 1965, a large fire devastated Motomachi, the largest town on [the island of] Izu Ōshima. The national government invoked the Disaster Relief Act. When, in the immediate aftermath of the disaster, the government presented various emergency support measures, the island's residents said that the assistance measures arbitrarily decided upon [by the government] were not what they wanted most. When all the residents got together and discussed the matter, what they came up with first was the plan "to use their own abilities to construct a reservoir (what the residents referred to as a *mizutoriyama*) in the [volcanic] desert zone of Mount Mihara.
>
> Of course, the reservoir would alleviate the island's chronic water shortage, but what the townspeople wanted, more than any other sort of recovery measure, was to use their own abilities to build a reservoir for the use of the whole town. They said they would first come up with their own plan for the reservoir, and then come together to construct it themselves. Building a reservoir themselves was likely based on the philosophy that you should help build up your region with your own hands, and there was probably also the idea of rediscovering the fundamental power of a place like a reservoir. It was also a rejection of an outsider's recovery plan in favor of putting community development in the hands of the island's residents. This was about putting the miscellaneous skills of a large group of people before mechanized power, wisdom before knowledge, durability before speed, and passion before reason. In other words, by asking for a reservoir, what the island's residents really sought was something with the symbolic power to unify the islanders. Government officials and disaster experts never dreamed that the islanders would ask to build a reservoir. On the contrary, what was so amazing about these islanders was that at such a time they didn't instantly swarm around the government 'menu' of support [measures], but prioritized coming together.
>
> So, what would the people of Otaru see as their reservoir? Was…. the canal not the *mizutoriyama* for Otaru's residents? Was that…. not what the Port Festival had revealed? Rather than transforming an outdated port facility into a modern road, shouldn't the foundation for the future of community development in Otaru be Otaru's own *mizutoriyama*—the Otaru canal….?
>
> At any rate, this was the story I heard thirty years ago. {…} And that's the story of the *mizutoriyama*. It made a deep impression on the young Port Festival organizers who were there to hear about it. It was that precise phrase ["the Otaru canal is our reservoir"] that led to the formation of the Yume-machi Committee. I was also fascinated by [the tale].
>
> "'Otaru Yume-machi, Soshite Mizutoriyama,' Mōhitori no Otaru Sobaya Oyaji no Hitorigoto 09/11/21," viewed on July 28, 2017. Extended dots are in the original, information in parentheses has been added by the author, information in brackets by the translator.

There is little need to expand upon Ogawara's account. As the Yume-machi Committee began to take shape, its members discussed the Izu Ōshima as an example of the "rejection of an outsider's recovery plan in favor of putting community development in the hands of the island's residents." The wisdom of Motomachi's residents, who asserted their need for the reservoir as a "symbolic" project to "unite the islanders" left a "deep impression" on Ogawara and his comrades.[72] Convinced the canal represented their own version of the islanders' reservoir, Ogawara and his comrades launched the Yume-machi Committee, with Ogawara himself penning an inaugural declaration.[73] After the first Port Festival demonstrated, almost by accident, that the canal was a symbol capable of uniting Otaru residents, Ogawara went on to chair the planning committee for the second Port Festival.

As we know, the canal was not preserved in its entirety, and the campaign to preserve it fissured over time. We have also seen how Otaru went on to witness a boom in tourism, welcoming nine million visitors a year at its peak. Ogawara offers an unreserved assessment of the quality of tourism in Otaru:

> Sakaimachi, the canal...six million tourists came to this area alone, and it's been completely transformed into a kind of tourist Dejima. It's like the island of Dejima in Nagasaki, a tourist Dejima.[74] And there is absolutely no interaction with residents. And no decent businesses... {...}Now of course a few excellent, expensive facilities (such as the Orugōrudō music box museum and the Ishihara Yūjirō Memorial Hall) also opened up, but a bunch of souvenir shops of the type you can find anywhere in Japan came in on their coattails. Above all, these [travel] agencies and [air] carriers get together [to offer cheap travel packages to Otaru], and these agencies keep funding these tourist shops. And we had talked about tourism as a means of achieving Otaru's economic revival, but we asked ourselves whether this kind of cheap, souvenir-based tourism was what we had really wanted. It really brought these questions to the fore. And from our perspective, the level of tourism was approaching that of Expo or Nōkyō Tourist Corporation tours [large group tours organized around the purchase of souvenirs]. And the Port Festival was fully incorporated into the event schedule.
>
> Interview with Ogawara Tadashi in Otaru on August 31, 1999. Information in parentheses has been added by the author, information in brackets has been added by the translator.

In Ogawara's assessment, Otaru's tourism had devolved into a "cheap souvenir-based tourism." This forced him and his friends to ask themselves "whether this...was

[72] The author visited Izu Ōshima on two occasions (February 24–26, 2010 and March 10–12, 2011) to research the reconstruction process in Motomachi. This survey was conducted under the guidance of Dr. Jun Ōyane, the foremost authority in the field of the sociology of disasters, and with the assistance of two students of Hosei University, Yamashita Takako and Sasaki Kenta. It appears that Ogawara's account of the *mizutoriyama* episode is not entirely accurate. However, the important question here is not the truth of this episode, but rather how their understanding of it (mistaken or otherwise) served to unify the young people of Otaru. In this sense, there should be further discussion of the potential "power of mistakes" and "power of misunderstandings."

[73] "We are not 'preservationists.' We do not want the historical and cultural environment of our town, Otaru, to be subjected to the conventional preservation method for national cultural properties, and simply turned into museum pieces, a form of preservation that leaves them frozen in time. What we want is to bring back to life this environment—the living environment of generation after generation of people, and to restore its value. In other words, we want to fill this old vessel with a new energy. {...} We hereby announce our campaign to fill this good old vessel (the city) with something new (community and human development) {...}. *Fiesta Otaru*, inaugural edition, December 1978. Cited in Otaru Unga Mondai o Kangaeru Kai, ed. (1986a: 222–223).

[74] See Footnote 30 for an explanation of "Dejima tourism."

what we had really wanted." Ogawara also identified a significant change in the Port Festival:

> After seventeen years of the Port Festival, you have these old hands producing the stage show, old hands managing the rock performance space... {...} And you become stuck within a framework in which these people are more interested in how many meetings you've attended, and how much scaffolding you've put up [in other words, your seniority] than in your ideals for the town and community development. I mean, it's not a conversation about your hopes or attachment to your town, or about what should be done for the town, or your own role within that process. Everything became about organizing the festival, and staff tended to be evaluated on how large a role they had played, and things like that. And I think, ultimately {...} the clincher was the weakening of the community development orientation. You know, in the old days, at the beginning, we would start preparing for the next year in the autumn, as soon as the festival was over ... {...}
>
> [We used to talk about] what we should do for the town. Or about a building that seemed to be on the verge of collapse. Or a good building that was going to be sold or something. All sorts of topics would come up, one after another. Even if we didn't (end up) doing anything. We'd talk about getting everyone together to paint an old building, for example. Even if we didn't end up doing it, we would always discuss it. If nothing else, we would always talk about these things. And in doing so, we'd take a look at our town, piece by piece. That's all over now. That's what happens when fifteen people run the festival and as soon as it's over, they don't want to do anything for a good six months. I can understand (why they feel that way). And ultimately, the process between one festival and the next, that continuity, disappeared. And all these different factors combined to [make us question] the significance of holding the festival at all. And this was a surprise to us—the first generation behind the Port Festival. That it had come to this {...} And, well, it raised the question: after all of our hard work, what did the event really mean?
>
> Interview with Ogawara Tadashi in Otaru, August 31, 1999. Information in parentheses has been added by the author. Information in brackets has been added by the translator.

As roles became fixed and the "old hands" came to wield the greatest authority, "everything became about organizing the festival." The event had become an end in itself. The Port Festival's community development orientation, which "look(ed) at our town, piece by piece," dissipated, prompting Ogawara to propose putting an end to the annual festival:

> (We held discussions about the future of the Port Festival) For about six months, from around November up until the Golden Week [holiday in late April and early May]. And well, right before Golden Week we somehow arrived at a unanimous decision to make it [that year's festival] the final Port Festival. It was then that I told everyone to "go back to being an individual." For a while. Before that, you would earn instant respect if you introduced yourself as "so-and-so from the Port Festival." But I said it [was time to] go back to being Ogawara, the individual. Yamaguchi, the individual. So-and-so, the individual. And I said that we could mobilize again if another event, like a new Port Festival, truly became necessary. That is, we could tout a new ideal, a new slogan. And that this could be the start of a 'second chapter in community development.' And having assigned everyone this 'homework,' we disbanded. Now how many years has it been since then...?
>
> Interview with Ogawara Tadashi in Otaru, August 31, 1999. Information in parentheses has been added by the author, information in brackets by the translator.

With its "community development orientation" "weakened," the Port Festival was held for the final time in July 1994. Ogawara's exhortation to the festival staff to "go back to being an individual," and tout "a new ideal, a new slogan" and thus "begin a second chapter in community development" can be described as the epitaph of the first chapter of the canal preservation movement. It could equally be described as the preface to the movement's second phase.

How, then, did this second phase unfold? Ogawara first uses the expression *otoshi mae*, or "settling debts."

> [The word] revitalization immediately signifies profits—it can be equated with profits, but I, for one, don't think that the word "bustle" means profit. When we were fighting to save the canal, we used tourism as an opening for a new kind of community development within our campaign. [People had] many different inclinations, but when we talked about the concept of "tourism-based community development" we didn't mean 'profit,' but 'bustle.' So even after the canal preservation movement was defeated we carried on as a way of 'settling our debts' so to speak. Well, that was the expression used by young people back then—at sixty-two (years old), I'm embarrassed to use it.
>
> Author interview with Ogawara Tadashi in Otaru, September 12, 2011. Information in parentheses has been added by the author, information in brackets by the translator.

Even after the "defeat" of the preservation campaign Ogawara remained involved in community development efforts because he had never sought "profit," but rather a revival of the prosperous, bustling town where people once gathered. He continued his efforts in an attempt to take responsibility for the unsatisfactory state of Otaru tourism—in other words, to "settle his debts." It is this self-awareness that informed the development of the second phase:

> If you really think about it, doesn't (the old Temiya line of the national rail) rival the canal as a legacy of [Japan's early] modernization? A great legacy of modernization that Japan can take pride in? {…} All right then, in that case we decided that citizen community development activists would attempt another community development effort, carrying on the currents of the canal campaign to address the question of how to use the old Temiya line.
>
> Interview with Ogawara Tadashi in Otaru, August 31, 1999. Information in parentheses has been added by the author.

Ogawara and his comrades decided that the preservation and reuse of the Temiya line—just the third railway line to be constructed in all of Japan—would become the target of this "second chapter" in community development. In this new phase, however, they adopted a more strategic approach, abandoning confrontation in favor of collaboration:

> The canal preservation movement, which gave rise to the Urban Design Section [at Otaru City Hall], changed its position and with regard to the [future of the] Temiya line is competing with the Community Development Council and the Urban Design Section, for example. I can't say it went well, but at least unlike the old administration, they are good at providing information, in a certain sense. {…} We can finally hold a meeting at city hall to criticize 'Dejima tourism.' {…} We can have these really vigorous discussions. And we can talk about how Otaru will be able to survive in the future. And city officials, who were convinced that the preservationists were inflexible and determined to oppose anything, well, they used to speak carefully and respond very cautiously. Especially when Yamaguchi (Tamotsu) and I would go [to city hall]. They didn't slip up, even when [everyone got overexcited]. Now,

> when we are in these debates, the fact that they used to choose their words so carefully...well, that's just not the case at all anymore.
>
> Interview with Ogawara Tadashi in Otaru, August 31, 1999. Information in parentheses has been added by the author, information in brackets by the translator.

Not only did the canal preservation movement lead to the establishment of an "Urban Design Section" at Otaru City Hall, but Ogawara and his fellow community development activists could engage in heated discussions of tourism strategy during meetings with members of the administration. City officials who had seen "preservationists" as "inflexible," and ready to "oppose anything" came to realize that was "not the case at all," enabling a real dialog.[75]

Rather than delve too deeply into this second chapter [in the preservation movement], let us examine instead the focal point of the first chapter: the meaning Ogawara gave to save the canal and his logic of preservation. To begin, Ogawara describes the significance of the canal to the people of Otaru as follows:

> At that time [the city would] take subsidies from the national government and use them to build public buildings, or tinker with [urban] hardware. This would make money for the construction industry, and the idea was that enriching the construction industry would in turn enrich other, fringe industries. That was the idea.
>
> Every town adopted this approach. In the case of Otaru, it just so happened that this took the form of a new road. That's all. Because it [took the form of a] road, a citizen's movement sprouted up. I don't know if a citizens' movement would have coalesced around a dam or some such thing. It had to do with the significance of the canal, and its location. Nowadays there's a beautiful word for it: *shinsuikūkan* [literally, a space where people have easy access to water and can enjoy it freely], but back in our day it was more difficult for people living at the port, in the town, to have a relationship [with the water]. Up until then this hadn't been a "canal town." This was the "port town" of Otaru, with a working port for distribution. It was the "port town of Otaru;" the "canal town of Otaru" was just a name created later on, after the preservation movement. {...} So the people of Otaru have a lot of sentiment for the port. {...} (Otaru) has no future prospects as a port, but there's a kind of nostalgia, or sentiment, about the port, and for the days when the town flourished as a working distribution port. This is why [we feel] different from that upstart, Sapporo, and why everyone at the public bathhouses goes on about how 'Sapporo's putting on airs now, but it only became [what it is today] because of the Otaru port.' And about how Otaru has a lot of things that Sapporo doesn't, and how people will return to it in the future.
>
> But you know, this is something you can only realize if you've left Otaru for the outside world. It's not easy to remain in Otaru and work hard knowing what a great place it is—after all, isn't that a matter of atmosphere? The port and historic buildings have been here a long time, but they were simply part of the atmosphere [of the place]—(there was no need to) objectify them, or become aware of them. To put it in stronger terms, to discover their value you have to first forsake Otaru and go out into the world. It's just like when you arrive at university, and recognize just how provincial your hometown is—but also [recognize] its good points.
>
> Interview with Ogawara Tadashi in Otaru, September 12, 2011. Information in parentheses has been added by the author, information in brackets by the translator.

[75]One might suggest that Ogawara's POA orientation, which had been negative during the battle over the canal, gradually grew more positive in subsequent years. In either case, one cannot dispute the fact that this new relationship with city bureaucrats was one clear result of the "canal war."

"The people of Otaru" harbor "enormous sentiment" for "the port of Otaru," and according to Ogawara, the Otaru canal, positioned at the interface between the port and the town center, was of no small "significance." According to Ogawara, the preservation movement formed because it was this canal that was threatened with destruction. He wonders whether a similar movement would have formed in opposition to a different type of development project, such as the construction of a dam ("I don't know if a community building citizens' movement would have coalesced if it had been a dam or some such thing"). Given the emotional significance of the "Otaru port" area and "the canal" to Ogawara, who describes "a feeling of nostalgia, or maybe sentiment," the NOS nostalgia factor here is positive.

As the interview continues, however, Ogawara begins to discuss the logic behind the preservation movement, and here we find instances in which he sets nostalgia aside. Having once "left Otaru for the outside" world, the "hometown" Ogawara later rediscovered was no longer an object of nostalgia, but the chosen site of his livelihood, the place where he worked to provide for his family:

> Railway buffs are called '*Tecchan*' ['train geeks'] but (following that logic) I never wanted to be a "historic building buff." Sure there's value to historic buildings, but you can't bring back a town's...what would you say—spirit? energy? You can't bring back a town's vitality unless you put these old buildings to practical use. On the other hand, if a town doesn't regain some vitality, these types of old buildings won't survive. So it's a 'chicken-or-egg?' type of argument. Without increasing a town's potential, there won't be any reserve funds. And without those funds, a town won't have the financial capacity to preserve its old buildings or environment. But saying that you can seek economic development by leaving these buildings intact? You see what a messy debate this was, a kind of Mebius circle. We went around and around in circles for eleven years... {...} I'm not a history geek and I don't have some kind of mania for old buildings. In short, I wanted this city to become a place where the people who were raised here, in every age, could continue to lead lively and energetic lives, here in this same city. The canal—and the idea of tourism—just provided [us with] the opening [to elaborate this goal].
>
> I would say that every citizen has his own vision for the city's future. Citizens don't want to entrust these visions to legislators or the business world—they have the right to self-determination when it comes to their own city. It's when you have citizens like this that citizen's autonomy can be established, without reliance upon the [local] administration or business. So the movement started amid doubts about whether there really was a 'civil society' in Japan. From that perspective, I think the idea that a bustling commercial district could transmit vitality throughout the city is consistent [with our goals]. Now Yamaguchi (Tamotsu) is a city council member and Sasaki Kyōjirō heads the (Otaru) Tourism Association, where they offer the best [local] information in the tourist industry. We don't have a mania for old buildings, or for the canal, but perhaps we have a kind of mania for this city—we all wondered how we could invigorate this city, make it prosper, and ensure its continued existence in the future. And that's why we all concentrated on the term *machizukuri,* or 'community development.'"
>
> Interview with Ogawara Tadashi in Otaru, September 12, 2011. Information in parentheses has been added by the author, information in brackets by the translator.

"Historic building buff." "History geek." "Canal mania." Ogawara's successive variations on this term convey his robust denial of a particular aesthetic disposition, which stems from a nostalgic obsession with historic buildings or a particular love for obsolete, unfashionable landscapes. A chief refrain is Ogawara's insistence that he is not some sort of historic building buff, but rather someone determined to increase the potential of his city, for which historic preservation, and tourism, represent a new approach. Moreover, because the movement sought "the right to self-determination when it came to their own city," "without reliance upon the [local] administration or business," the lofty goal of community development, was one "we all concentrated on," and gave consistency to everyone's actions. Thus, Ogawara's interview reveals both positive and negative NOS factors.

How do we explain Ogawara's emphasis upon preservation? What was his logic for preservation? Given the paramount importance of this question, let us turn to Ogawara's own words on the subject:

> You know, there's something called the 'Yufuin model.' With a reception area in the center, and guest rooms here—detached rooms—all around, like this (*Ogawara uses his hands to demonstrate the layout*). [Once that model of inn was developed] any hot spring resort across Japan with a bit of money abruptly tore down its huge main building and built an inn in the Yufuin style on the same site. They did this because they would be able to charge forty thousand yen per guest. Because the 'big three' resorts in Yufuin, places like Tama no Yu and Kamenoi Bessō, were charging around forty thousand yen a person. So for a pair, for a couple, it would be eighty thousand. {...} You can't imitate a philosophy, although you can copy (the architectural plan of) a central office and detached rooms.
>
> It's easy to understand this if you take a look at the gift shops [at these places]. At [inns like] like Tama no Yu or Kamenoi Bessō, the gift shops feature the very best local products of Ōita, from Yufuin—or from the neighboring city of Beppu at the very farthest. They have been packaged to convey a sense of luxury, a sense that Yufuin is the only place where you could buy such a thing. It's not one of these disorganized, muddled-together kind of souvenir corners. {...} Everyone believes they can improve their own brand if their own product were to be included, and so everyone, including the local people, work hard to make their packaging, and of course whatever is inside, look like the things at Tama no Yu or Kamenoi Bessō. This can elevate [the work of] local industries by leaps and bounds.
>
> In other words, leaders don't simply make money by leading—in this way they also push locals to improve the level of their industries [by acting as curators]. But [not everyone imitates] this system, so you might go to a souvenir shop and find something with a great design, but think 'didn't I see this at some roadside kiosk [selling local products]?' {...} In other words, Japanese tourism would copy anything to survive. {...}
>
> I'm not saying there is anything wrong with imitation. After all, in order to cultivate yourself in any way, you have to imitate. But you have to fully understand the essence of what you are imitating, and translate it—so if it's me, Ogawara, doing this, then I would have to translate it into something [distinctively] 'Ogawara's.' {...} I've seen rural [villages] and I've seen the lives of city-dwellers, and having seen such things, I think the town of Otaru has the right foundation to become such a town (a town that can take pride in its originality, like Yufuin). But everyone sees [Otaru's features] as drawbacks. A canal that no longer plays the role it did in a different age. A meaningless port. Old buildings, when there are now [functional but dull] modern high rises. I'd tell them: 'Don't be such idiots! Wait a minute. Take a look at the details.' I'd say: 'You can't find these things anywhere else, and you wouldn't be able to recreate them if you tried.' And: 'You couldn't imitate these places in Sapporo!'"
>
> Interview with Ogawara Tadashi in Otaru, September 12, 2011. Information in parentheses has been added by the author.

Ogawara's detailed explanation of the "Yufuin model" expands his argument that even if one is to copy an original, one must first understand the underlying philosophy, and refine its essence—thus transforming it into something of one's own. Ogawara is also advocating an understanding of Otaru as an authentic town that cannot be found anywhere else—a town that even a place like Sapporo, with all of its resources, would be unable to imitate.

In a conversation with the author and a student, Ogawara also referred to a famous concept developed by the "Hokudai trio": the "educational power of the environment" (*kankyō no kyōiku ryoku*) and touched upon some of the fundamental reasons for his continued interest in the canal, and continued advocacy for its preservation:

> Student: You mentioned "the educational power of the environment" just now. Mr. Ogawara, you've remained involved with [the canal] ever since the time of the preservation movement, and you remember the canal from that time. If you were to compare the canal of that time with the canal as it is today, would you say that there has been a change to its "educational power" or the effect it has on people?
>
> Ogawara: "Oh, that's astute. That's the question I dislike most. What I mean is: the canal today isn't [really] a canal. It's a waterway. And you know, I think the Hokudai (trio) who came up with that term would also hesitate if you were to ask them if what you all came to observe—one half of the old canal, that's to say the canal today with the road alongside it—really fits the description of an 'environment with the power to educate.' The canal that you see now—well, Professor Iida taught architecture in Hokkaido University's engineering department at the time, and this Professor Iida proposed creating the canal you see today as a kind of compromise between the preservation camp and the administration. As I look back now, we pursued total preservation and we lost. But the final result was not the full reclamation of the canal—[we] settled on an intermediate solution. So, for a time after the preservation campaign came to an end, I wrote all sorts of pieces on the subject, before I could really [put my thoughts] in order. I gave one piece a cool title: 'The Eleven Year Campaign to Save the Canal was a Splendid Nothing." I also once used the phrase 'partial-victory, partial-defeat' to describe the movement. But if you ask me if the canal today has the power to educate [people about history], my answer to that is no, not at all. What I mean is that there's no philosophy behind it; it simply represents a compromise between the preservationist faction and the faction who wanted to pave it over. {…}
>
> Student: With the passage of time, can it return? Can the educational power of the environment return?
>
> Ogawara: No. Never. The raw materials are different and all that. You know, I'm not one for writing poetry, but when I [first] returned to Otaru I did write one poem. It was a poem about how 'the canal and historic buildings have grown up from the ground, from the roots.' The point is, I'm not sure how [closely] you looked at the canal or the stone warehouses, but they were not created with a bulldozer, or a crane. Everything was built by humans, with human strength alone. That's the only reason that the stone warehouses deserve [their name]. That's why the space, the buildings, feel as if it they grew up [naturally] from underground roots. We refused to tolerate modern construction, construction tools; we rejected them. Although that may be putting it a bit too strongly. {…} An environment doesn't have "historical educational power" [just] because it has granite kerbstones—the kind of kerbstones that high heels can get caught in.[76] {…} That is, I don't believe that construction today is the kind of construction that can last hundreds of years. That's the significance of Otaru as a town that can't be copied, as I was talking about earlier. A city with energy and assets, like Sapporo, would have no problem constructing stone warehouses like those in Otaru. But they would be mere copies, not the stone warehouses that line the canal. And even if they

[76]Ogawara's point is that high heeled shoes would not get caught between the old stones of the original streets and alleyways, but only when the streets are new, and made of imitation materials.

> created stand-alone replicas, they could never replicate the way that these [buildings] have existed together as a group, like I was talking about—as an area. {…} The canal would have to return to its original state to have that kind of [authentic] status. That is to say, it would have to measure forty meters [in width], not the twenty meters that remain today.
>
> Author: You're saying that it was forty meters out of necessity.
>
> Ogawara: Yes.
>
> Author: To allow barges to pull up, loaded with cargo.
>
> Ogawara: Loaded [with cargo]. Of course, this made economic sense for those working at the port. {…} A waterway of forty meters allowed barges to change direction, and approach the quay perpendicularly while still leaving enough space for other barges to pass by. In terms of the scenery, historically that width of forty meters meant both sides of the canal were reflected in the water. The buildings on both sides. The warehouses along the mountain side, and the warehouses along the sea side—both sides were reflected in the water. But [now that the canal is only] twenty meters, the buildings on the mountain side are [no longer] reflected. Because that side [of the canal] is now a road. What this means is that the ridgeline of the mountains is no longer visible. From the canal, you can no longer see the landscape of the city extending toward the mountains, or the ridgeline of the mountains. I mean, you can't see the town [of Otaru]. So architecturally—in terms of grand design—there are also these problems. Taking all these things into account, I do think there is this concept of the educational power of the environment.
>
> Interview with Ogawara Tadashi in Otaru, September 12, 2011. Information in parentheses has been added by the author, information in brackets by the translator.

Ogawara sees "a space…that feels like it grew up naturally, from underground roots," as the source of "the significance of Otaru…a town that can't be copied." His lack of hesitation in replying "No, never" when asked if the canal today retains the educational power of its specific environment identifies him as LOC positive.

What is Ogawara's "logic of preservation"—in other words, what is Ogawara's tolerance for change (the CNG variable)? Ogawara spoke fervently on the subject while referencing the issues of the launch of Mycal Otaru and the reuse of the Temiya Line:

> [Speaking not] merely as Ogawara the soba shop proprietor, or the activist, or the crusader, but as the Ogawara who lives in Otaru, and as the Ogawara who fought to save the canal—although I don't really make any internal distinction between these different roles anymore—but anyway, speaking as this Ogawara, I don't think it [the supermarket chain Mycal Otaru] suits Otaru at all. It will ruin the image of Otaru. You know, I'm not opposed to everything new. I don't think there's any need to go out of our way to give the light rail that will run on the Temiya line any kind of "retro" appearance. The popular aerodynamic model with really low floors, all glass windows… {…} the most modern, popular model that's prevalent in Europe—that's the light rail I think we should bring to Otaru. Cities will absolutely change, and cities with a future, cities with energy, will be able to accept this fusion of old and new.
>
> Interview with Ogawara Tadashi in Otaru, August 31, 1999. Information in parentheses has been added by the author.

Ogawara insists that "I'm not opposed to everything new." Not only did he lack "any inclination" to "go out of our way" to give a "retro look" to the light rail he hoped to bring to the Temiya line, but he wanted to introduce the type of light rail that was "the most modern" and "prevalent in Europe." Clearly, Ogawara's approach to preservation is not rooted in a nostalgia for the good old days. But how exactly did Ogawara think old and new should come together?

> Yes, cities change. I think cities are living things. They can't be suspended [in time]. Both in a good sense and a bad sense. Cities change steadily, and people change as well with each generation. So change in itself is not the problem. The more important [question] is how actors get involved to bring about change. If no one is involved, would [the city] just degrade? Does resistance maintain the status quo, even if it can't really change anything? I'm not sure. And then there are cases where people are still working to achieve change. I don't think there's such a thing as an unchanging city. Not even Kyoto. {...} Cities keep on changing. They have to change. But can they really be changed according to a certain philosophy? I mean, going back to what I was talking about earlier, can individuals change a place according to their vision of the city's future? I wish people would have that kind of debate. It's good to have both opposition and agreement. But when there's just a fierce debate between opposing sides, the two camps put the future of their city on the line. I think this was the case with the canal—perhaps at a different level. So, usually—well, take the case of siting a nuclear plant. You have a group opposing the plant and another camp that wants the plant. And it [always] ends in a resounding victory for the camp that wants the nuclear plant. And the people who opposed it can never rise to the top in that town again. They're crushed. But the movement to save the canal wasn't destroyed. I believe we campaigned in a certain fashion so that, even if we lost, we couldn't be wiped out entirely. And that's why community development lives on, even today.
>
> Interview with Ogawara Tadashi in Otaru, September 12, 2011. Information in parentheses has been added by the author, information in brackets by the translator.

Ogawara states: "Cities keep on changing. They have to change. But can they really be changed according to a certain philosophy?" And: "Change in itself is not the problem. The more important [question] is how actors get involved to bring about change." By focusing on methods, Ogawara suggests that even inevitable changes need not be accepted unconditionally; the ideas informing change and the role of various actors in its execution should also be scrutinized.

It is this type of attitude to change that gives the community development group both a positive and negative CNG orientation. Ogawara saw the space surrounding the diminished canal as "just a compromise between the preservationist faction and the faction who wanted to pave it over," without "any philosophy behind it." His sustained involvement with the canal represents a continued interrogation of "how actors get involved to bring about change." It is also how Ogawara settles his debt to Otaru's "community development" efforts, which "survive even today."

The community development group thus retained a certain nostalgic outlook. Ogawara's nostalgia centers upon the town in which he was born and raised. Yamaguchi's nostalgia is for that "first impression" of Otaru on a snowy winter evening, which prompted his realization that "there were still some good sights to be seen in Japan." In that sense, their NOS orientation can be described as positive. What distinguished the community development group, however, was their ability to shake off this nostalgia in order to pursue regional revitalization—and their keen sense of strategy. Ogawara's refusal to be identified as a "building buff" with a "mania for the canal" underlines the fact that the group's NOS orientation could also be identified as negative.

The same applies to the CNG and LOC variables. While the community development group sought to preserve the canal in its original form, this desire did not stem from an attachment to old things, as it did for the aesthetic group. Rather, it was rooted in their understanding of "the canal as an original place that seemed to grow from underground roots," and their goal of self-determination. Their conception of self-determination emphasized locality and a flexible approach to coping with change. Their LOC orientation is thus positive, while their CNG orientation is both positive and negative. Their POA orientation is negative, as evidenced by the community development group's use of political negotiation tactics to realize their goals, which included the threat of a recall effort against Mayor Shimura.

5.4.4 The Traditional Leftists

The fourth and final group within the movement to save the canal can be described as the traditional left-wing activists. As readers may well consider the terms "traditional" and "left-wing" to be antithetical, this description demands some explanation. Put simply, these activists adopted the traditional approach of postwar leftist movements when fighting to save the canal. The term "traditional" does not connote the deep respect for tradition typically associated with a more conservative value system. Rather, while drawing freely from the leftist lexicon, these activists nevertheless pursued their goals within the pre-established framework of left-wing activism. It is in this sense that they can be described as "traditional."

More specifically, the thoughts and actions of this fourth group are distinguished by their identification of the canal issue as just one front in a larger "battle" against the government and bureaucracy. Put differently, this group did not ascribe any particular significance to the preservation of the canal in and of itself. The battle over the canal's fate simply presented them with a concrete opportunity to challenge established authority. This explanation is supported by their subsequent actions: as the canal's fate was replaced by more controversial issues, such as the presence of U.S. aircraft carriers carrying nuclear warheads in Japanese ports and the question of nuclear power plants in neighboring regions, these activists shifted their focus accordingly.

Given that they saw the canal issue as nothing more than an opportunity to display an anti-establishment stance, it was not surprising that this group remained outside the mainstream of the canal preservation campaign. After all, their motivations proved entirely incompatible with the determination of other activists to avoid the establishment/anti-establishment framework and rely entirely upon the intrinsic and unique landscape of Otaru to make their case for saving the canal. The general aversion to a "red threat" in the conservative, provincial society of Otaru also sidelined the preservation movement's "traditional leftists." In the final stages of the fight to save the canal, their decision to skip an internal vote and decision-making protocol and unilaterally pursue an unsuccessful mayoral recall effort represented a last-ditch effort to wrest back control of the movement.

Anti-Authority Community Development: A Life History of Noro Sadao

Noro Sadao was born in Otaru, in September 1950. He grew up in the Masakae-chō neighborhood in the southeastern corner of the city, an area lined with small workshops and factories. Noro's father worked as a laborer in one such factory. Noro began working at city hall in 1974 and remained a city employee until his retirement. There appear to have been just two city employees who crossed lines to join the canal preservation movement. Noro was one of them. While this fact alone makes him an intriguing figure, the manner in which Noro involved himself with the preservation movement, and his stance on the canal issue, also demand close attention.

Let us begin with a significant turning point in Noro's life—well before Noro became involved with the canal:

> It was in my third year of high school that I first developed a way of thinking about and looking [critically] at society… Up until then I was at a local…university preparatory high school, and in my first two years there I belonged to the tennis club, playing hard tennis. {…} And because this was a prep school, all the second year students quit the tennis club after the national [tennis] championships, to give way to the underclassmen [and focus on their university entrance examinations]. So from the autumn of my second year in high school I had nothing to do…I became quite aimless. A classmate of mine became involved with protests against the war in Vietnam through the Beheiren [Citizen's League for Peace in Vietnam], which was this national-level movement completely unrelated to the more sectarian groups—to factions like the Chūkaku or the Kakumaru. It was a nonsectarian citizens' movement. Well, you could blame this on my youth, but I had nothing to do, so I accepted an invitation to attend a rally being held that day in Otaru. I didn't attend with any purpose in mind, but simply because the opportunity presented itself. I was shocked by the gap between the consciousness of other rally participants and my own view of society. To be honest, I still can't remember what they talked about [at that rally], but what I do remember, even now, is my feeling that even though I was living through the same times, I wasn't discerning [what was going on]. You could say the rally revealed to me the shallowness of my understanding. [It revealed] new truths, a new reality. And after that my view of life and society changed completely. Even though I had been a huge sports nut up until that point with no occasion to read books—I'm embarrassed to admit it, but up to that point, I had only read things like *The Gratitude of the Crane* and *Old Tales of Japan*. {…} So from the latter half of my second year of high school and into my third, I changed completely. You know, even now I get chills thinking about what I would have been doing all this time if it had not been for that chance encounter.
>
> Interview with Noro Sadao in Otaru, September 4, 1999. Information in parentheses has been added by the author, information in brackets by the translator.

As a student at an elite Otaru high school, Noro was "a sports nut" devoted to tennis, with "no occasion to read." In an unexpected turn of events, he happened to attend a Beheiren rally "held…in Otaru." There Noro felt keenly the inadequacy of his understanding of the world. The youthful Noro was shocked to discover that "even though" he was "living through the same times," his "view of society" could be so different. His view of life, and society, was "changed completely" by the experience. Noro quickly became a bookish young man:

> After that, the time to study for university entrance exams approached and I was motivated by my friends, but I ignored those studies and instead read everything I could lay my hands

> on—all the books I had never read before. I was on my way toward becoming a kind of pseudo young literary enthusiast. And as a result, although I did study for the exams, I failed on my first attempt… {…} So I had to wait a year to take the examinations again.
>
> Interview with Noro Sadao in Otaru, September 4, 1999. Information in parentheses has been added by the author, information in brackets by the translator.

Following the footsteps of many, a bookish youth, Noro failed his first attempt to gain acceptance to the university and embarked upon an additional year of study. This year of additional preparation, however, was not entirely devoted to his studies:

> (Now that you mention) my *rōnin* year, well, (ultimately) I entered Hokkaido University, but {…} during the year before I entered [as a university student]—the year after I graduated from high school—I spent more time commuting to Hokkaido University for (Beheiren activities) than I did attending cram school. {…} At that time, first and second year university students belonged to the liberal arts faculty. This was located on the north end of the Hokkaido University campus, and it was commonplace for certain sectarian groups [within the student movement]…. to occupy many of the classrooms there. {…} I didn't really have any desire to affiliate with one of these established sectarian groups, or wear those helmets and colored hats… {…} There was an "International Anti-War Day…" This was a big deal for the new left across the country—a big event for that era, and for the Beheiren, it was the day of a big demonstration. Even in Japan it was referred to as 10.21. And on that day…I collided with someone while being chased by the riot police, and fractured a bone. After that I withdrew from the front lines, so to speak, and finally started preparing for my entrance examinations.
>
> Interview with Noro Sadao in Otaru, September 4, 1999. Information in parentheses has been added by the author, information in brackets by the translator.

During Noro's additional year of study, which was ostensibly devoted to preparing to retake his university entrance examinations, he nevertheless frequented the base of the student movement within the Hokkaido University campus. It was only after he sustained a bone fracture while fleeing the riot police that he "finally started preparing for my entrance examinations." The following spring, Noro gained admission to Hokkaido University. He reports that he did not involve himself with the student movement while at university. Noro had hoped to work at a publishing house after graduation, but was disappointed both in this aspiration and his attempt to attend graduate school. Noro ended up taking the examination to work at Otaru City Hall, and in 1974 became a municipal civil servant. Noro joined the Otaru city government just as the nascent movement to save the Otaru canal began to draw notice.

Noro says that he initially found labor union activities within city hall more interesting than the canal:

> After starting university, I stayed away from political movements—and from the student movement in particular. I graduated from university and immediately joined the Otaru city government, and that is when my interests turned to union activities, which I continued all the way through [my time there]. Some old die-hards from the student movement started working at Otaru City Hall—people I'd met before. Some had started before I did, and others came later. And they started a "modern history research society."[77] This was a cobbled-together sort of name but…in spirit it opposed the enemies of the Japan Socialist Party…and the

[77] Within the leftist lexicon of the late 1970s, the term "research group" (*kenkyūkai)* was intended to give a neutral guise to what is in fact a minority group dedicated to developing a highly politicized

> Japan Communist Party, and the anti-union forces at the time. Back then the Japan Socialist Party controlled the unions, so it also opposed the union leadership. {...} It was that kind of thing, handing out flyers and naturally...we went after both, we went after the powers that be in city hall. We saw everyone as enemies, and put out a newspaper to share our opinions with union members {...} There were only six or seven of us and, yes, everyone gave us the cold shoulder {...}"
>
> Interview with Noro Sadao in Otaru, September 4, 1999. Information in parentheses has been added by the author, information in brackets by the translator.

As a newly minted city employee, Noro joined the "modern history research society" and "went after" both the labor union leadership and the "powers that be" within city hall. His group "saw everyone as enemies and put out a newspaper to share our opinions with union members." Looking back at that time, Noro offers the following assessment of his younger self:

> Well, (I) was very far from being an earnest city employee. Although that's a strange thing to say about oneself.
>
> Interview with Noro Sadao in Otaru, September 12, 2007. Information in parentheses has been added by the author.

Noro recalls overtly conducting union business during working hours and seeking compensation from management when his eyeglasses were broken by a door that had not been properly maintained (Interview with Noro Sadao on September 12, 2007). Noro's lack of aversion to conflict and self-advocacy suggest a negative POA orientation. In Noro's own words:

> Hmm. Well, I suppose I knew I wasn't doing anything wrong. Or I had this strange strength. I was full of confidence.
>
> Interview with Noro Sadao in Otaru on September 4, 1999.

"Full of confidence" that "I wasn't doing anything wrong," Noro showed little concern for the repercussions of his behavior or their impact on his chances for promotion. Noro did not try to disguise his actions, but undertook them in plain sight—much as one would undertake a crime of conscience. In that sense, he was "very far from being an earnest [city] employee." Unsurprisingly, Noro's career trajectory at city hall was not the typical one.

> I spent eight years in the property tax division. {...} I started in 1974, in a group of twenty new hires, but no one else spent eight years in the same office. Usually one would spend between four or five years in one place {...} From the point of view of the authorities at city hall, my involvement in union activities or the modern history study group meant that I was a really outrageous sort. {...} That's probably why I wasn't transferred [to other divisions within city hall]. I was too dangerous.
>
> Interview with Noro Sadao in Otaru, September 4, 1999. Information in parentheses has been added by the author, information in brackets by the translator.

---

critique of the status quo, including the city government and the unions, which at this time were aligned with the JSP.

Noro was not bothered in the least at having been branded a "dangerous" sort who could not be safely "transferred" to other divisions; on the contrary, it made him all the more inclined to do as he pleased:

> Over time you could say that I became dissatisfied with both union activities and the modern history study group, and the anti-nuclear movement that started up around 1980 felt completely different from my own instincts. For me, this was right in the middle of my twenties…and, well, at that time there was something called the 'ten feet campaign…' {…} in which individuals donated money to buy [back from the U.S. the footage of Hiroshima and Nagasaki after the atomic bombings] to make their own anti-nuclear film. {…} This 'ten feet campaign' emerged, and their actions resonated with me. But the other members of the modern history study group disavowed the 'ten feet campaign' as a kind of 'sham citizens' movement.' I was a member of the study group, but after that I distanced myself from it somewhat, and grew more involved with the anti-nuclear movement, which was at a remove from both the unions and my workplace. I started attending anti-nuclear campaigns in town, and viewings of [anti-nuclear themed] movies.
>
> Interview with Noro Sadao in Otaru, September 4, 1999. Information in parentheses has been added by the author, information in brackets by the translator.

Noro now felt unease about not only union activities, but also the study group. He took part in organizing the "buy ten feet" campaign, and in order to organize a viewing of the film he grew more involved with the regional community, "which was at a remove from the unions and my workplace." This coincided with the impending arrival of the U.S.S. warship *Blue Ridge* at a Japanese port, an increasingly controversial development for Japanese society. Noro threw himself into the movement with fervor:

> I had just become involved with the [ten feet movement], and in the midst of that, the *Blue Ridge*, which was the flagship of the Seventh Fleet, an American warship that was suspected—strongly suspected—of carrying a nuclear arsenal on board, was set to arrive at the port of Otaru. Now, I was challenging the status quo right and left, but at that time [Japanese] citizens also had a very strong level of interest in the nuclear issue {…} We held a protest right in front of the warship that was written up in the newspapers. Apparently this was very poorly received within city hall.
>
> Interview with Noro Sadao in Otaru, September 12, 2007. Information in parentheses has been added by the author, information in brackets by the translator.

This episode underscores Noro's indifference to his reputation at city hall. It also led to his introduction to the Otaru canal preservation campaign:

> After I'd been doing [this kind of protest] for about a year or so, I happened to meet someone involved in the campaign to save the canal. Now, I'm an Otaru native, but to be honest I had very little opportunity to visit, or look at, the Otaru canal. And my birthplace was a place called Masakae-chō, in what you could describe as the suburbs of Otaru. And, well, if you take a look at a map of Otaru later…in terms of north, east, south, and west [*gesturing with his hands*], it's further away from here, or in terms of another distant place, it's at the foot of Chōryō High School.
>
> Interview with Noro Sadao in Otaru, September 4, 1999. Information in parentheses has been added by the author, information in brackets by the translator.

Noro became involved with the canal preservation effort because he "happened to meet someone from the campaign to save the canal," although, surprisingly enough, he says he had virtually no previous encounters with the canal. To be sure, Masakae-chō is faraway from central Otaru, where the canal is located. However, this physical distance was not the only reason for Noro's lack of familiarity with the canal:

> To be honest, in the beginning I didn't have much concern, or interest, in preserving the canal. {...} It wasn't just me. The majority of Otaru residents saw the canal as a nuisance—as an out-of-date relic.
>
> Interview with Noro Sadao in Otaru, September 12, 2007. Information in parentheses has been added by the author, information in brackets by the translator.

Because he "saw the canal as a nuisance" and "an out-of-date relic," the preservation issue initially aroused little "interest or concern" in Noro or most of his fellow citizens. He only became involved with the issue in the final five years of the canal war:

> ...I became involved with the canal preservation movement later on. I think I was involved with the movement for around five years in total. {...} The city issued a policy to reclaim the entire canal {...} and build a new road called the Rinkōsen. Well, it was the administration's duty to promote this [plan]. That someone who belonged to the side promoting this policy would try to save the canal—well, no matter how [good an employee I was, that raised eyebrows]. My immediate colleagues didn't seem to take much notice, but the city leadership at that time apparently took quite a bit of interest in just what kind of person (this Noro) was. I myself was relatively unconcerned about it.
>
> Interview with Noro Sadao in Otaru, September 12, 2007. Information in parentheses has been added by the author, information in brackets by the translator.

True to form, Noro continued to act as a sort of "conscientious objector" within the city administration, and candidly acknowledges that he remained "unconcerned" by the increasingly vigilant gaze of his superiors. Noro also says that he did not face any overt discrimination because of his affiliation with the canal campaign:

> Through the movement, I became involved with the canal. And there was a woman whose name came up a minute ago—Ms. Kumagai (Yōko). In fact, this Ms. Kumagai [started] working at city hall in the spring of that year. But I had no idea that she was working for the city. It was only later, at some gathering, that I found out she worked at city hall. I was a bit surprised, because I had always figured that I was the only [city employee]. It was reassuring to learn there happened to be someone else from the city [within the movement]. And {...} regarding [your] question about my position inside city hall, and how I saw my position and how I was seen by others {...} There wasn't any backbiting. The fact that I was part of (the preservation movement) didn't mean that people around me at work refused to speak to me or anything like that. But, as you know, the mayor himself was pushing for the canal's reclamation, and naturally he was backed by the entire management [level]. So there was no way I could discuss the canal issue at work. I never made it a conversation topic, or—how can I put it? While there wasn't any sort of gag order in place, there was just an atmosphere within city hall [that prevented me from talking about it].
>
> Interview with Noro Sadao in Otaru, September 4, 1999. Information in parentheses has been added by the author, information in brackets by the translator.

Noro, who "had always figured that I was the only" city employee involved in the canal preservation movement, admits he was "surprised" to learn that Kumagai Yōko also worked at city hall, and found it "reassuring to learn that there happened to be someone else from city government" within the preservation movement.

But why should Noro, who admits he was "indifferent" to the fate of the Otaru canal, become a regular participant in the effort to save it? Put differently, what rationale did Noro find for saving the canal? Noro's recollection of one particular episode provides us with the answer:

> At that time we launched a clean-up effort in the area around the canal. Now, the city government wanted to get rid of the canal, so of course no effort had been made to keep the area clean. There were weeds growing everywhere, and mounds of garbage. And since we were the ones mounting this appeal to save the canal, we decided we should start off by doing whatever we could, straight away. A clean-up was something we could do, and of course the general public was invited to join in. {...}
>
> (The clean-up) started in the afternoon, and we kept at it for quite a long time. It was a Sunday. {...} Now, by October or November, the sun starts setting pretty early here (in Otaru). And so, at around 4 p.m., the setting sun starts to give off this evening glow. I had been invited to join [the clean-up] and I was there almost entirely out of a sense of obligation—because I had been asked, and because I knew they didn't have many people [helping]. It felt like the right thing to do. [Members of the preservation movement] had cooperated with me in the anti-nuclear campaign, and so it simply felt like a kind of reciprocal duty. Anyway, I was working away pretty zealously—I liked cutting back the grass, although picking up trash was less enjoyable—and suddenly, at around 4 p.m., from behind the Hokkaiseikan [warehouse] I could see the sun starting to sink behind the mountains in the direction of Temiya. It wasn't quite an afterglow, but the sky was reddish and dusky, like twilight. I glanced up and took in this view. Around me, everyone was working away. I was just taking a short rest. I can still remember that view now. This might sound affected, but it touched my heart. {...} At that moment, the canal was right in front of me, with the Hokkaiseikan building to my right and the mountains behind. And the mountains were covered in this red glow, and even the Hokkaiseikan building was tinged with red light.
>
> As I looked out over all this, I decided that this type of scene was worth saving. There was something sacred about it, and I felt that Otaru absolutely had to retain this kind of scenery. It wasn't that I had these impure [economic] motivations, thinking that [this kind of scene] could be used later to promote tourism. No, it was more that I felt we should be able to live in a town with the slow pace of life that this scene conveyed. {...} And that was the real reason why I grew more deeply involved with the canal issue—more so than with the anti-nuclear movement... Nowadays we talk about "the educational power (of the environment)," but it was then that I first realized—or was made to realize—the impact of that landscape.
>
> Interview with Noro Sadao in Otaru, September 4, 1999. Information in parentheses has been added by the author, information in brackets by the translator.

Noro joined the canal clean-up effort out of a sense of "reciprocal obligation" to those who had "cooperated with me in the anti-nuclear campaign," joining in "because I was asked, and because I knew they didn't have many people." When he took a brief rest from his labors, he looked up—and discovered a scene that "touched my heart," and was "worth saving." Indeed, Noro goes even further, declaring "Otaru absolutely had to retain this kind of scenery." In this passage, Noro recalls the precise moment at which the "educational power of the environment" touched a previously "unconcerned" Otaru resident. Yet Noro's recollections do not reflect a nostalgic sensibility. Let us turn to another portion of his narrative:

> ...You could blame this on my youth, but I was late (to the preservation movement), and didn't necessarily join in with a clear understanding of preservation, or the value of the canal. I had only shifted to a softer, more impressionistic feeling, as I just described, that this type of scenery should be saved. So I acquired the theory behind the movement, shall we say, or a clearer understanding [of the issues] one by one, in the course of conversation. A lecture series on the Otaru canal was held just around then, and even though I didn't attend all that often, when I did I learned [from the various speakers] that the canal issue was also an environmental issue, and that there were many different themes to the movement. And the [speakers] took up these themes one by one and simplified them to explain why we had to fight to save the canal—I think being able to make sense of these issues gave us the encouragement, or confidence [to fight for the canal].
>
> Interview with Noro Sadao in Otaru, September 12, 2007. Information in parentheses has been added by the author, information in brackets by the translator.

As Noro "didn't necessarily join in with a clear understanding," he strived to simplify the various issues "one by one," and master the argument for preserving the canal. This is quite different from Ogawara Tadashi's perception of the canal as a space that seemed "to grow up from [underground] roots." Meanwhile, Noro's explanation of his participation in [the movement] with reference to "my youth," and a shift to a "softer, more impressionistic" [response to the canal] lacks the sense of urgency that defined Mineyama Fumi's view of the canal as a place of decisive importance entity for Otaru. For these reasons, Noro's LOC and NOS orientations are both negative.

The most important thing for Noro was rather the enjoyment he derived from involving himself with current issues:

> To put it simply, I felt very strongly that 'time was running out,' and that 'the canal will definitely disappear if we don't act now.' Also—while this doesn't express it adequately—I also felt the excitement of enlisting myself in this cause, where things changed hour to hour. Of course I wasn't changing the world all by myself, but for a brief moment I had what you might call a sense of being in the moment—of being part of something at that moment.
>
> Interview with Noro Sadao in Otaru, September 12, 2007. Information in parentheses has been added by the author, information in brackets by the translator.

Naturally enough, Noro's involvement in the preservation movement convinced him that "if we don't act now, the canal will definitely disappear," and inspired him to fight all the harder. Significantly, Noro also describes "the excitement of enlisting myself in this cause." Noro's enjoyment of being a "part of" current issues and "sense of being in the moment" led him to participate in the final five years of the effort to save the Otaru canal, but also in protests against the docking of the U.S.S. *Blue Ridge* in Otaru, and other anti-nuclear demonstrations (Interview with Noro Sadao in Otaru, September 12, 2007). Noro's disposition—namely, his ability to derive pleasure from participating in the issues of the day—stands in striking contrast to that of Mineyama, who remained single-mindedly fixated on the canal and its preservation. Noro's remarks that "Otaru absolutely had to retain this sort of scenery," and his "strong sense" that "I was doing nothing wrong," indicates his low tolerance for change. In other words, his CNG orientation is negative.

As internal differences fissured the movement in its final stages, Noro distanced himself from the preservation effort:

> When I was thirty-six or thirty-seven, the canal preservation campaign had largely come to an end, and I myself had lost any desire to participate in a movement that had become so divided. And the history group, (I mean) the modern history study group, had clearly differentiated its activities from those of the canal preservation movement... {...} Because I intentionally avoided joining the union leadership, there wasn't anyone I was particularly close to there. At that time, both the study group and the union refrained from commenting on the canal issue, although the union changed a bit at the end. The modern history study group saw the canal preservation campaign in the same light as the anti-nuclear movement—as a pseudo citizens' movement. I was even told they would kick out anyone who was got too involved with it. But the study group had only six or seven members, so any talk of expulsion was ridiculous.
>
> Interview with Noro Sadao in Otaru, September 4, 1999. Information in parentheses has been added by the author, information in brackets by the translator.

By Noro's account, not only had he distanced himself from the canal issue, but he had been told by his study group that expulsion awaited "anyone actively involved with the 'sham citizens' movement" to save the canal. All that remained for Noro was his work at city hall.

> By the time I was thirty-six or seven, I had calmed down about both the anti-nuclear movement and the canal, and when I wondered what else I could do, it turned out I had nothing left but my own work. {...} It was then...that I started thinking about how, if I was bound to my work for eight hours [a day], I could work in my own way. {...} For me, personally, I was entering my late thirties, with three children... So, in my own way, I was thinking about the future. I had to ask myself what I was doing, and whether it was okay to simply carry on as before.
>
> Interview with Noro Sadao in Otaru, September 4, 1999. Information in parentheses has been added by the author, information in brackets by the translator.

Leaving the preservation movement behind, Noro confronted the reality of his own work at city hall. He decided to set his own professional goals, and proceeded to execute them with success:

> You didn't really have to feel any great amount of motivation in order to get by at work. But that didn't feel like any way to live, so I set my own goal: as of that April, there were six hundred cases of arrears [for tax payment]. I decided I would do my best to reduce that number by half. This (*laughing*) was a really tough job, if I say so myself. {...} It wasn't a task that I announced from the start. Without saying anything to my supervisor, I set my own goals and started working to meet them. I decided I would try my best for a year, with the goal of reducing the [case number] by half. I'm the stubborn type when it comes to this kind of thing, so once I set my own target, I stubbornly pursued it. I really worked hard. Even during the summer heat, if someone invited me to take a break at a coffee shop, I'd put them off and say I'd join later. I went about it in a sort of disagreeable way at times, but I was determined. And at the end of the fiscal year, the following March, the books are closed. And when my manager looked over [the accounts], he knew without being told what [I'd done]. {...} One day, after everyone else had left and I happened to be the last to leave the office, he called me over and acknowledged how hard I had worked. That one brief comment [from him] had a big effect. I realized that he had really seen [what I had achieved]. {...} I realized that my job could be pretty interesting, if I pursued it in this way. I don't think I would have worked in the same way on someone else's orders, because I was told to do it. But I realized the tax section could be an interesting place to work if I set my own goals, worked toward

> them in my own way, and produced those kinds of results. And my whole world changed. My approach to work had been so negative up until that point, but…I became someone who [really] worked.
>
> Interview with Noro Sadao in Otaru, September 4, 1999. Information in parentheses has been added by the author, information in brackets by the translator.

Noro was subsequently promoted to head his subsection of the tax department and later rose through the ranks of city government, serving as vice-chairman of the Otaru Literary Museum, section chief within the finance division and the tourism promotion office, and finally, deputy director of the finance division.

Given how easily Noro could have remained a lifelong rank and file employee, this professional transformation was nothing short of miraculous. It stemmed from Noro's realization that his work at the tax division could actually prove interesting. When asked about his involvement with the canal movement during his time in management-level positions with the city, Noro shared the following reflection:

> Ironically, when I arrived at the [tourism promotion office], people by the name of Ogawara (Tadashi) of the soba restaurant Yabuhan and Yamaguchi (Tamotsu) were already getting involved, and had started their own tourism organization. This was called the Tourism Promotion Council, and it no longer exists today. This year (2007) it merged with the Otaru Tourism Association, so it no longer exists [as an independent entity]. This group was started as a kind of extension of the canal preservation movement, although they didn't put it in those terms. They didn't announce that they had created the group as an extension of the canal movement, but in terms of the consciousness—the consciousness of people like Ogawara and Yamaguchi—it was the same. They said the Tourist Promotion Association wasn't addressing the current situation of tourism in Otaru, so they decided to create a group to do so themselves. They decided that the new, future Otaru had no other option but to depend on tourism. It wasn't that they had any inherent interest in tourism. And they weren't happy about the new canal—the canal that they had fought to save. The canal that had been sliced in half. So in a sense, it was ironic that the very people who had campaigned to save the canal later got involved with tourism. And it was another irony of history that I ended supporting them in my position as the chief of the city's tourism section. So now that we are promoting what remains of the canal as a symbol of [Otaru] tourism, well, when we get together the conversation can turn bitter. We ask each other: 'How is it that we're doing this kind of thing?'
>
> Interview with Noro Sadao in Otaru, September 12, 2007. Information in parentheses has been added by the author, information in brackets by the translator.

Did Noro abstain from further activism? During his tenure as vice-chairman of the Otaru Literary Museum, he responded to a question from a student of the author as follows:

> Noro: The relationships I forged with people I met in the course of my activism. These are assets, even today.
>
> {…}
>
> Student: Mr. Noro, are you involved in any sort of movement now? Community development or some different (type of activism), outside of your work?
>
> Noro: No, I'm not. Definitely not. But I'm still inextricably linked to the people who are involved, and I still talk with them. We're still connected. I'm not directly involved with

> community development efforts, to be sure, but I'm interested in them. And this might make me sound self-important, but even though it was only by chance that I was transferred to this Literary Museum and art museum, I do think of whatever progress I can achieve here as my own contribution to community development.
>
> Interview with Noro Sadao in Otaru, September 4, 1999. Information in parentheses has been added by the author, information in brackets by the translator.

Noro may have prized the relationships he formed with people during the course of the canal preservation campaign, but he did not participate in any subsequent citizens' movement. On the contrary, he saw his work at the Otaru Literary Museum and, later, at the economic revitalization unit within the city's finance division, as "my own contribution to community development."

Inspired by the social movements he first encountered as a high school student, Noro went on to participate in the labor union movement and other forms of activism that addressed the issues of the day. His participation in the fight to save the Otaru canal can be understood as just one more fight that Noro might explain with a shrug and the suggestion that we "blame it on my youth."

5.5 The Logic of Preservation and the Structure of Activism

We have reviewed the life histories of representative activists and their various logics for protecting the canal. What conclusions can we now draw from their efforts to save the canal, and from the actual results of the preservation movement? Let us conclude this chapter by reviewing several key points.

First, how did the canal preservation movement unfold?

The campaign to save the Otaru canal was sparked by the demolition of the Arihoro warehouses. The original aim of the preservation movement was to save the canal in its entirety.

This goal was only natural. After all, the movement was the product of residents' deep affection for their hometown, and their sense of crisis that "Otaru would no longer be Otaru" if the canal was reclaimed and paved over.

Yet the goal of saving the canal in its entirety was also a refutation of change. If this goal was realized, the canal would remain frozen in time. Nothing would change. Nothing *could* change. There could be no new economic activity, and the full potential of the surrounding area would remain unrealized. The earliest efforts to save the canal did not elicit support from Otaru's business community or most local residents precisely because "you can't make a living off of a canal" suspended in time.

Enter a new approach to preservation: the idea of "canal-based community development." According to advocates of this approach, the canal was not a mere relic from Otaru's past, but also a guide to its future. They sought to break the impasse created by the earlier goal of "cryopreservation," and reimagined the canal as a community resource with the potential to stimulate local tourism. As we have seen, this new "community development" approach, which remained predicated upon preservation but tolerated some forms of change and economic activity, elicited tremendous support from Otaru residents. Ultimately, advocates of the community development approach prevailed within the preservation movement. Briefly summarized, the four central tenets of their approach are as follows:

1. The canal should be preserved because it is essential to the identity of Otaru residents.
2. Consequently, changes to the Rinkōsen plan—rather than the complete suspension of the project—should be considered.
3. The canal should be understood as the cornerstone of tourism development, not the site of a new road.
4. The preservation of the canal was a form of "community development" increasingly commonplace in Europe and the U.S. It would not signal Otaru's transformation "into a history museum."

Significantly, this ideological shift "from cryopreservation to community development" was accompanied by changes to the power structure of the preservation movement. In the early days, the movement was led by members of the "aesthetic school" and by "traditional leftists," whose only strategy was to submit petitions and collect signatures in the street. Ultimately, they proved unable to rally the public to their cause. The latter group relied upon the traditional vocabulary of left-wing politics to passionately defend their own righteousness and protest the unjust suppression of their views, and pursued their own form of justice through litigation. Indeed, they perfectly embodied the spirit of Japan's postwar left by embracing a form of direct confrontation that could only end in one of two ways: in "total victory or honorable death."

Later on, however, the "pure preservationists" and "community development" groups assumed leadership of the movement, gaining extensive public support and political bargaining power along the way. The "pure preservationists" successfully appealed to Otaru residents' genuine love for their hometown. Meanwhile, members of the community development group won the support of the local business community by making a more strategic argument: "saving the canal could be an effective vehicle for economic revitalization." We have already seen how these two groups, operating in tandem, brought the preservation campaign to a position of unprecedented strength. What demands our attention here is the fact that while the community development group's insistence upon "preservation" remained constant, their definition of this goal did not. Specifically, over time their idea of preservation changed from the type of "cryopreservation" favored by the aesthetic school and the pure preservationists to something else: namely, a form of "preservation with development" that tolerated a certain level of development while giving local citizens control over change.[78] More concisely, they made an appeal for "social control over change,"[79] which gradually evolved into the paradoxical argument that "to preserve is to change."

[78]We should recall Ogawara Tadashi's description of the preservation movement's goal: "for individual citizens the right to self-determination regarding their own town" September 12, 2011.

[79]Emphasizing *social* control over change here differentiates the control wielded by civil society from the systemic regulations of "the state," "the bureaucracy," "the law," or "the plan."

Despite the confusion created by the continued use of the single word "preservation," this change in definition revolutionized the ideology of preservation. This ideological revolution and the accompanying expansion of the preservation movement, ended up making bitter enemies out of fellow activists, and led to the breakup and collapse of the preservation movement (Table 5.3).

Next, who participated in the campaign to save the Otaru canal? What can be learned from an analysis of the various life histories of prominent activists?

Far from being monolithic, the canal preservation movement was composed of four distinct types of activists who, as we have seen, often jockeyed for power. One thing that is clear from a close examination of the activists associated with each of the four groups is that the preservation campaign cannot be explained solely by activists' "love for their hometown" (Table 5.4).

The desire to save the canal was common to all activists. However, they differed in their reasons for wanting to protect the canal, in their ways of thinking, and their modes of activism. To be sure, representatives of all but the traditional left-wing group displayed a pronounced locality orientation and nostalgic understanding of the canal, demonstrating that love of one's hometown, or similar nostalgic attachments, was a powerful motivation for participating in the movement. However, while Fujimori and Mineyama spoke frequently of their deep attachment to the canal and their irrational desire to save it, Ogawara comes across quite differently. Without disavowing his affection for the canal—a place known to him since childhood—Ogawara nevertheless refutes the label of an "historic building nut" with a "mania" for old ruins, and instead advocates tourism development as a means of realizing Otaru's latent potential. Lumping these various perspectives together under a shared "love for one's hometown" obscures crucial differences. This is also the reason why, despite the importance of the LOC and NOS variables, we must also assess the CNG orientation of each activist to further articulate this concept of "love for one's hometown." The detailed life histories presented in the fourth section of this chapter demonstrate that while all activists loved the city of Otaru, the differing extents to which they tolerated change brought about a significant transition in the internal hegemony of the movement.

These rich narratives also underscore the enormous influence of postwar social movements and the Japanese student movement in particular. Fujimori's insistence that the preservation campaign be carried out as a citizens' movement stemmed from his sense of alienation from existing forms of activism, and his consequent desire to avoid "being used as a tool in a political battle." Meanwhile, Yamaguchi, Ogawara, and Noro deliberately sought out a different course from that of the student movement of which they had all once been a part. The "pure preservationist" Mineyama made no mention of politics or economics and was careful to keep her distance from all the political parties. Even former student activists within the preservation movement were determined to make a fresh start and disavowed any nostalgic attachment to their days as student radicals. Nevertheless, the influence of the Japanese student movement was considerable, if only in a negative sense: Otaru preservation activists took enormous care to forge a different path from that of Japan's radical student movement. Indeed, new modes of activism such as the Port Festival emerged in

Table 5.3 Three stages of the canal issue

	First phase (1973–1976)	Second phase (1977–1984)		Third phase (1985–)
Plan targeted by activists	Initial (Inoue) road plan (Ministry of Construction Bulletin #2912)		Revised (Iida) plan (Hokkaido Bulletin #2361)	
Logic of road construction	The canal as the site of a new road		The canal as the site of a new road and city park	
Participants within the movement	The "aesthetic school" and the "traditional leftists" (a)	The "pure preservationists" and the community development (*machizukuri*) activists (b)	Resistance of (a) to the leadership of (b)	–
Central entity within the preservation movement	APOC	APOC, Yume-machi, Port Festival	APOC, Yume-machi, Committee of One Hundred, Port Festival	Revitalization Committee, Otaru Rebirth Forum
Goal of the preservation movement	Total preservation of the canal ("cryopreservation")	Community development (preservation through adaptive reuse)	(a): total preservation (b): community development	(a): attempt to rebuild a total preservation effort (b): participate in the Revitalization Committee
Method of activism	Petitions and discussions with city officials	Presentation of sophisticated counter-proposals and a debate surrounding the goals of redevelopment	Signature-gathering campaign and mayoral recall effort	–
Logic of preservation	The canal as a symbol of local identity	The canal as a core resource for community development	Canal-based tourism development	Waterfront-based community development
Relationship to political parties	Rejected the involvement of political parties	Rejected the involvement of political parties	Some of the (a) group attempt to use the Japan Communist Party (JCP) to further their cause	–

Table 5.4 The canal preservation movement: four categories of activist

	Locality-oriented (LOC)	Nostalgia-oriented (NOS)	Tolerance of change (CNG)	Political attitude (POA)
The aesthetic school	+	++	–	–
Pure preservationists	+	+	–	+
Community development activists	+	+ –	+ –	–
Traditional leftists	–	–	–	–

Otaru because they remained focused upon the canal as a highly specific, distinctive place, and avoided the abstracted political language of the student movement.[80] In that sense, the Otaru canal preservation movement occupies a precise position within the various historical layers of postwar social movements (Horikawa, 2021). It is only through the various life histories that we gain this perspective.

What was the logic for preservation put forth by activists?

"If the canal is paved over, Otaru will no longer be Otaru." This refrain underscores the fact that the canal was the object of considerable nostalgia. In Fujimori's words, at some irrational level, "Otaru natives just love [the canal]." A symbol of the region, the canal had borne witness to the lives of generations of Otaru residents. It is thus unsurprising that the campaign to save the canal continued for over a decade. Unsurprising that, even in her sixties, Mineyama Fumi threw herself into the preservation effort. Unsurprising that Fujimori Shigeo wept tears of despair as he painted the "Red Canal." For local residents, preserving the canal was a way of protecting a "place" intrinsic to Otaru.

Nevertheless, a cool-headed recognition that change is constant in urban environments prevailed within the preservation movement. As Ogawara Tadashi acknowledged, "cities are always changing. They have to." But unregulated change need not be accepted without question. The preservation movement thus developed a more centrist proposition: Otaru's redevelopment should be rooted in its distinctive history and scenery. The real problem, activists contended, was not change but the chipping away of the distinctive "placeness" of urban environments. Yet urban regeneration *must* stem from such "placeness." We should recall Yamaguchi's assertion that "redevelopment has to be led by citizens," and Ogawara's argument that "change in itself is not the problem. The more important [question] is how actors get involved to bring about change." The ultimate goal of Otaru's preservation activists was thus something very different from what the word "preservation" suggests: namely, "social control over change." Their beloved hometown of Otaru would change. That much

[80]We should recall Ogawara's insistence that the canal and historic buildings "grew up from the ground, from underground roots."

was inevitable. What activists really wanted was to realize this change in the manner, and at the speed, of their own choosing.

References

Akashi, T. (2002). *Jichitai Esunogurafī: Chihō Jichitai ni Okeru Soshiki Henyō to Aratana Shokuinzō* (Law and Society Debut Series 3). Shinshinsha.

Arisue, K. (1992). Seikatsu kenkyū to raifu hisutorī: seikatsushi kenkyū kara. In N. Kawazoe (Ed.), *Seikatsugaku e no Apurōchi* (Seikatsugaku Sōsho), pp. 49–68. Domes Shuppan.

Barthel, D. (1984). *Amana: From pietist sect to American community*. Lincoln, NE: University of Nebraska Press.

Barthel, D. (1996). *Historic preservation: Collective memory and historical identity*. New Brunswick, NJ: Rutgers University Press.

Castells, M. (1983). *The city and the grassroots: A cross-cultural theory of urban social movements*. Berkeley, CA: University of California Press.

Edagawa, K. (1985). Otaru no Merī gō Rando. *Teretopia, 269*, 10–14. NTT.

Horii, K. (2009). *Designare: Furusato otaru o aishita dezainā fujimori shigeo*. Sakuraishi, Nara: Yamatobito Henshūbu.

Horikawa, S. (1989). Otaru unga hozon undō no bunretsu katei: Undō rinen no hensen to tenkai. *Hōgaku Kenkyūka Ronbunshū, 30*, 111–128. Keio University Press.

Horikawa, S. (1990). Sengo nihon ni okeru 'Machinami Hozon' Undō no tenkai: Otaru unga hozon undō o jirei toshite. Keiogijuku Daigaku Daigakuin Hōgaku Kenkyūka 1989-nendo Shūshi Ronbun (A4, 216 pp.).

Horikawa, S. (1994) Chiiki Shakai no Saisei Toshite no Machinami Hozon: Otarushi Saikaihatsu Chiku o Meguru Undō to Gyōsei no Ronri Kōchiku Katei. In Shakai Undōron Kenkyūkai (Ed.), *Shakai Undō no Gendaiteki Isō,* pp. 95–143. Seibundō.

Horikawa, S. (1998). Rekishiteki Kankyō to Chiiki Saisei: Machinami Hozon ni Okeru 'Bashosei' no Sōtenka. In H. Funabashi, & N. Iijima (Eds.), *Kankyō* (Kōza Shakaigaku 12), pp. 103–132. University of Tokyo Press.

Horikawa, S. (2000). Unga Hozon to Kankō Kaihatsu: Otaru ni Okeru Toshi no Shisō. In S. Katagiri (Ed.), *Rekishiteki Kankyō no Shakaigaku* (Shirīzu Kankyō Shakaigaku 3), pp. 107–129. Shinyōsha.

Horikawa, S. (2007). Toshi o Kioku Suru no wa Dara ka: 'Rekishiteki Kankyō Hozon no Shakaigaku' e Mukete no Skecchi. *Testugaku, 117*, 177–218. Keiōgijuku Daigaku Mita Tetsugakukai.

Horikawa, S. (2009). Kichō Hōkoku: Hozon Undō no Keika to Seika. In S.J. Otaru (Ed.), *Otaru Unga to Sekizō Sōkogun no Hozon Undō Kara Nani o Uketsugu ka: Chiiki ni Iki, Chiiki o Mamoru...Machizukuri Undō no Senkusha Mineyama Fumi-shi ga Tsutaeru Koto Shinpojiumu Kaisai Hōkoku*, pp. 9–30. Sapporo: Nihon Kenchikugakkai Hokkaido Shibu (A4, 76 pp.).

Horikawa, S. (2010). Basho to Kūkan no Shakaigaku: Toshi Kūkan no Hozon Undō wa Nani o Imi Suru no ka. *Shakaigaku Hyōron, 60–4,* 517–534.

Horikawa, S. (2012). Ameniti. In M. Ōsawa, S. Yoshimi, & S. Washida (Eds.), *Gendai Shakaigaku Jiten*, p. 26L. Kōbundō.

Horikawa, S. (2021). Do as democracy demands: The irony of historic preservation movement and its relevance in popular sovereignty in postwar Japan. In H. Hardacre, T. George, K. Komamura, & F. Seraphim (Eds.), *Japanese Constitutional Revision and Civic Activism*, pp. 279–289. Lanham, MD: Lexington Books.

Horikawa, S., & Fukaya, N. (Eds.). (2013). *Toshi Gabanansu no Shakaigakuteki Jissho Kenshū (4): 2012-nendo Hosei Daigaku Shakaigakubu Shakai Chōsa Jisshū Hōkokusho.* Hosei Daigaku Shakaigakubu Shakai Chōsha Jisshūshitsu (A4, 194 pp.).

Ishizuka, M. (1975). (Shiryōhen) Intabyū (1974-nendo Hokkaido Daigaku Kōgakubu Sotsugyō Ronbun, Bessatsu Shiryōhen (A4 blueprint, 227 pp.).

Kihara, K. (1982). *Rekishiteki Kankyō: Hozon to Saisei* (Iwanami Shinsho Kiban 216). Iwanami Shoten.

Kihara, K. (1992). *Kurashi no Kankyō o Mamoru: Ameniti to Jūmin Undō* (Asahi Sensho 452). Asahi Shinbunsha.

Kikkawa, T. (2001). *Gakurekishakai no Rōkaru Torakku: Chihō kara no Daigaku Shingaku.* Kyoto: Sekai Shisōsha.

Kikkawa, T. (2001). 'Ryō' kara 'Shitsu' e: Hōhōron no hashi o wataru. *Sociologi, 46–2,* 105–110.

Kitō, S. (1996) *Shizen Hogo o Toinaosu: Kankyō Rinri to Nettowāku* (Chikuma Shinsho 68). Chikuma Shobō.

Kitō, S. (Ed.). (1999). *Kankyō no Yutakasa o Motomete: Rinen to Undō* (Kōza Ningen to Kankyō 12). Shōwadō.

Mineyama, F. (1985). *Haha no Omoide* (B5, 6 pp.). Printed and distributed privately by the author.

Mineyama, F. (1995). *Chiiki ni Ikiru: Otaru Unga to Tomo ni.* Sapporo: Hokkaido Shinbunsha Shuppankyoku.

Miyamaru, Y. (1983). Kokoro ni Kui wa Utarenai, Otaru Unga no 10-nen. *Kankyō Bunka, 60,* 49–59. Kankyō Bunka Kenkyūjo.

Miyano, M. (Ed.). (1975). *Dōdō Rinkōsen, Otaru Unga Fu Otarukō 100-nen no Ayumi.* Sapporo: Hokuensha.

Morihisa, S. (2002). Shōtengai ni yoru Machizukuri no Kokoromi: Machizukuri to Shōtengai Saisei no Gendaiteki Isō (2001nendo Shūshi Ronbun). Hōseidaigaku Daigakuin Shakaigaku Kenkyūka (A4, 69 pp.).

Myōen, K., Adachi, F., Morishita, M., Satō, K., Saeki, S., Yanagida, R., & Ishizuka, M. (1993). Otaru no Rekishiteki Kankyō no Henyō Kōzō ni Kan Suru Kenkyū: Unga, Ironai, Sakaimachi Chiku no Bāi. *Nihon Kenchikugakkai Hokkaido Shibu Kenkyū Hōkokushū, 66,* 529–532.

Nakazawa, H. (2005). *Jūmin Tōhyō Undō to Rōkaru Rejīmu: Niigataken Makimachi to Kongenteki Minshushugi no Hosomichi, 1994–2004.* Hābesutosha.

Nishimura, Y., & Rachi, M. (Eds.). (2007). *Shōgen, Machinami Hozon.* Kyoto: Gakugei Shuppansha.

Ogasawara, M. (1986). *Otaru Unga Sensō Shimatsu.* Asahi Shinbunsha.

Otaru Kankō Yūchi Sokushin Kyōgikai Chōsa Kenkyūbu, (Ed.). (1998). *Otaru Kankō o Kangaeru: Otaru Kankō no Genjō to Korekara.* Otaru Kankō Yūchi Sokushin Kyōgikai (A4, 22 pp.).

Otaru Shinpojiumu Jikkō Iinkai, (Ed.). (2009). *Otaru Unga to Sekizō Sōkogun no Hozon Undō Kara Nani o Uketsugu ka: Chiiki ni Iki, Chiiki o Mamoru...Machizukuri Undō no Senkusha Mineyama Fumi-shi ga Tsutaeru Koto Shinpojiumu Kaisai Hōkoku.* Sapporo: Nihon Kenchiku Gakkai Hokkaido Shibu (A4, 76 pp.).

Mondai, O. U., & o Kangaeru Kai, (Eds.). (1986). *Otaru Unga Hozon no Undō, Rekishihen.* Sapporo: Otaru Unga Hozon Undō Kankōkai.

Mondai, O. U., & o Kangaeru Kai, (Eds.). (1986). *Otaru Unga Hozon no Undō, Shiryōhen.* Sapporo: Otaru Unga Hozon Undō Kankōkai.

Otaru Unga o Mamoru Kai. (1974). *Otaru Unga o Mamorimashō! Undō no Tebiki.* Otaru Unga o Mamoru Kai (B5, 12 pp.).

Otaru Yuki Akari no Michi Jikkōiinkai. (1998). *Otaru Yuki Akari no Michi.* Otaru Yuki Akari no Michi Jikkōiinkai (Kikakusho, A4, 24 pp.).

Sasaki, K., (Ed.). (1983). Machizukuri ni Okeru Kankyō no Kyōikuryoku to Kankyō Ibentogata Shimin Undō no Tenkai ni Kan Suru Kenkyū: Otaru Unga Mondai o Tōshite" (Dai 2-kai Toyota Zaidan Kenkyū Konkūru Chūkan Kenkyū Hōkokusho). Toyota Zaidan.

Shibakōdai Tōsōshi o Katarukai. (2010). *Mōhitotsu no Zenkyōtō: Shibaura Kōdai Zenkakutō 1968–1972.* Tsuge Shobō Shinsha.

Suzuki, T. (2013). *'Kokoro no Yami' to Dōki no Goi: Hanzai Hōdō no 1990-nendai* (Seikyūsha Raiburarī 78). Seikyūsha.
Tamano, K. (1997). Toshi Shakai Undō to Ikirareta Kūkan: Seikatsushi Chōsa no Chiken Kara. *Ryūtsū Keizai Daigaku Shakaigakubu Ronsō, 7–2,* 71–107.
Tamura, Y. (2009). *Otaru Unga Monogatari.* Kajima Shuppankai.

Chapter 6
What Was Won? What Was Lost?: The Transformation of the Otaru Townscape in the Post "Canal War" Period

Abstract Based on twenty-two annual fixed point observation surveys, in this chapter I make an empirical assessment of the impact of tourism development on Otaru's townscape. I conclude that Otaru is losing the very scenery that made the city such a popular tourist destination, and describe the logical structure of this "paradox of tourism development." I also examine the structure of the "canal war," and identify how the two sides understood the canal in opposite ways. City officials saw the canal as an outdated port facility ready to perform an entirely new function as a road, while the preservation movement understood the Otaru canal not as a functional "space," but as a highly specific, irreplaceable, and deeply meaningful environment—in other words, as a "place." The canal was a "space" to administrative authorities, but a "place" to the movement. Moreover, while city officials saw the issue as a single layer, the preservation movement understood it as a composite of multiple layers.

Keywords Space/place · Tourism development · Layer · Fixed point observation survey

6.1 The Logic of Change and the Logic of Preservation

In the preceding chapters, we have reviewed Otaru's history and examined the respective logics informing the advocates of both change and preservation in Otaru. What has been revealed in the process? Let us begin with a brief summary.

6.1.1 The Logic of Change

The actors who pushed to reclaim land from the canal to construct the Rinkōsen road took the following approach to the issue:

1. They identified the modernization of Otaru's port facilities as a policy priority.
2. They saw the canal as a *space*.
3. They approached the Rinkōsen construction project as one of "reuse" or "repurposing."

S. Horikawa, *Why Place Matters*,
https://doi.org/10.1007/978-3-030-71600-4_6

4. They believed that public works projects have a ripple effect across the local economy.
5. They shared the expectation that any change to the Rinkōsen project would negatively impact the city's ability to obtain further subsidies from the national and prefectural governments.

As recounted in Chap. 3, city officials intent upon reviving the local economy saw the redevelopment of Otaru's port as an urgent policy priority. However, the relative lack of flat land adjacent to the Otaru coastline complicated the task of modernizing port facilities and facilitating cargo transport by truck. The pressing need to overcome these challenges, coupled with the city government's approach to the canal as a physical space, resulted in the Rinkōsen road construction plan. City officials saw the canal as a transparent cube—an abstracted space, which they could easily convert to new purposes. If one viewed the canal as nothing more than a physical space, there was nothing problematic with the notion that it could be reused—or repurposed—to better meet the demands of a new age. Repurposing the canal space was thus seen as the perfect solution to the imperative of port modernization and redevelopment.

This proposal received powerful support from the collective faith in the ripple effect of public works projects. The push for change was fueled by officials' conviction that the huge influx of national funding would ripple through the entire local economy. Coupled with the belief that space should always be optimized to meet the needs of the present, this "ripple effect of public works" paradigm created a powerful pro-development ideology.

The fifth and final point requires a bit more explanation. The word "expectation" refers to the predictions of Otaru city officials regarding the decision-making process at higher levels of government. If changes were made to the Rinkōsen construction plan halfway through its execution (and a portion of the associated budget thereby relinquished), national and prefectural authorities might be less inclined to allocate future subsidies to Otaru. This prediction functioned as a form of self-regulation that imposed rigidity on city policy. Without actually consulting the central bureaucracy, Otaru officials claimed that "reversing a previous decision would discredit the local administration, and impact future projects" (Shisei no Ayumi Henshū Iinkai, ed. 1988: 214). This "prediction" was a powerful factor informing local officials' inability to contemplate changes to the Rinkōsen plan.[1] Put differently, Otaru city officials operated within the framework of a national subsidy system that hardened their resolve to execute the original road plan at all costs. If the first four points affirmed change, then the fifth precluded any deviation from policy and perpetuated the pro-development orientation of city officials.

[1]In fact, the budget for the Rinkōsen project decreased year by year, due to the lack of local consensus, making city officials "impatient" to complete the construction as soon as possible. Readers may refer to comments made by former Otaru mayor Shimura Kazuo and Otaru's former director of civil engineering, Nishio Akira, for a 2002 NHK special program on the Otaru canal dispute. "Kita no Chōsensha-tachi, Dai-ikkai: Yomigaeru Unga no Machi: Otaru o Nibunshita Daironsō," NHK Sapporo (2002).

This comprised the logic informing the actions of the "agents of change"—primarily local politicians and bureaucrats.

6.1.2 The Logic of Preservation

The men and women who sought to preserve the Otaru canal were guided by an equally powerful but opposing logic. In summary:

1. They saw the canal as a *place*.
2. Their perspective on the canal shifted over time. While activists initially saw the canal as a cultural asset, they came to understand it as a resource for promoting local tourism.
3. Consequently, they argued that change should be spearheaded by local residents, and based upon the city's existing stock of buildings.

The canal preservation movement that sprung up in opposition to the Rinkōsen's construction initially demanded the canal be kept exactly as it was, "frozen in time." Yet describing the canal, which had once been the center of Otaru's economic life, as a "cultural property" was a dubious analogy that repudiated all change and failed to reflect the realities of daily city life. This outright rejection of change isolated the preservation movement. Subsequently, activists developed a new and more dynamic logic of preservation: "community development" (*machizukuri*). With this, their focus changed from "the canal as an important cultural property" to "the canal as a resource for tourism." Activists' insistence that the canal was an important resource for local tourism expanded public support for their cause and also justified a certain level of preservation in the canal district. Activists believed that their willingness to tolerate some forms of change would legitimize their demands for the canal's preservation.

One striking characteristic of this evolution in logic was the notion of "resident-led change." Because the new goal of "community development" conveyed a willingness to accept a certain degree of change, it was far more susceptible to attack or exploitation by the city government. Preservation activists thus took care to emphasize that it was actually slow change, instigated *by* residents and *for* residents, that was needed (here we should recall the words of Ogawara: "what matters is how actors get together to bring about change," and Yamaguchi: "we said that redevelopment had to make sense to citizens"), and developed a rationale for preservation that remained compatible with changes to the city. This new approach suggested that unchecked change would not be tolerated, residents should control other forms of change, and that even if residents decided to accept certain changes, these could not be overly abrupt or drastic. Underlying this evolution in approach was a recognition that the Otaru canal and its surroundings were a *place* of irreplaceable sentiment and meaning to the people of Otaru. This was the logic informing the actions of the local residents who resisted the city's efforts at development and fought to save "the irreplaceable Otaru canal."

6.2 The Structure of the "Canal War" Conflict: What Did Otaru Gain?

Our next task is to explain the extraordinary duration of Otaru's canal war. What structures perpetuated the conflict and what did the Otaru gain as a result? Both city authorities and the majority of Otaru residents recognized the need for a new road along the coast to alleviate the terrible congestion on National Highway 5, which runs through the city. Given this shared recognition, why did the battle over the canal persist for more than a decade, leaving behind bitter grudges that remain to this day? These questions remind us that attempting to understand the protracted canal controversy as a simple "road dispute" obscures the real dimensions of the conflict. There must have been another more essential issue on which the city administration and the preservation movement continued to disagree.

Let us review the following three points in our quest to explain the structure and length of the canal war.

6.2.1 "Space" Versus "Place:" Two Opposing Perspectives on the Otaru Canal

Differing understandings of the environment in question—the Otaru canal—were one reason for the grave social divisions occasioned by the canal war and the duration of the canal dispute. People saw the same canal in very different ways. A canal would seem to be an exceedingly straightforward public works installation, and yet the Otaru canal was assigned very different meanings. To explain this, let us return to the framework of "space" and "place" (Horikawa 1998b, 2010a).

Approaching an environment through the framework of "space" is to understand it as a generic cube—a transparent container, capable of becoming almost anything. City planning terminology reflects this concept of space. By contrast, approaching a particular environment as a "place" is to understand it as governed by the values of the humans who interact with it and by the meanings they assign to it. In the latter approach, the environment is not a colorless, transparent "space," but a highly specific, individual, and heterogeneous *place*, which has been imbued with a particular significance. Naturally enough, strategic visions for the future of the Otaru canal district varied drastically depending on whether the area was understood as a "space" or as a "place."

If we apply these definitions of space and place to our review of the canal war, it is abundantly clear that the Otaru city government approached the canal as a space. City officials saw the canal as an outdated port facility ready to perform an entirely new function as *the site of a new road*. In other words, while in a previous era the transparent "container" of Otaru's port district had been filled by the canal, city officials believed it was now time to fill it with something new: a road.

Meanwhile, preservation activists understood the Otaru canal not as a functional "space," but as a highly specific, irreplaceable, and deeply meaningful environment—in other words, as a "place." The refrain "Otaru will no longer be Otaru without the canal" underlines this understanding. To the men and women who joined the preservation campaign, the canal and its surroundings *symbolized their lives* in Otaru. One might say they experienced the comprehensive transformation of the canal district as the destruction of not only a place, but also their very identity. Activists claimed that city planners' understanding of the canal as a "space" stripped away the memories and sentiments of local residents—in other words, it chipped away at the "placeness" of the urban environment. Moreover, they argued that it was precisely this "placeness" that ought to form the foundation of the city's rebirth. Finally, activists saw the scenery of the canal district and the local community as inseparable: the canal must be saved because the loss of that physical built environment would alter the character of the local community (Nishimura et al. 2003: 121–155).

The fundamental disagreement between municipal authorities and preservation activists stemmed from their conflicting approaches to the canal (as space versus place). The divide between the forces of change and preservation aligns perfectly with the fault line between their respective forms of spatial awareness. This helps to explain why the two camps continued to "talk past each other" in the protracted dispute over the canal's fate.[2]

6.2.2 "To Preserve Is to Change:" A New Understanding of Preservation

A second factor pertains to how the word "preservation" was understood. Activists demanded the preservation of the canal as a "place" beloved by local residents. Yet city planners cannot fully address or incorporate each individual interpretation of the position, significance, or weight of "place." For this reason, activists did not try to save everything. Paradoxically enough, preservation activists actually argued that the area around the canal *should* change. But why would the preservation movement make the case for change?

We know that the preservation movement abandoned the early goal of "cryopreservation" in favor of a "community development" (*machizukuri*) approach. This community development approach anticipated a certain degree of change to the existing townscape of the canal district. Refashioning historic buildings as art galleries or cafes would not have been possible under the earlier goal of total preservation. After all, it would have been impossible to install so much as an air-conditioning

[2]Differences in spatial awareness are accompanied by differences in temporal awareness. The understanding of an environment as a "space" is inevitably related to a modern understanding of time as the irreversible progress forward along the number line. By contrast, in a "place," time accumulates in drifts and piles, while certain points in time and place are selectively recalled (Halbwachs [1950] 1968 = 1989; Lynch 1972 = 1974; Hama 2010).

unit in a historic building along the canal if changes to accommodate the needs of modern life were prohibited. The central concern of the preservation movement was thus not the wholesale rejection of change, but rather an argument for "giving residents—not city planners or outside experts—control over the extent and pace of change." This is reflected in the words of the Yume-machi executive committee: "We are not 'preservationists;'" "it is essential 'to put new energy into old vessels'" (Otaru Unga Mondai o Kangaeru Kai 1986a: 222–223). Significantly, despite this evolution in approach, from beginning to end activists used the same word to describe their goal: preservation (*hozon*). Activists continued to use this rather retrospective word because they were opposing a form of "space"-based redevelopment (Otaru Shinpojiumu Jikkō Iinkai, ed. 2009: 59). City officials failed to recognize that the movement's continued use of the word "preservation" belied a surprising change in definition: over time, activists came to accept that "to preserve is to change." This is another reason for the protracted standoff between the administration and the preservation movement.

6.2.3 Understanding the Multilayered "Canal Issue"

We have not yet exhausted the reasons behind the lengthy canal war. For it was not only the terminology employed, or its substance, that kept the two sides from finding common ground, but also the register in which the issue was discussed. What does this mean? In this section, I identify the different layers—the registers of debate—within the canal issue (Horikawa 2010a, 2011).[3]

The pros and cons of the new highway construction were fiercely contested, so naturally the "road dispute" was one layer. Indeed, the assertion that "the canal issue was really a road issue" is not correct—but neither is it entirely mistaken. This is because the road construction issue remained the central layer of debate during the early days of the canal war.

The questions posed by the preservation movement uncovered still other layers. The canal issue originated as a road construction project that had already been approved and funded. By the time preservation activists had collected and submitted the signatures of 98,000 local residents, however, different questions were being asked: Who really represented the will of the people? Would the city government's economic stimulus policies actually prove effective? And on a more philosophical level, how should people live in a city? In other words, the register in which the movement addressed the issue had three layers, which went far beyond the merits

[3]As I have explained in Chap. 1, the concept of "layers" refers to the elements and "registers" of dispute within a particular social conflict. Articulating these different layers allows us to grasp the various societal levels at which a single element (for example, the pros and cons of the new road project) is contested. As social conflicts typically involve diverse and complex disputes between multiple actors and emerge as the composite of various actors, registers, and situations, the "layer" is a conceptual apparatus for understanding the shape of these multiple conflicts and the substance of debates through a focus on the "register" in which they are discussed.

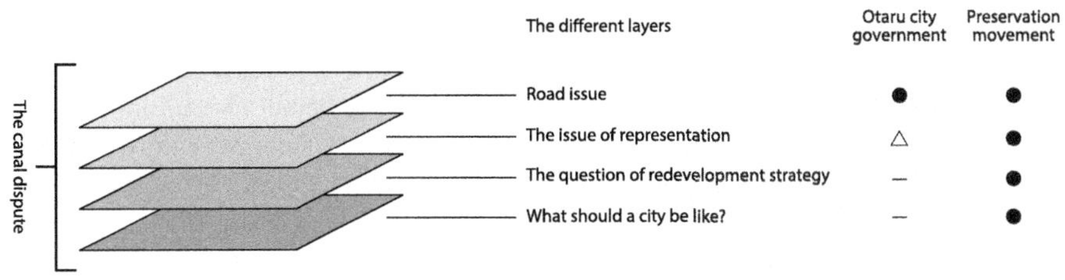

Fig. 6.1 The structure of the multilayered canal dispute (Author's illustration)

of the proposed road construction project to other questions of representation and legitimacy with regard to decision-making, the validity of the city's strategy for redevelopment, and the image of the ideal city. The "canal issue" must be understood as the composite of these many layers (Fig. 6.1).

The Otaru city government continued to understand the canal problem as a single-layer[4] issue involving the construction of a new road. The preservation movement understood the issue as a multilayered one. This is yet another way in which the preservation movement, whose members sought to reverse a pre-approved public works project and place control over change in the hands of local residents, remained out of step with the city government. The inability of the two sides to understand each other—across three different layers—was yet another reason for the drawn-out fight over the fate of the Otaru canal.

6.2.4 From Port City to Tourist Destination: What Did Otaru Gain?

What did the structure of the canal war conflict produce, and what did Otaru gain as a result?

First, the protracted fight over the canal made Otaru and its canal household names throughout Japan. The number of visitors to Otaru grew steadily in the years following the end of the canal war in 1984. At the peak of its tourism boom, Otaru welcomed nine million visitors in a single year. This was a truly remarkable number for a provincial city with a local population of just 150,000. Looking back, Otaru's rise to national prominence began when a handful of local residents raised their voices to "save the canal." This pushed the canal issue beyond the city council to the prefectural assembly, and finally all the way to the National Diet. All the while, APOC chairwoman Mineyama Fumi made frequent appearances on NHK television

[4]The Otaru city government's response to the submission of the petition signed by 98,000 local residents in support of the canal's reservation was to "inspect" the authenticity of the signatures submitted. This was an effort to avoid, at all costs, the introduction of a new layer: the issue of representation in the decision-making process. The triangle symbol is a representation of this in Fig. 6.1.

news and in such national publications as the *Asahi Shinbun*. Well-known novelists and commentators (Natsubori 1980, 1992; Muramatsu 1986, Maruya and Yamazaki 1987) and artists (Chiba Shichirō) were also inspired to depict Otaru in their work, drawing further attention to the city. Otaru's newfound fame was an unintended consequence of the preservation campaign, which allowed the fading port city to make a comeback as an extraordinarily popular tourist destination.

Second, the canal war helped to solidify the direction of local community development efforts. The register in which the issue was discussed deepened over the course of the canal dispute. As the stalemate between the city government and preservation movement continued, a simple "road dispute" prompted a much deeper debate. Who should speak for the citizens of Otaru? How should local residents seek to revive the fortunes of a city in decline? And what kind of city should we try to live in? The "canal issue" was a composite of these various layers of debate, with the real issue becoming that of the direction of change in Otaru, and mechanisms for controlling this change.

The multilayered canal issue posed the question: should Otaru adopt a "scrap and build" approach, destroying the existing townscape to create something new? Or should the city adopt a "rehabilitation" strategy and utilize the existing townscape toward new ends? *In essence, this was a debate about the design of the future Otaru.* Should changes to the city be entrusted to the redevelopment policies of the local administration or controlled according to a consensus that incorporates the wishes of local citizens? Because internal divisions brought the preservation movement to an abrupt end, the city's "scrap and build" approach was implemented, and the Rinkōsen was constructed—albeit only after city officials adopted a "compromise plan" that made some concessions to the concerns of the preservationist camp and saved a narrow portion of the canal. Thus, at the conclusion of the "canal war," there was still no clear prospect for the future design of Otaru.

This task was left to the Otaru Revitalization Committee (established in 1984 as a result of the five-party talks) and the administration of Shimura Kazuo's successor, Araya Masaaki. After winning the election in 1987, Mayor Araya and his new administration rapidly adopted a series of measures pertaining to Otaru's scenery and townscape. A Citizens' Colloquium on Community Development (Otaru Machizukuri Shimin Kondankai) was established in January 1988, and a "Basic Vision for the City's Scenery" was formulated in March 1988. Between December 1989 and December 1990, the city installed gas lamps and a footpath along the banks of the northern portion of the canal, which was preserved at its original width. Finally, in April 1992, a "Landscape Ordinance for Community Development through Otaru's History and Natural Environment" was adopted, and an Urban Design Section was established within Otaru City Hall. The Araya administration pursued a form of community development that drew upon Otaru's local history as a resource and was based upon the traditional scenery of the city. That this policy course was adopted and pursued, however imperfectly, was a direct result of the questions raised by the canal dispute, including that of the future design of Otaru.

Meanwhile, the social divisions created by the canal dispute remained. This could be described as a third product of the canal war. An insurmountable rift developed between Fujimori Shigeo and Mineyama Fumi, two of the earliest and most dedicated champions of the canal's preservation. The hardliners behind the premature mayoral recall effort were subsequently shunned by other activists, and the divisions created by their mutual outrage and resentment remain to this day. One could say that these wounds and the destruction of longstanding personal relationships were the heavy price of Otaru's rebirth as a city of tourism.[5]

6.3 A Framework for Making Sense of Urban Landscape Change

We have seen what the canal war brought to Otaru. Now let us turn to another question: *what type of local landscape did the long canal war produce*? Issues and disputes do not emerge out of thin air. On the contrary, they emerge as a result of the meanings residents ascribe to their everyday life, within the particular historical context of a city or town—a unique and highly specific built environment. The preservation movement originated out of residents' sense of crisis at the impending destruction of the canal, which was a cornerstone of their identity. Their preservation campaign has brought about the new landscape of the canal district, and this new landscape has in turn created a new identity. If we accept this premise, then we must also examine the specific ways in which the landscape of Otaru has been changed by the city's transformation into a popular tourist destination. What has been the impact of local tourism on Otaru? And what has the city lost as a consequence? In this section, I draw upon the fields of architecture and urban planning to address these questions.

6.3.1 The Unintended Consequences of Tourism Development

The transformation of historic buildings into tourist cafes and offices into souvenir shops was pursued with energy and enthusiasm in the newly minted tourist destination of Otaru. In fact, the district of Sakaimachi is now lined with souvenir shops. The wholesale transformation of the scenery of Otaru's port and canal district has prompted a collective sense of loss among local residents. Indeed, this profound sense of loss was a theme repeated by a number of interview subjects. One can understand the long fight between the forces of "change" and "preservation" as one

[5]Yamaguchi Tamotsu and Ogawara Tadashi have described the persistence of this damage and likened it to the feeling of "something sharp that's still stuck in my throat." To be sure, the canal fight casts a long shadow over the relationship between local citizens and the Otaru city government, but it is important to recognize that this was not an entirely negative development; it created a new "check" upon municipal authorities by imposing a new standard for transparency with regard to administrative policy and demanding a higher level of accountability.

that resulted—in *substance* if not in name—with the preservation and repurposing of Otaru's stock of historic buildings for use as local tourist resources, transforming a "city in decline" (*shayō no Otaru*) into a booming tourist town. Why, then, do so many local residents mourn the transformation of the Otaru townscape? How has the landscape of Otaru changed as a result of the city's unexpected and precipitous transformation into a tourist destination? What has been the impact of tourism development upon a distinctive regional "place?" These are the questions that guide our examination in this section.

6.3.2 Grasping Change: From Architecture to Sociology

What type of research has already been guided by such questions? What have been the concerns, analytical tools, and conclusions of such research? Let us begin with a review of the literature.

6.3.2.1 Previous Architectural Surveys of Otaru

Sociologists themselves do not study the physical environment or architectural structures. In order to understand changes to the built environment, sociologists must refer to existing research in the field of architecture. The questions guiding this section led me to a series of architectural surveys undertaken by students of city planning within the Architecture and Engineering Section of Hokkaido University's Engineering Department. Conducted in 1981, 1986, and 1992, these surveys examined changes to more than two hundred historic buildings in the canal, Ironai, and Sakaimachi districts of Otaru. The five surveys (hereafter referred to as the Hokkaido University surveys) were conducted in the course of research for three undergraduate theses, one master's thesis, and one conference paper developed from the master's thesis.

The Hokkaido University surveys were panel studies of the same buildings in the same (port and canal) district of Otaru, from the 1980s to the 1990s, as shown in Table 6.1.

If we avail ourselves of earlier surveys conducted in the course of undergraduate and master's thesis research, we can trace changes to the district back even further in time, to the early 1970s, when the canal preservation campaign began.[6] Although these surveys were conducted by undergraduate and graduate students, they offer first-rate and extremely detailed data on the scenery of the canal district and neighboring districts of Otaru before the construction of the Rinkōsen—scenery that has since been lost.

[6]The graduation theses dating from the late 1970s are the work of the "Hokudai trio." These are indispensable resources for charting the development and evolving ideals of the preservation movement. However, the approach of the Hokudai trio, which centers on interviews with participants in the preservation movement, is fundamentally different from the fixed point observation method. For this reason, a discussion of their work is not included in this section's review of existing research.

Table 6.1 The Hokkaido University surveys: titles and preservation status of individual survey sheets

Year	Type	Author and title	Survey sheets
1981	Graduation thesis	Hamada Takenori, "Rekishiteki Kankyō no Seibi ni Kan Suru Kenkyū: Otarushi Ironai Chiku o Kēsu Sutadi Toshite"	×
1986	Graduation thesis	Matsushita Shigeo, "Rekishiteki Machinami ni Okeru Keikan Seibi Shuhō ni Kan Suru Kenkyū: Otarushi Ironai-dōri, Midoriyamate-dōri no Bāi"	×
1992	Graduation thesis	Satō Kuniaki and Saeki Satoshi, "Otarushi Unga Shūhen Chiku no Machinami Henka ni Kan Suru Kenkyū"	△
	MA thesis	Myōen Kōichi, "Otaru ni Okeru Keikan no Henyō Kōzō ni Kan Suru Kenkyū: Unga Shūhen Chiku o Megutte"	△
1993	Journal article	Myōen et al., "Otaru no Rekishiteki Keikan no Henyō Kōzō ni Kan Suru Kenkyū: Unga, Ironai, Sakaimachi Chiku no Bāi"	NA

Note An × indicates that the individual survey sheets were lost and could not be reviewed by the author. An △ indicates that only a portion of the survey sheets was available for review by the author

In these surveys, researchers assigned numbers to buildings located within the greater port district of Otaru, where the characteristic townscape of Otaru was best represented (in Ironai 1 chō-me, Ironai 2 chō-me, Ironai 3 chō-me, Minatomachi, and Sakaimachi). A detailed questionnaire sheet was created for each building. Researchers supplemented this questionnaire with door-to-door interviews with the building's occupants and firsthand assessments of the building's current condition (including photographic records). Researchers divided the survey's target area into ten different districts and identified them with the letters "A" through "J." In total, 272 buildings were included in the survey (Fig. 6.2). From the 1980s onward, the same buildings, identified by the same number, were examined by each successive survey. The questionnaires capture a vivid picture of each individual building at a certain point in time: each begins with a color photograph of the building in question

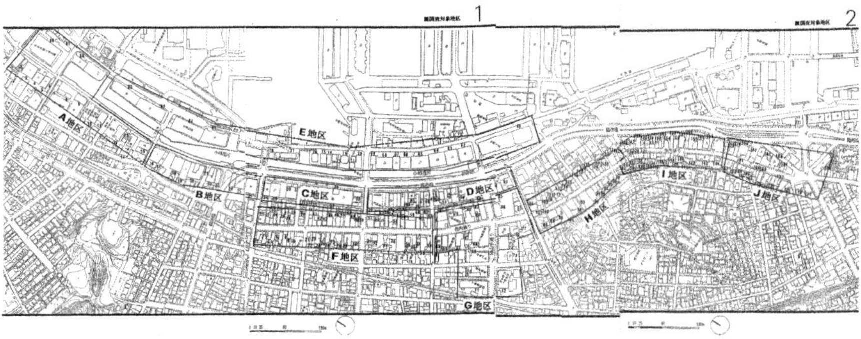

Fig. 6.2 The target survey area of Myōen (1992). *Source* The author has scanned and reduced the figure presented in Myōen (1992: 8–9)

and includes highly detailed questions regarding the building's structure, the nature of any changes or renovations, architectural style, and materials. Direct transcriptions of statements from the building's residents are also included.

Unfortunately, these individual survey sheets cannot be replicated in their entirety; not all were preserved at the university, and some of them have been scattered or lost. Given this, our examination of the Hokkaido University survey data must be limited to the aggregate data presented in the final product (in other words, the final research theses) rather than the contents of each building-specific questionnaire.

The Hokkaido University survey data is extremely valuable in spite of these limitations. There is very little precedent for this type of fixed point observation of individual buildings, over such a long period of time. Because the first 1981 thesis incorporates data from the late 1970s, the Hokkaido University surveys span almost full two decades—a remarkable length of time for such a study. Although the color photographs atop each individual building questionnaire have faded with time, the value of these questionnaires actually increases every day.

Of the five Hokkaido University surveys, the research conducted by Myōen Kōichi for his 1992 master's thesis is of particular note. Indeed, Myōen's work represents a remarkable achievement, as he captures the changes to the historic buildings of Otaru's port district. By comparing the results of his 1992 survey with the earlier research of Hamada (1981) and Matsushita (1986), Myōen compiled a crucial data set regarding changes to Otaru's built environment, making his research of foundational importance to this volume. Although we cannot examine each individual questionnaire collected by Myōen, his inclusion of a clear and detailed table of aggregate data and his replication and explanation of this data at academic conferences and in subsequent journal publications (Myōen et al. 1993) allows us to refer to his work as a principal example of prior research on the subject.

At the same time, there is a certain "indecisiveness" to Myōen's work, which stems from his inability to fully synthesize the concepts of change and urban landscape transformation. One could also say that Myōen was confused in his definition of change. Let us take a moment to examine Myōen's 1992 thesis in more detail.

The stated objective of Myōen's research thesis, entitled "A Study on Structure of Change in Historic District of Otaru: a case of Otaru Canal, Ironai, Sakaimachi districts," was to "seek out the reality and context of change and identify the structure of urban landscape transformation" (Myōen 1992: 2). In his first chapter, Myōen analyzes "the actual conditions of changes to the external appearance" of buildings. In his second chapter, Myōen analyzes changes in the ways these buildings were used.

The four "changes" examined in Myōen's first chapter are listed in Table 6.2.

Taken together, these four changes and Myōen's own assertion that "changes to the interior of the buildings were not taken into account" (Myōen 1992: 15) allow us to conclude that his study of what we might call "external changes" was exactly that: an assessment of changes (or non-change) based on the extent of modifications to the external façades of the buildings surveyed. Consequently, Myōen defined "change" as the extensive remodeling of a building's exterior. Structural alterations to a building's interior did not qualify as "changes" if they did affect the external

Table 6.2 The definition of "Changes to External Appearance" in Myōen (1992)

Status	Category	Definitions and specific examples
No change	No change, repairs	• Buildings left untouched
		• Small-scale repairs for building maintenance (E.g., repainting a roof, putting up or removing a billboard)
Change	Substantial repairs or remodeling	• Large-scale repairs or alterations, repainting or repair of the entire exterior (E.g., changes to the building façade, large-scale repairs)
	Rebuilding	• Demolition of the existing building followed by new construction
	Demolition	• Demolition of the existing building and no new construction (E.g., vacant lots, parking lots)

Note In this table, the author summarizes the definitions used by Myōen Kōichi (1992: 15)

appearance of the building. In other words, among all possible physical changes, only major modifications to the exterior of a building were defined as "changes" by Myōen.

In the second chapter, the focus is different: Myōen "puts aside, for the time being, changes to external appearance in order to discuss how the uses of these buildings have changed" (Myōen 1992: 42). Myōen creates three broad categories to capture changes in usage, as depicted in Table 6.3.

The first category is rather cryptic: "buildings that cannot be freely utilized by the general public." Myōen's meaning becomes a bit clearer when we compare this category with the second: "buildings that can be freely utilized by the general public." It would seem that Myōen's first category refers to buildings where the authorized activity is limited to the building's owner or tenant, or where other forms of activity

Table 6.3 The categorization of building "Uses" in Myōen (1992)

Category	Examples
Buildings that cannot be freely utilized by the general public	• Warehouses, factories
	• Offices, hospitals, banks
Buildings that can be freely utilized by the general public	• Retail stores, souvenir shops
	• Restaurants
	• Hotels
	• Museums, galleries, public halls
Other	• Highways, roads, open spaces
	• Parking lots and garages
	• Vacant lots
	• Other

Note In this table, the author summarizes the categorization of building use developed by Myōen Kōichi (1992: 42)

are heavily restricted. The main point of this distinction thus seems to revolve around the question of public access.

However, Myōen's next assertion makes this very interesting categorization problematic—and it is this problem that introduces the aforementioned "indecisive" element into Myōen's argument. Myōen defines "actual changes to a building as a consequence of a change in its use" as the "remodeling of shape" (Myōen 1992: 42) and classifies these as "conversion," "reconstruction," and "demolition." "Conversion" occurs when an existing building is put to a different use. "Reconstruction" occurs when an existing building is torn down and replaced with a new building. "Demolition" refers to instances in which a building is torn down and the site is left vacant or simply turned into a parking lot, for instance.

There is a fundamental contradiction between Myōen's stated goal in the second chapter ("put(ting) aside, for the time being, changes in external appearance in order to discuss how the uses of these buildings have changed") and his assertion that "actual changes to a building as a consequence of a change in its use" should be defined as the "remodeling of shape." There is a very clear contradiction when a procedural assertion that sets external appearance to the side is followed by a stated focus on the remodeled appearance of the building. This is not all. Myōen's premise in Table 6.3 (regarding the standard of public access to a building) is hardly necessary to differentiate between the three forms of change in usage: diversion, reconstruction, and demolition. As classificatory criteria, there is no logical relationship between the categories presented in Table 6.3 and the "remodeling of shape." In short, Myōen's decision to examine the ways in which the use of buildings has changed in his second chapter lacks theoretical coherence.

Given this lack of theoretical coherence, why not consign Myōen's study to the dustbin of long-forgotten masters' theses? It is important to recognize the theoretical deficiencies of Myōen's argument, but we should also ask *why he chose to focus on the changing uses of buildings covered by his survey*? Myōen's decision to analyze changes to the way buildings were used was a result of his assessment that "more than general improvements, repairs, or remodeling, it is 'changes in usage' that bring about the more influential types of 'landscape change,' giving us a deeper understanding of the district's transformation and one avenue for exploring the broader context [of change]" (Myōen 1992: 42). This single sentence contains several key points:

1. Myōen differentiates the factors behind changes to architectural features of buildings: some are the result of "general improvements and repairs" while others are the consequence of a "conversion in usage."
2. He places more relative importance on conversion in use as a causal factor in change.
3. He proposes an analysis of the pattern behind the conversion in building usage.

When we recall that Myōen conducted his survey in the early 1990s, when Otaru was newly overrun with tourists, it seems clear that the massive overhaul of buildings converted for use in the tourist trade was the primary factor behind the transformation of Otaru's townscape. The Otaru tourism boom led to a conversion in the use of many buildings. It seems likely that Myōen's analytical framework was deeply influenced by the concurrent tourism boom, even if Myōen himself was unaware of this influence.

This would explain Myōen's unique methodology of examining how the use of specific buildings changed from one year to the next. This approach allowed Myōen to seek out the types of conversion in building use that were behind Otaru's rapid transformation into a tourist destination and enabled him to identify which areas had made the greatest "contributions" (so to speak) to this transformation. If we move past the theoretical gaps of his enquiry, this is the real thrust of Myōen's analysis.

Later, Myōen (Myōen et al. 1993) joined the other members of the Hokkaido University research team to update his study of changes in land use and scrupulously avoided the theoretical pitfalls of his original 1992 master's thesis. Instead, Myōen lists changes to the external appearance of buildings and changes in land use as equal factors behind landscape change and refrains from positing any causal relationship between the two in this comprehensive analysis of Otaru's landscape transformation.

Specifically, Myōen and his colleagues enumerate six "realities of the transformation of Otaru's historic environment" and analyze their context and causes to delineate the "structure" of environmental transformation. Changes to a building's external appearance and changes to the ways a building is used are treated as components of the "reality of historic environment transformation." Changes to the external appearance of buildings are referenced in the first "reality" (the loss of townscape continuity once created by the stone warehouses), and changes to building use are referenced in the third (the hollowing out of the historic district) (Myōen et al. 1993: 529–532). A summary of the analyses developed by the Hokkaido University surveys of Otaru's transformed landscape appears in Table 6.4.

Table 6.4 The issues presented by Myōen et al. (1993)

Specific transformation
1. The loss of continuity in the townscape of stone warehouses
2. Civic life cut off from the waterfront
3. Hollowing out of the historic district
4. Deterioration of pedestrian zone amenities
5. An increase in tourism
6. Growing reuse of historic buildings
Background/factors
1. Destabilization of the district due to a decline in major industries
2. Construction and subsequent impact of the Rinkōsen (both direct and indirect)
3. Pressures for development, including growing numbers of tourists and external capital investment
4. Rapid functional conversion to tourism
5. Various major changes to the overall environment

Note This is the author's summary of the argument presented in Myōen et al. (1993)

As shown in Table 6.4, Myōen and his colleagues identify the Rinkōsen's destruction of the continuity of the Otaru townscape, the decline of principal industries within the historic district, and their subsequent replacement with tourism-related businesses as the real conditions of local landscape transformation. They identified two primary causes: the "pressure for development" exerted by the vast number of tourists and an influx of outside investment, and the "rapid functional conversion of buildings to tourism." The fatal "indecision" embedded in Myōen's 1992 thesis was thus resolved in his 1993 analysis of the factors contributing to the transformation of the Otaru landscape.

The Hokkaido University surveys were informed by a shared concern that rapid tourism development was destroying the traditional landscape of Otaru. Successive panel surveys traced the transformation of the port district while focusing on conversions in building use. The key analytical tools (or concepts) were the focus on the rate of survival of historic buildings (Table 6.5) and a "use conversion matrix" for buildings (Table 6.6).

The first tool simply measures the continued existence of buildings within the relevant district, with no reference to their external transformation. The simple survival of a historic building is the only measure, regardless of how much its façade has been altered. The second tool allows us to understand the real ways in which buildings are put to use. It enables this type of analysis: "Stone warehouses, which numbered *n* in the year *x*, were reduced to *o* by the year *y*, while the *p* (number of) buildings with a change in use have been converted into parking lots" (Table 6.6). This simple cross-tabulation depicts the reality of Otaru's tourism explosion and represents a significant contribution by Myōen (1993) and his colleagues at Hokkaido University.

Table 6.5 The survival rate of surveyed buildings: 1986/1992 (Myōen et al. 1993)

	All buildings			Historic buildings		
	Number of buildings		Survival rate (%)	Number of buildings		Survival rate (%)
	1986	1992		1986	1992	
Canal district	86	57	66	45	26	58
Ironai district	85	72	85	25	23	92
Sakaimachi district	99	90	91	24	23	96
Total	270	219	81	94	72	77

Source Table 1 in Myōen et al. (1993: 529), with supplementary data added by the author
Note The "survival rate" is calculated based on a building's continued existence, regardless of the extent of changes to its external appearance through remodeling or repairs. See the main text for further explanation

Table 6.6 Changes in building usage between 1986–1992 (Myōen et al. 1993)

									(Unit: Buildings)
		Before (Use in 1986)							1992 total (Horizontal total)
		Warehouses, factories	Offices	Retail	Restaurants	Museums, etc.	Parking and vacant lots	Other	
Present (Use in 1992)	Warehouses, factories	**3**	0	0	0	0	0	0	3
	Offices	1	**8**	0	0	0	0	0	9
	Retail, souvenir shops	10	8	**1**	1	1	1	1	23
	Restaurants	5	9	0	**0**	0	0	0	14
	Museums, hotels, etc.	4	2	0	0	**0**	0	2	8
	Roads, sidewalks	2	2	0	1	0	**0**	1	6
	Parking lots, vacant lots	13	7	0	0	0	**1**	1	22
	Other	1	3	0	0	0	0	**0**	4
1986 total (Vertical total)		39	39	1	2	1	2	5	

Source Table 2 in Myōen et al. (1993: 531), with supplementary data added by the author

6.3.2.2 "Change:" A Sociological Expansion of the Concept and Its Application

I have continued the analytical approach and methods employed in the Hokkaido University surveys to trace subsequent changes to the Otaru townscape. Since the summer of 1997, I have continued with the fixed point observation approach, visiting and examining the same set of buildings targeted by the earlier Hokkaido University surveys. I have used the same ten-district organization in the survey and the same identifying number for each building (Fig. 6.3, Table 6.7).

These surveys were undertaken with the dedicated assistance of students enrolled in the social research practicum offered by the Department of Sociology at Hosei University. Together we created a questionnaire almost identical to the one used in the Hokkaido University surveys. When it came to taking photographs, the differences between camera types and the division of labor made it difficult to adopt the same approach as the earlier surveys. Nevertheless, the absence of any other follow-up to the Hokkaido University surveys means that our investigation is the sole source of data on the subject, which is far from insignificant. Together, the data collected in the course of the Hokkaido University surveys and the later surveys conducted by my research team allow us to trace changes to the Otaru townscape over almost four decades.

While labor-intensive, the specific methodology of these surveys was quite simple; either alone or in the company of my students, I conducted a fixed point observation survey of 276 historic buildings in the port district and another 116 stores in Otaru's commercial center.[7] There were two components to this very simple survey structure: on-site interviews and the firsthand observation of each and every building. Through these methods, researchers can use the built environment to uncover the history of a particular small district (Horikawa 2000a, 2000c, 2003; Horikawa, ed: 1998a, 1999, 2000; Horikawa and Morihisa, eds: 2008a, 2009; Horikawa and Fukaya, eds: 2012, 2013; Horikawa and Matsuyama, eds: 2016, 2017).

Let us first examine the significance of conducting interviews on-site, while looking at the actual building under discussion. The building anchors the narrative of residents, for whom the physical presence of the building lends a smoothness or fluidity to their discussion. It allows researchers to hear specific and richly varied narratives that could not have been anticipated back in a seminar room, rather than abstracted discussion or popularized local histories. Interviewing the residents of a specific district also avoids the pitfall of listening exclusively to participants in the preservation movement. It is extremely important to take into account not only the voices of those who actively joined the fight to save the canal, but also the ordinary people who simply go about their daily lives in the area. Ideas or subjects that

[7]The Hokkaido University surveys did not include Otaru's commercial districts. I designed and undertook this portion of the survey in response to the controversy surrounding the opening of the national supermarket chain MyCal in Otaru. However, the full results of my commercial district survey do not warrant more discussion here.

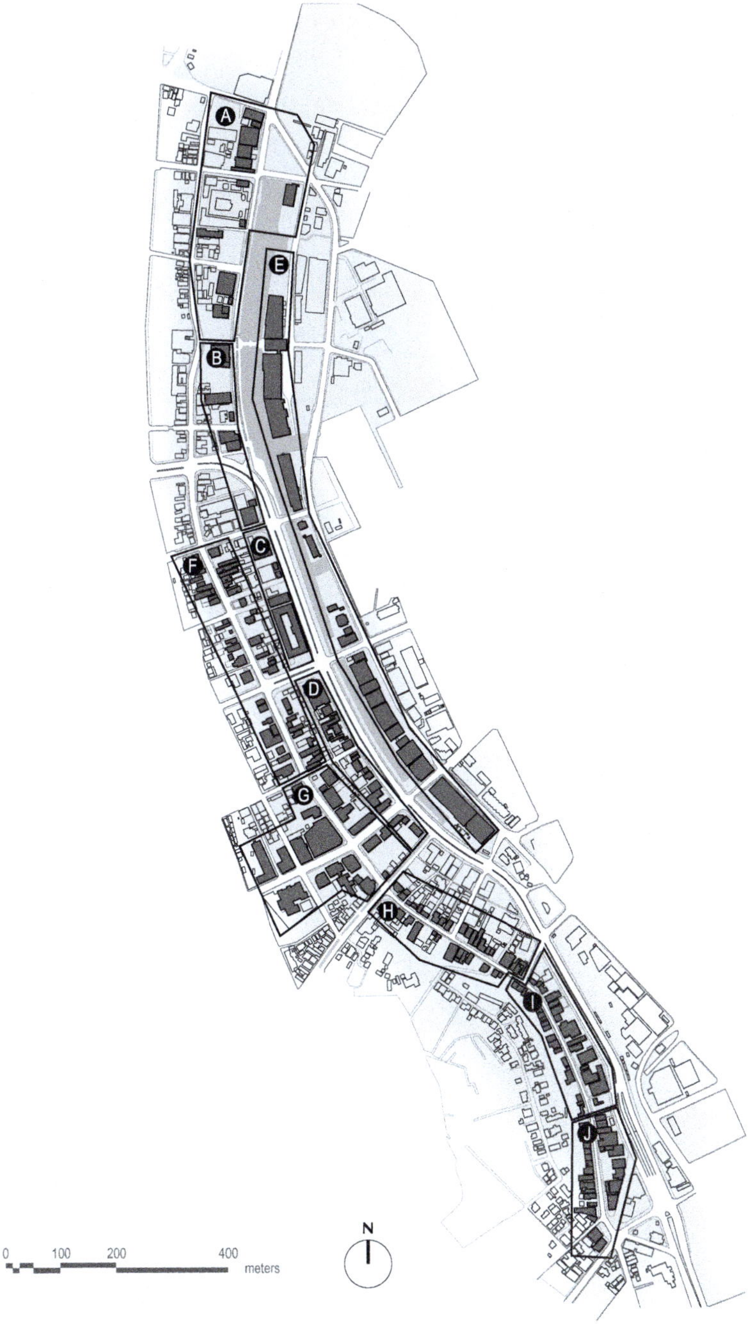

Fig. 6.3 A map of the total survey area. Drawing by the author

Table 6.7 An overview of the author's fixed point observation surveys (1997–2016)

Survey type	District	1997	1998	1999	2000	2001	2002	2003	2004¶	2005¶	2006	2007	2008	2009	2010	2011	2012	2013	2014	2015	2016
Building survey	A	○	○	○	□	□	□	□	..	..	□	○	○	□	□	○	○	□	□	○	○
	B	○	○	○	□	□	□	□	..	..	□	○	○	□	□	○	○	□	□	○	○
	C	○	○	○	□	□	□	□	..	..	□	○	○	□	□	○	○	□	□	○	○
	D	○	○	○	□	□	□	□	..	..	□	○	○	□	□	○	○	□	□	○	○
	E	○	○	○	□	□	□	□	..	..	□	○	○	□	□	○	○	□	□	○	○
	F	..*	○	○	□	□	□	□	..	..	□	○	○	□	□	○	○	□	□	○	○
	G	..*	○	○	□	□	□	□	..	..	□	○	○	□	□	○	○	□	□	○	○
	H	○	○	○	□	□	□	□	..	..	□	○	○	□	□	○	○	□	□	○	○
	I	○	○	○	□	□	□	□	..	..	□	○	○	□	□	○	○	□	□	○	○
	J	○	○	○	□	□	□	□	..	..	□	○	○	□	□	○	○	□	□	○	○
Commercial district survey	Miyako-dōri	—#	○	○	□	□	□	□	..	..	□	○	○	□	□	○	○	□	□	○	○
	Sun Mall	—#	○	○	□	□	□	□	..	..	□	○	○	□	□	○	○	□	□	○	○

Notes

□ Independent survey conducted by the author

○ Survey conducted with students participating in the Hosei University social research practicum

* Districts that could not be surveyed due to a lack of manpower

The commercial district survey began in 1998

¶ Years in which the author's overseas research precluded survey trips to Otaru

emerged when we compared and contrasted the narratives transcribed for each individual questionnaire often helped us to identify important new topics for subsequent interviews. These annual surveys, conducted with the help of my social research students, would involve nightly meetings at which the various narratives collected each day were presented to the group for discussion and debate. This helped us to identify potential meanings and implications, or themes for further exploration.

The second component of our surveys was the physical observation of every single building. In order to replicate the structure of the Hokkaido University surveys, this approach was adopted out of necessity—but it is important to stress the enormous *sociological potential* of such an approach. Personally observing buildings obviously allows one to glean far more information regarding the building's shape or changes to the building than simply relying on a Zenrin residential map. The greatest merit to this approach, however, is the potential for collecting "micro-level" information: news regarding a change in ownership, for instance, or the financial situation of local businesses. While this could be dismissed as "local gossip," this type of information actually illuminates a number of "changes" that might otherwise remain undetected. For example, one might learn that a change in management at a local café drove away all of its regular patrons, and while the name of the café remains unchanged and it now serves a very different clientele and has an entirely new atmosphere, making it more akin to a new business than an existing one. It also offers an opportunity to learn how the circumstances leading up to a building's construction or the origins of the builder may influence how residents evaluate the structure. In other words, their assessment of a building is often powerfully regulated by social factors.[8] Identifying these types of changes—which could not be captured through the strictly architectural concept of "change" employed in the Hokkaido University surveys—is central to this sociological survey of "change." Put differently, the sociological expansion of the concept of change allows us to analyze what was measured according to the architectural concept of change employed by the Hokkaido University surveys. Consequently, results and conclusions may vary from earlier reports on the same survey data (Horikawa, ed. 1998, 1999, 2000; Horikawa 2000a, 2000c, 2001; Horikawa and Egami 2002; Horikawa and Morihisa, eds. 2008a, 2009; Horikawa and Fukaya, eds. 2012, 2013).

Our analysis of survey data will proceed as follows.

[8]Hosei University student Kuramoto Yurie has written about her interview with a Sapporo confectioner who set up a shop along the Otaru canal out of a desire "to pitch in and help Otaru." This confectionery maker erected a faux Edo-period "fire watchtower" that did not conform to the architectural style of Otaru and had never existed in the city. Nevertheless, the building won favor with local residents. By contrast, a national clothing chain that opened next door met with a far harsher local reception. Kuramoto discusses an interesting contradiction: a national clothing chain that changed its shop design to conform with a local ordinance regulating the appearance of local buildings nevertheless met with local censure, while the confectionery maker that erected a kitschy "watchtower" won favor. This is a good example of the "sociological potential" of firsthand observation. See Horikawa and Morihisa, eds. (2009: 44–46).

First, a straightforward aggregation of the data collected during the fixed point observation surveys can be used to analyze the special characteristics of each particular area within Otaru's larger port district. This enables us to identify key trends that illuminate changes to the spatial construction of the area targeted by the preservation movement. This analysis also serves to refine and supplement the local history of Otaru presented earlier in this volume.

We will then consider how the impact of tourism development has been manifested by comparing rates of changed external appearance and rates of converted usage among the historic buildings targeted in the survey. We will continue the "use conversion matrix" developed by the Hokkaido University studies to assess the reality of change in Otaru.

Let us begin, however, by confirming the special characteristics of this line of analysis.

First, this study does not make use of the Hokkaido University survey's "building survival rate" for historic buildings. At the risk of repetition, this "building survival rate" indicated the number of buildings within a given district that remained unchanged from the year before. The building survival rate for year *xx*/*yy* was computed by asking how many buildings standing in year *xx* were still standing in year *yy*, taking no account of repairs or modifications to the external appearance of the buildings. Within the study of architecture, which takes buildings themselves as the subject of investigation, this is a concept of fundamental importance. However, it does not suit the objective of the current investigation, which is more interested in how changes to the urban landscape are understood by local residents, and the social processes that emerge in response to such changes. Moreover, as we shall see, a high "building survival rate" should not be equated with a low "rate of change." For someone attempting to spotlight change, the architectural question of a building's continued existence is of little concern and is thus not taken up here.[9]

The second characteristic of this line of analysis pertains to the substance of change. Nobody knows a particular area better than the people who live and work there. Unlike surveys that seek to capture nothing more than a particular moment in the target location, local residents know that conditions on the same road differ from morning to night; they can recognize the sound of people breathing in the next house and can tell you the volume of garbage their neighbor produces each week. They know who gathers nightly at the local watering hole, the inventory of all the local shops, and street parking conditions, day and night. They possess all these intimate forms of local knowledge (Horikawa 1998a). In fact, after years of interviewing Otaru residents, it is very clear that local residents are extraordinarily sensitive to changes in their area and highly knowledgeable about the family structure of local households and questions of local land ownership. In some cases, interviews revealed that respondents did not view architectural changes to a building as real changes. In such cases, the

[9]Of course, I have no intention of negating the concept's validity to the objectives and context of Myōen et al. (1993). On the contrary, the "building survival rate" concept employed by Myōen and his colleagues effectively demonstrates that even Otaru's "historic" buildings are being destroyed. The concept simply does not fit with the purpose of this investigation.

sociological concerns of this book dictate that the residents' perception of non-change prevail over the factual occurrence of architectural change. The concept of change employed in this volume is thus a sociological expansion of the architectural concept of change.

With these methodological concerns established, the following section uses landscape survey data to assess the impact of tourism development on the city of Otaru.

6.4 The Impact of Tourism Development: What Did Otaru Lose?

6.4.1 Distinctions Between Districts Revealed by Land Use

Let us begin by ascertaining the spatial structure of Otaru's port district, how this structure has changed over time, and the key trends associated with these changes by using the data regarding categories of use presented in Tables 6.8 and 6.9.

Table 6.8 traces changes to the most common forms of building use within individual districts, allowing us to identify both the prevailing forms of use in each district in any given year and potential trends.

In District A, the most common form of building use changed over time from "(1) warehouses or factories" to "(7) parking lots, vacant lots, and unoccupied buildings," and then to "(8) other." In other words, District A changed from a district of warehouses and factories to one dominated by vacant lots and parks. The increase in parks (the creation of the new "canal park") and vacant land indicates the hollowing out of the district.

Table 6.8 Most common forms of building usage by district (1986–2016)

District	1986	1992	1998	2008	2016
A	1	7	7	8	8
B	1	1	1	7	7
C	1	3	3, 7	7	7
D	2	2	2, 5	3, 4, 5	5
E	1	1	1	7	7
F	2	2	2	3	7
G	2	2	2	2, 7	2, 7
H	2	2	3	3	3
I	2	1	3	3	3
J	2	2	3	3	3

Table 6.9 Shifts in building "Usage" in Otaru's canal and port district (1986–2016)

Block name	District name	Most common category of use	Trends and nature of change		Notes
			Before (1986)	After (2016)	
Canal district	A	1 → 7 → 8	Warehouses, factories	Vacant lots, parks	Hollowing out of the district
	B	1 → 7	Warehouses, factories		Decrease in warehouses and factories and a striking number of empty buildings
	C	1 → 3 → 3 + 7 → 7	Warehouses, factories	Souvenir shops	Increasing number of parking lots and vacant lots
	D	2 → 2 + 5 → 3 + 4 + 5 → 5	Offices	Museums, hotels	Dramatic drop in the number of offices, and an increase in retail shops targeting tourists
	E	1 → 4 → 7	Warehouses, factories, and offices	Vacant lots, restaurants	In the southern part of the district, a group of warehouses remains in active use
Ironai district	F	2 → 3 → 7	Offices	Vacant lots, tourism-related retail	Increase in number of parking lots, vacant lots, retail, and apartment buildings
	G	2 → 2, 7	Offices		Museums, hotels, retail, increase in parking lots and vacant lots
Sakaimachi district	H	2 → 3	Offices	Souvenir shops	Sharp increase in restaurants
	I	2 → 1 → 3	Offices	Souvenir shops	Sharp increase in retail shops targeting tourists, an increase in restaurants

(continued)

Table 6.9 (continued)

Block name	District name	Most common category of use	Trends and nature of change		Notes
			Before (1986)	After (2016)	
	J	2 → 3	Offices	Souvenir shops	Sharp rise in retail shops targeting tourists, an increase in restaurants

1: Warehouses, factories
2: Offices
3: Retail
4: Restaurants
5: Museums, hotels
6. Roads, sidewalks
7: Parking lots, vacant lots, vacant buildings
8: Other

In contrast to District A, District B continues to be dominated by "(1) warehouses or factories," although the total number of such buildings decreased by half over time. Since 2000, however, the number of "parking lots, vacant lots, and unoccupied buildings" has increased, suggesting that this district is also hollowing out. In District C, the warehouse quarter has been transformed into an area full of small souvenir shops. However, the growing ascendancy of "parking lots, vacant lots, and unoccupied buildings" in this district indicates that the same hollowing out phenomenon is at work. This district is home to a famous music box museum and other well-known souvenir purveyors, but it also includes many shuttered warehouses. As such, it lacks the energy of districts with working factories or a continued residential presence. This transition began in 1998.

In District D, the most common form of building use changed from "(2) offices" to "(4) restaurants" and "(5) museums and hotels." This transition began around 2005. The long and narrow District E was originally comprised primarily of "(1) warehouses or factories" and "(2) offices," giving it the distinct character of a working area. Over time, however, these numbers were halved. The late 1990s witnessed the emergence of "(4) restaurants" in this area, but many have since closed and today there are a growing number of vacant lots and parking lots where buildings used to stand. Some warehouses remain in active use in the southern portion of this district, but this could also be an indication that the trend of refashioning old warehouses into restaurants is ending. In all cases, areas of the city that were formerly identified as working areas have been transformed into tourism-oriented districts. This trend is corroborated by the rapid proliferation of parking lots in these areas.

Taken together, Districts A through E could be referred to as the "canal block" of Otaru. This "canal block" of districts was formerly the very center of the commercial port of Otaru: a focal point for trade to and from mainland Japan and a major distribution center for goods arriving from overseas. Today it incorporates both empty, hollowed out areas and areas transformed by the tourist trade, which are now lined with restaurants and shops.

Let us turn our attention to Districts F and G, collectively known as the "Ironai block," where the wholesalers and financial institutions were concentrated during Otaru's heyday as a commercial port city. District F was traditionally comprised of wholesale offices. More recently, retail shops began to open in the district and retain a presence to this day. These retailers most likely opened in this district in order to target the growing number of tourists to neighboring areas. Since around 2005, the number of "(7) parking lots, vacant lots, and unoccupied buildings" has increased, as have the number of high-rise apartment buildings. While these apartment buildings increased the population of the district, we should note that since 2008 the number of vacated houses and parking lots exceeds that increase in population. District G, meanwhile, is the traditional financial center of Otaru—the area once known as "the Wall Street of the North." The survey data shows this district has consistently been dominated by "(2) offices," yet the data also reveals a marked increase in "(5) museums and hotels," "(3) retail shops," and "(7) parking lots, vacant lots, and unoccupied buildings."

Districts H, I, and J are contiguous and occupy long narrow strips of land along both sides of a road. Together, they can be referred to as the "Sakaimachi block." Survey data shows that while District H began as an area dominated by "(2) offices" it now has more "(3) retail shops." While the former comprised 54% of the district in 1986, by 2003 it comprised just 12%. The latter, meanwhile, steadily increased in ratio from 37% in 1997 to 47% in 2003. If we include "(4) restaurants" in the count, a full 70% of the district is comprised of retailers or restaurants. When we also consider the increase in the number of parking lots, it is clear that the entire district has been transformed in order to target tourists. The same process is apparent in District I. In 1986, this district was still predominantly a place of work: together "(1) warehouses and factories and "(2) offices" comprised 75% of total buildings usage. By 2003, this percentage had dropped to just 15%, with "(3) retail shops" and "(4) restaurants" comprising 73% of the total. In the first half of the 1990s, "(1) warehouses and factories" maintained their ascendance in the district, but since then "(3) retail shops" have held the dominant position. The surge in tourism-related businesses has largely been the result of redevelopment led by parking lot conversions.

District J, like the other two districts in the Sakaimachi block, has come to revolve around the local tourist trade. District J was also once a "place of work," where "(2) offices" comprised the vast majority of local buildings. Since 1998, however, the most common form of building usage in the district is "(3) retail shops," followed by "(4) restaurants." Together the two comprise 77% of usage in the district (as of 2003). Considering the fact that in 1986, "(1) warehouses and factories" and "(2) offices" comprised 85% of the district, the extent of change is as astonishing as that in Districts H and I.

To review, the canal block displayed two distinct trends: certain districts hollowed out but continue to be dominated by warehouses and factories, while other districts were entirely given over to the tourist trade. The Ironai block remains a working area, but the signs of hollowing out (the growing number of parking lots and vacant lots) are apparent. The Sakaimachi block, finally, has been entirely transformed by the predominance of retail shops and restaurants, largely aimed at tourists (Table 6.9).

6.4.2 *What Has Changed? an Analysis of the Changes to the Rate of Use and the Rate of Landscape Change*

To grasp the realities behind these changes, let us begin by comparing the rates of conversion in building use and rates of townscape change between 1997 and 2003—the pinnacle of Otaru's tourism boom. Looking at Tables 6.10 and 6.11, we can see that there is little difference between the rate of use conversion and the rate of townscape change.

In other words, we can see that both rates progress within the same range. This seems to support the implicit hypothesis in Myōen (1992): renovations undertaken as part of conversion in usage are a primary cause of townscape transformation. Of course, this simple tabulation is not enough to discuss statistical causality. If we consider the issue with specific regard to Otaru, however, it seems reasonable to assume that conversions in building use have led to changes in the appearance of the physical townscape.

Nevertheless, it is important to consider minute differences between the two. In other words, are there circumstances in which changes to building use do not correspond to changes in the townscape (use conversion ≠ townscape change)? A close examination of the survey results allows us to identify such instances. For example, a retail shop may be converted into a storage building for remaining stock but not remodeled in any way. There are also cases in which one restaurant becomes another (sushi restaurant A → sushi restaurant B, or set meal (*teishoku*) restaurant → sushi restaurant) or is subdivided to allow for additional uses. Such changes may not involve any external change in the architectural sense, yet they do signify another form of change: the hollowing out of the district. In short, Otaru has witnessed similar levels of change to building usage and its townscape; changes in use appear to be a primary cause of townscape change, and there is a striking trend toward the conversion of buildings into parking lots or vacant lots.

Let us now examine this block by block. In the Sakaimachi area, the rate of use conversion (Fig. 6.4) is consistently higher than the overall average; conversely, the canal district has almost always remained lower than this average. In the Ironai area, conversions in use have been pronounced since the beginning of the 2000s. If we look at the rate of townscape change by block (Fig. 6.5), the results are similar to those for the use conversion rates: Sakaimachi is almost always above average, and the canal district is below. Change in the Ironai district has fluctuated repeatedly. What does this suggest?

The industrial structure of the canal district changed dramatically following the end of barge-based cargo transport, the rise of truck-based distribution, the demolition of warehouses for the construction of the Rinkōsen, and the transfer of warehouse

Table 6.10 A six-interval comparison of changes in building usage rates by survey block (1997–2003)

			Year to year comparison					
			1997/1998	1998/1999	1999/2000	2000/2001	2001/2002	2002/2003
Canal district	Total number of buildings (a1)		87	89	89	89	89	89
	Number of buildings with a change in use	A	1	2	0	0	0	0
		B	1	2	1	0	1	1
		C	1	1	0	0	0	1
		D	1	1	0	2	2	1
		E	0	3	0	0	1	1
		Total (b1)	4	9	1	2	4	4
	Rate of use conversion (b1/a1)		**4.6%**	**10.1%**	**1.1%**	**2.2%**	**4.5%**	**4.5%**
Ironai district	Total number of buildings (a2)		..	..	85	85	87	85
	Number of buildings with a change in use	F	..	..	6	2	7	9
		G	..	..	2	1	1	4
		Total (b2)	..	..	8	3	8	13
	Rate of use conversion (b2/a2)		..	..	**9.4%**	**3.5%**	**9.2%**	**15.3%**
Sakaimachi district	Total number of buildings (a3)		102	103	103	103	104	103
	Number of buildings with a change in use	H	8	4	3	4	6	4
		I	1	1	3	4	2	5
		J	3	4	2	1	1	1
		Total (b3)	12	9	8	9	9	10
	Rate of use conversion (b3/a3)		**11.8%**	**8.7%**	**7.8%**	**8.7%**	**8.7%**	**9.7%**
Overall rate of use conversion ([b1 + b2 + b3]/[a1 + a2 + b3])			8.5%	9.4%	6.1%	5.1%	7.5%	9.7%

Notes (1) See the main text for the definition of the "rate of use conversion."
(2) Two dots (..) indicate areas outside the survey scope, for which data is unavailable (Ironai district Blocks F and G, from 1997–1998)
(3) Blocks F and G in 1997–1998 were excluded from the calculation of the overall rate of use conversion ([b1 + b3]/[a1 + a3])
Sources (1) Data for 1997–1999: Survey data collected as part of the Hosei University social research practicum in 1997, 1998, and 1999 (Horikawa, ed. *Otarushi ni Okeru Rekishiteki Kankyō Hozon to Kankō Kaihatsu 1–3: Hosei Daigaku Shakaigakubu Shakai Chōsa Jisshū Hōkokusho, 1998–2000*)
(2) Data for 2000–2003: Survey data from supplementary and follow-up surveys to (1) above, conducted by the author and his research team

Table 6.11 A six-interval comparison of landscape transformation rates by survey block (1997–2003)

			Year to year comparison					
			97/98	98/99	99/00	00/01	01/02	02/03
Canal district	Total number of buildings (a1)		87	89	89	89	89	89
	Number of buildings with external changes	A	1	2	0	0	1	0
		B	1	0	1	0	1	1
		C	1	0	0	1	1	0
		D	1	0	0	2	2	1
		E	0	3	4	4	1	1
		Total (b1)	4	5	5	7	6	3
	Rate of townscape change (b1/a1)		**4.6%**	**5.6%**	**5.6%**	**7.9%**	**6.7%**	**3.4%**
Ironai district	Total number of buildings (a2)		..	..	85	85	87	85
	Number of buildings with external changes	F	..	..	7	3	8	6
		G	..	..	1	1	1	1
		Total (b2)	..	..	8	4	9	7
	Rate of townscape change (b2/a2)		..	..	**9.4%**	**4.7%**	**10.3%**	**8.2%**
Sakaimachi district	Total number of buildings (a3)		102	103	103	103	104	103
	Number of buildings with external changes	H	5	3	4	3	3	7
		I	2	2	5	4	2	5
		J	1	5	1	1	2	1
		Total (b3)	8	10	10	8	7	13
	Rate of townscape change (b3/a3)		**7.8%**	**9.7%**	**9.7%**	**7.8%**	**6.7%**	**12.6%**
Overall rate of change ([b1 + b2 + b3]/[a1 + a2 + b3])			6.3%	7.8%	8.3%	6.9%	7.9%	8.3%

Notes (1) See the main text for the definition of the "rate of townscape change."
(2) Two dots (..) indicate areas outside the survey scope, for which data is unavailable (Ironai district Blocks F and G, from 1997–1998)
(3) Blocks F and G in 1997–1998 were excluded from the calculation of the overall rate of use conversion ([b1 + b3]/[a1 + a3])
Sources (1) Data for 1997–1999: Survey data collected as part of the Hosei University social research practicum in 1997, 1998, and 1999 (Horikawa, ed. *Otarushi ni Okeru Rekishiteki Kankyō Hozon to Kankō Kaihatsu 1–3: Hosei Daigaku Shakaigakubu Shakai Chōsa Jisshū Hōkokusho, 1998–2000*)
(2) Data for 2000–2003: Survey data from supplementary and follow-up surveys to (1) above, conducted by the author and his research team

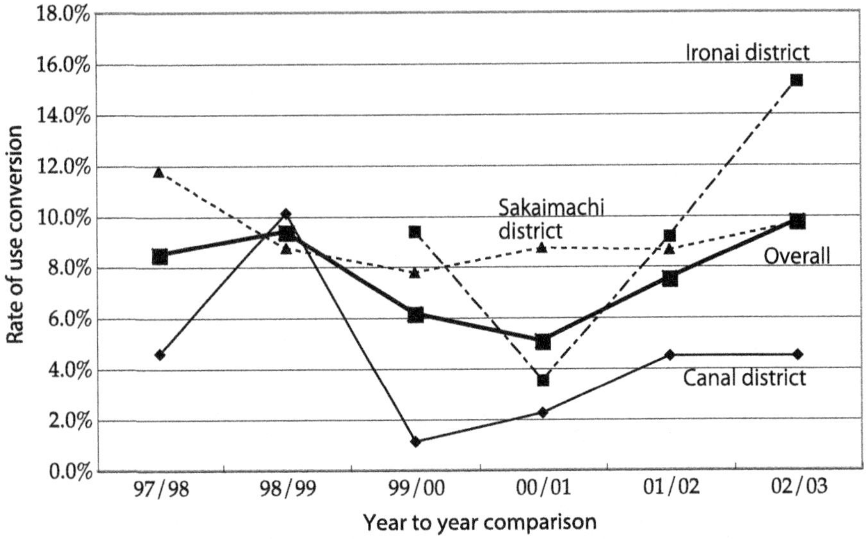

Fig. 6.4 Shifts in the rates of change to building usage by survey block (1997–2003)

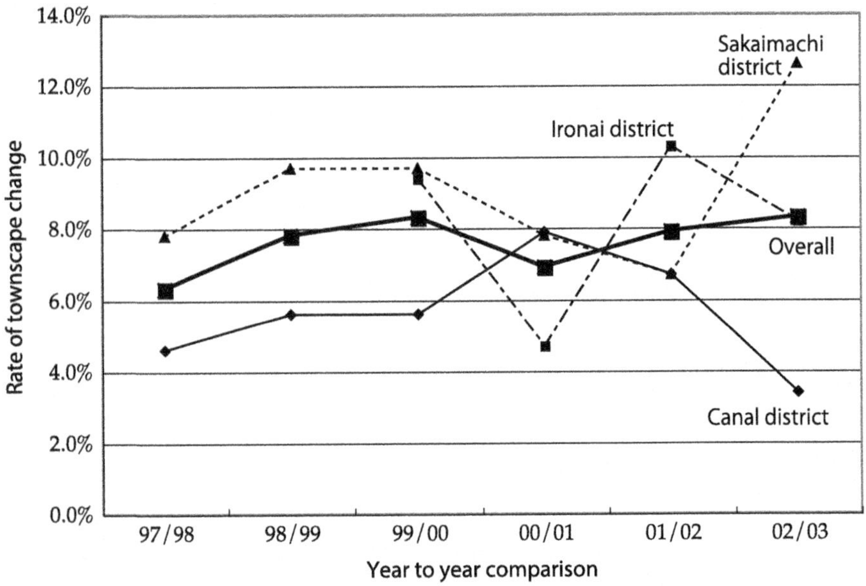

Fig. 6.5 Shifts in the rates of landscape change by survey block (1997–2003)

functions to the piers. A cycle emerged in which warehouses no longer in use were left vacant and eventually destroyed. The area lost approximately one-third of its warehouses to this vicious cycle. Indeed, the distinctive townscape of Otaru, with its lines of gabled-roof warehouses, was lost in the process (Myōen et al. 1993).

The Kitahama portion of the canal (Districts A–C) does not attract many tourists and has no souvenir shops. Nor is there much evidence of discontinued warehouses being converted to new uses. The new canal park was completed at the northern tip of the canal in the early 2000s, but this did not necessitate any major changes to the landscape. This is reflected in the rates of townscape change (3.4%) and use change (4.5%).

The Ironai block (F–G districts) was originally comprised of wholesalers and banks. The fact that these districts have maintained their character can be attributed to the renovation of large buildings (such as former banks), funded by large corporations or the government. Since 2001, the rate of change in building use has continued to increase, suggesting that some buildings are being converted for tourism-related purposes. Given that many buildings have a historic designation from the city of Otaru, however, changes to external appearance as a result of conversion have been kept low.

The Sakaimachi area is home to stores with national reputations, including Kitaichi Glass and the confectioner LeTAO, and has been a focal point for tourism since 1986. The area has experienced enormous change as a result. Historically, Sakaimachi was the center of the wholesale trade, and its clusters of small, wooden buildings were easily converted into souvenir shops with just minor alterations to their exterior facades. The area was thus engulfed by the growing wave of tourism. The data indicates that Sakaimachi has been the site of much of Otaru's townscape change in the 2000s. Indeed, almost all of the comments made by local residents in the course of fieldwork or door-to-door interviews referred to the Sakaimachi district ("lines of souvenir shops," cheap, "souvenir-based tourism," "Dejima tourism"). In the words of former preservation activist Ogawara Tadashi: "Six million tourists (in 1999) came just to Sakaimachi and the canal…it's become an entirely 'Dejima-style' style of tourism" (Interview with Ogawara Tadashi in Otaru, August 31, 1999. Information in parentheses has been added by the author). Ogawara is referring to the hordes of tourists who arrive in Sakaimachi to quickly eat sushi and purchase souvenirs and then leave again, having had no interaction with local residents. He is likening Sakaimachi tourism to that on the island of Dejima in Nagasaki, where a similar phenomenon occurs, and casting it in a very negative light.

This is not merely Ogawara's personal impression. In the words of Ōno Tomonobu, a former vice chairman of the Otaru Chamber of Commerce and Industry and a supporter of the preservation movement:

> If you go down this street you'll see there's a new conveyor-belt sushi restaurant opposite the Orugōrudō [music box museum], run by someone from Honshu. Locals who try it once would never return. Tourists don't know any better and will end up eating and drinking there because it looks good from the outside. And if the sushi is just of that [mediocre] level, they'll never bother coming back. So if we don't think more about tourism in Otaru, we'll end up

with the kind of tourism, the souvenir-buying kind of tourism, that happens anywhere. We have such a unique town, so I wish we could come up with a unique way of using it.

Interview with Ōno Tomonobu in Otaru on September 3, 1997. Information in brackets has been added by the translator.

Ōno's reference to "souvenir-buying tourism" carries the same negative connotation as Ogawara's disparaging reference to "Dejima tourism."

This sense of discomfort with "Dejima tourism" and "souvenir-buying tourism" is also evident in the following narrative:

In early (19)90, when tourism was taking off, I had the feeling that something was wrong…I had a strong sense of unease at [what was going on in places like] Sakaimachi. (Former APOC chairwoman) Mrs. Mineyama would often say 'there's something so strange about all of this, this isn't the way it should be…' There was a real feeling of discomfort about it.

Interview with Yanagida Ryōzō, Ishizuka Masaaki, and Morishita Mitsuru in Sapporo on September 1, 1999. The words spoken here are Yanagida's. Information in parentheses has been added by the author, information in brackets by the translator.

We should note that amid all the discussions of general discomfort with the state of tourism in Otaru, Sakaimachi is the only place referenced by name. According to Yamaguchi Tamotsu, a former preservation activist who later continued his community development efforts through a new tourism council:

"Tourism in Otaru was left uncontrolled (by any concrete tourism policy) for ten years. The number of tourists increased from one million to six million last year (1997). So [Otaru] has become a major tourist city. And if you stop and look closely, the area around Kitaichi Glass, in Sakaimachi, is entirely covered by vulgar souvenir shops that are riding on Kitaichi's coattails and will sell anything. They'll sell crabs. They'll sell melons. So we tried to (limit this with a landscape ordinance regulating scenery), so they can't put up some weird building, but they're still allowed to sell anything they like, which I don't agree with at all… That's the problem. It's 'Dejima tourism' through and through. I want to expand the area for tourism and make Otaru a tourist city with real atmosphere."

Interview with Yamaguchi Tamotsu in Otaru, September 4, 1998. Information in parentheses has been added by the author, information in brackets by the translator.

Here again we encounter the term "Dejima tourism" and a specific reference to Sakaimachi. Hardly any mention is made of the Ironai district. This is probably because many of the old buildings along the former "Wall Street of the North" carry historic designations, restricting changes to the townscape.

Let us turn to the question of how the use of buildings has changed. Following Myōen et al. (1993), I have created a use conversion matrix to identify the specific changes at work.

Table 6.12a depicts Hokkaido University survey data that has been updated by the author.

Of 138 office buildings recorded in 1986, 97 were still being used by the same office in 1992. The majority of the remaining 41 buildings had been converted into retail stores (14 buildings) or parking lots or vacant lots (15 buildings). Similarly, of the 88 warehouses and factories that once stood in the district, the majority were

Table 6.12a A matrix of changes to building usage 1986/1992 (Across all survey blocks)

		Building uses in 1986								1992 total
		Warehouses, factories	Offices	Retail	Restaurants	Museums, hotels	Roads, sidewalks	Parking lots, vacant lots	Other	(horizontal total)
Building uses in 1992	Warehouses, factories	**51**	0	0	0	0	0	0	0	51
	Office	2	**97**	0	0	0	0	0	3	102
	Retail	11	14	**22**	0	0	0	1	3	51
	Restaurants	5	5	0	**4**	0	0	0	1	15
	Museum, hotel	2	3	0	0	**3**	0	1	1	10
	Roads, sidewalks	2	2	0	1	0	**0**	0	1	6
	Parking lot, vacant lot	14	15	1	0	0	0	**4**	0	34
	Other	1	2	1	0	0	0	0	**15**	19
1986 total (vertical total)		88	138	24	5	3	0	6	24	***288***

Source The data for 1986 and 1992 is from the Hokkaido University surveys (Myōen 1992; Myōen et al. 1993), with some corrections by the author
Notes (1) Due to certain corrections and recalculations, the totals (vertical and horizontal) may not match those for 1992/1998
(2) Cases of multiple uses have been tabulated as follows
A case of "restaurant → restaurant and parking lot" is counted as one instance each of "restaurant → restaurant" and "restaurant → parking lot."
A case of "restaurant and parking lot → restaurant and parking lot" is counted as one instance each of "restaurant → restaurant" and "parking lot →parking lot."
A case of "restaurant and parking lot → "parking lot" is counted as one instance each of "restaurant → parking lot" and "parking lot → parking lot."
(3) Buildings that did not yet exist in 1986 were excluded from the totals, but cases of new construction within existing structures are included

converted into retail stores (11 buildings) and parking lots or vacant lots (14 buildings). These changes indicate that one quarter of the offices and warehouses that were once accessible only to concerned parties have been transformed into retail stores and parking or vacant lots accessible to the general public.

Table 6.12b depicts the use conversion matrix developed using the final data set from Hokkaido University and subsequent data collected by the author (Hosei University data).

Fluctuations in the number of buildings have led to some small inconsistencies between the figures presented here and those in the previous table. Of the 104 offices in existence in 1992, just 65 remained as the same offices at the time of the 1998 survey. Of the 39 buildings that had changed, most had been converted into retail stores (19 buildings) or parking (or otherwise vacant) lots (13 buildings). Of the 54 warehouses and factories, nine buildings were converted into retail stores, and eight buildings were converted into restaurants or bars. When we look more closely at the specific buildings that were converted, it is immediately clear that these buildings were intentionally converted into tourist-oriented facilities.

Table 6.12c shows the use conversion matrix developed with Hosei University survey data for 1998 and 2008. Of the 76 offices in existence in 1998, only 43 remained ten years later, while the others had been converted to other uses: retail, restaurants, and parking lots. The same overall trend of warehouses and offices converted into retail stores and tourist-oriented restaurants, with the remainder put to use as parking lots, is apparent—but the increase in the number of retail shops is striking. Finally, the use conversion matrix in Table 6.12d depicts changes between 2008 and 2016. There was little change to the 21 remaining warehouses and factories in 2008: in 2016, 17 remained in operation, while the remainder had been converted into retail stores, restaurants, or offices.

Taken together, the interview data, the data presented in Myōen et al. (1993), data on the rate of townscape change, and the use conversion matrixes enable us to summarize our findings as follows:

Changes to Otaru's townscape during the 1980s occurred primarily in the canal and Sakaimachi districts. There was relatively little change to the appearance of buildings in the Ironai district. In the canal and Sakaimachi districts, groups of stone warehouses were demolished and replaced with parking lots, while other buildings were overhauled and transformed into souvenir shops or restaurants. In the late 1990s, the change occurred primarily in the canal district, particularly in Districts D and E. Since the 2000s, Sakaimachi has been the "epicenter" of townscape transformation; this has been characterized by a remarkable increase in the number of souvenir shops targeting tourists, and—despite the fact that Otaru's tourist boom stemmed from its reputation as the "canal town"—changes to both the appearance and use of historic buildings that have dramatically transformed the Otaru townscape. The use conversion matrix tables reveal the bipolar nature of this change: the trend toward the creation of ever more souvenir shops and restaurants aimed at tourists has been accompanied by an increase in vacant land and parking lots. In some cases, the latter has created space for the construction of new stores. The revival of large-scale tourism since 2012 has only obscured the steady destruction of Otaru's historic townscape.

Table 6.12b A matrix of changes to building usage 1992/1998 (Across all survey blocks)

		Building uses in 1992								1998 total (horizontal total)
		Warehouses, factories	Offices	Retail	Restaurants	Museums, hotels	Roads, sidewalks	Parking lots, vacant lots	Other	
Building uses in 1998	Warehouses, factories	**30**	2	3	0	0	0	0	0	35
	Offices	1	**65**	2	0	0	1	4	1	74
	Retail	9	19	**37**	0	0	0	1	2	68
	Restaurants	8	4	3	**13**	0	0	3	3	34
	Museums, hotels	1	0	2	0	**10**	0	2	0	15
	Roads, sidewalks	0	0	0	0	0	**5**	0	0	5
	Parking lots, vacant lots	3	13	4	0	0	0	**19**	2	41
	Other	2	1	0	1	0	0	0	**11**	15

(continued)

Table 6.12b (continued)

	Building uses in 1992								1998 total (horizontal total)
	Warehouses, factories	Offices	Retail	Restaurants	Museums, hotels	Roads, sidewalks	Parking lots, vacant lots	Other	
1992 total (vertical total)	54	104	51	14	10	6	29	19	***287***

Sources The data for 1992 is from the Hokkaido University surveys (Myōen 1992; Myōen et al. 1993), with some corrections by the author. The 1998 data is from a survey conducted by the members of the Hosei University social research practicum (1997–1999) under the direction of the author (Horikawa, ed. 1999), with subsequent corrections and recalculations by the author. The data for 1992 is from the Hokkaido University surveys (Myōen 1992; Myōen et al. 1993), with some corrections by the author. The 1998 data is from a survey conducted by the members of the Hosei University social research practicum (1997–1999) under the direction of the author (Horikawa, ed. 1999), with subsequent corrections and recalculations by the author

Notes (1) Due to certain corrections and an increase in the number of buildings, the 1992/1998 totals (vertical and horizontal) may not match those for 1998/2008

(2) Cases of multiple uses have been tabulated as follows

A case of "restaurant → restaurant and parking lot" is counted as one instance each of "restaurant → restaurant" and "restaurant → parking lot."

A case of "restaurant and parking lot → restaurant and parking lot" is counted as one instance each of "restaurant → restaurant" and "parking lot →parking lot."

(3) Buildings that did not yet exist in 1992 were excluded from the totals, but cases of new construction within existing structures are included

Table 6.12c A matrix of changes to building usage 1998/2008 (Across all survey blocks)

		Building uses in 1998								2008 total
		Warehouses, factories	Offices	Retail	Restaurants	Museums, hotels	Roads, sidewalks	Parking lots, vacant lots	Other	(horizontal total)
Building uses in 2008	Warehouses, factories	**17**	0	1	0	0	0	0	1	19
	Offices	1	**43**	1	0	0	0	0	1	46
	Retail	7	13	**58**	6	0	0	8	2	94
	Restaurants	5	8	3	**27**	2	0	8	3	56
	Museums, hotels	1	2	1	0	**13**	0	0	0	17
	Roads, sidewalks	0	0	0	0	0	**4**	0	0	4
	Parking lots, vacant lots	8	7	6	5	2	2	**24**	1	55
	Other	0	3	1	0	0	0	8	**11**	23
1998 total (vertical total)		39	76	71	38	17	6	48	19	***314***

Source The 1998 data is from a survey conducted by the members of the Hosei University social research practicum under the direction of the author (1997–1999) (Horikawa, ed. 1999), with subsequent corrections and recalculations by the author. The 2008 data is from a survey conducted by the author and his Hosei University research team

Source (1) Due to certain corrections and recalculations, the totals (vertical and horizontal) may not match those for 1992/1998

(2) Buildings that did not yet exist in 1998 were excluded from the totals, but cases of new construction within existing structures are included

Table 6.12d A matrix of changes to building usage 2008/2016 (Across all survey blocks)

		Building uses in 2008								2008 total
		Warehouses, factories	Offices	Retail	Restaurants	Museums, hotels	Roads, sidewalks	Parking lots, vacant lots	Other	(horizontal total)
Building uses in 2016	Warehouses, factories	**17**	0	0	0	0	0	0	0	17
	Offices	1	**32**	1	0	0	0	3	0	37
	Retail	2	2	**80**	4	1	0	7	1	97
	Restaurants	0	1	7	**50**	2	0	4	0	64
	Museums, hotels	0	0	1	0	**13**	0	1	0	15
	Roads, sidewalks	0	0	0	0	0	**4**	0	0	4
	Parking lots, vacant lots	1	8	5	4	1	0	**37**	2	58
	Other	0	4	2	0	0	0	4	**20**	30
2008 total (vertical total)		21	47	96	58	17	4	56	23	***322***

Source The 2008 and 2016 data are from a survey conducted by the author and his Hosei University research team

Notes (1) Due to certain corrections and an increase in the number of buildings, the totals (vertical and horizontal) may not match those for 1998/2008

(2) Buildings that did not yet exist in 2008 were excluded from the totals, but cases of new construction within existing structures are included

The townscape of Otaru has undergone a dramatic transformation. While "the canal and stone warehouses" continue to define the city's image, the number of warehouses has declined steadily. Quietly but surely, the survey data tells us that Otaru's unique townscape is disappearing.

6.4.3 A History of the Land: Transitions in Tourism Development and Land Ownership

Having examined the trend of tourism development in Otaru's port area, we must now turn to the history of the land used for tourism development. After all, a comprehensive understanding of the confrontation between the Otaru city government and the canal preservation movement must consider how the land in question was owned and disposed of (Horikawa 2008a).

6.4.3.1 Previous Studies and Sociological Possibilities

Given their necessity, remarkably few attempts have been made to analyze urban issues, city policies, or urban history through an investigation of the structure of urban land ownership. In the words of Natake Natsuki: "Research into the issue of land ownership in urban areas of Japan has primarily focused on the early modern period and the period of high economic growth and after, while only a few works focus on developments between the Meiji Restoration and the postwar reconstruction period" (Natake 2007: 4). There are several reasons for this. First, research on Japanese economic history has traditionally focused on agricultural areas, not cities. Second, academic interest in postwar land issues has focused upon the land transactions of large corporations, not the overall structure of land ownership. Finally, research has been constrained by limited materials (Natake 2007: 4–5).

Within this limited number of studies, several share my focus upon changes to landscape and related issues of governance within a particular city. These include studies conducted by the University of Tokyo Institute of Social Science (Tokyo Daigaku Shakaikagaku Kenkyūjo, ed. 1952, 1968), Okamoto Satoshi (2003), the Hosei University Graduate School of Eco-Design and the Japanese Minatomachi Research Association (Hōsei Daigaku Daigakuin Eko Dezain Kenkyūsho, Nippon no Minatomachi Kenkyūkai, eds. 2005), and Natake (2007). Although the scope and target of these studies vary, they represent valuable scholarly contributions for their accurate delineation of land ownership structures through fieldwork and primary sources, thereby providing basic data on how cities have changed. These works paint a vivid picture of a particular city through the exhaustive compilation of the various forms of data inherent to the system of land ownership, which underpins the physical existence of buildings and land.

Let us now turn to a more detailed examination of this research.

Building upon earlier (1952) research by the same institution, the University of Tokyo Institute of Social Science (1968) conducted exhaustive surveys of land, buildings, and households in small sub-districts of Takemachi, a neighborhood in Tokyo's Taito Ward, at four junctures between 1948 and 1964. These studies carefully documented the structure of absentee landlords and the transition from owned to rented homes and bring into quiet relief the form of redevelopment that resulted from this structure of landlords and renters (Tokyo Daigaku Shakaikagaku Kenkyūjo, ed. 1968: 3–24).

Okamoto Satoshi's study of the Ginza district of Tokyo starts by focusing upon the differences between the Ginza townscape and the ordering of European townscapes. Compared to Western cities, where the relationship between the site and the building is relatively clear, Okamoto confronts the space of Ginza, which appears to be far more irregular and suggests that: "In Japan, unlike the West, land defines the space of cities" (Okamoto 2003: 5) and proceeds to trace transitions in land ownership. Beginning his analysis from this perspective, Okamoto (2003: 89) writes that around the year 1912, "the urban structure of Ginza underwent a major transition from a samurai residential quarter oriented toward the waterfront and the *daimyō* mansions to one centered upon the Ginza Street." It is important to remember that this type of account is only possible through Okamoto's painstaking chronological review of specific cases of land ownership and transfers over time (Okamoto 2003: 89–92). Given that "today, even the fundamental facts regarding the significance of advancing concrete, individual-level analysis of the development of urban land ownership in Japan remain unclear" (Natake 2007: 2) we should recognize the pioneering nature of the research conducted by Tokyo Daigaku Shakaikagaku Kenkyūjo, ed. (1968) and Okamoto (2003).

I have attempted to learn from the research conducted by Tokyo Daigaku Shakaikagaku Kenkyūjo, ed. (1952, 1968), Okamoto (2003), and Natake (2007) in my effort to trace changes to land in Otaru. From a sociological perspective, the significance of this research lies in its potential to identify gaps between the reality of land change and local residents' consciousness of such change. The fixed point observation of 277 buildings in Otaru complements extensive interviews with local residents, creating a comprehensive picture of the changes to Otaru's landscape. However, there are subtle differences between the land ownership data that clearly trace the pace of tourism development in Otaru and the consciousness of local residents. These differences should *not* be regarded as an error or as something to be corrected. Rather, we should ask *what the discrepancies between the data and residents' consciousness can tell us*. The strategy of decoding this gap between the data and local consciousness is one form of sociological research, which allows for discoveries that would not have been possible through the approaches of architecture or economic history.

6.4.3.2 The Target Districts and Data

Here our target area is the Sakaimachi neighborhood of Otaru's port district (formerly known as Minatomachi). As we have seen, this area was originally a mix of workplaces and private residences and played an important role within the port city of Otaru. Fixed point observation surveys confirm that buildings in this district have experienced the greatest rate of change, and the pressures of tourism development have dramatically altered the urban landscape. We will now use the structure of land ownership to explore how this area has changed from the business space of the wholesale trade to the commercial space of the tourist trade.

I have collected historical data for twenty-two parcels of land located along the intersection in front of the Orugōrudō Music Box Museum building, at the southernmost tip of the target area (Figs. 6.6 and 6.7). This data was obtained from copies and inspection of registers, closed registers, land use ledgers, and cadastral maps[10] housed at the Otaru branch office of the Sapporo Regional Legal Affairs Bureau. I also referred to three historical resources that offered valuable clues for tracking land ownership and transfers of ownership in the district: Nagano and Marufuji, eds. (1916), Marufuji and Wakamatsu, eds. (1920), and Satō, ed. (1927). As a result, I have been able to ascertain the status of land ownership in this area from 1889 through the 1990s. The results are summarized in Tables 6.13, 6.14 (Part 1) and 6.14 (Part 2).

6.4.3.3 The Gap Between Data and Oral Accounts: The Myth of "Outside Capital"

What is immediately evident from Table 6.13 is the transition from individual to corporate ownership in the target area. At present, almost all of the twenty-two parcels of land in this area are owned by corporations (cf. shaded area in Tables 6.14 (Part 1) and 6.14 (Part 2)). If we examine the timing of the transfer of land from individuals to corporations, we can identify one peak in the 1920s and a second peak between the 1940s and the 1960s. These likely involved the transfer of land from small, privately owned businesses and wholesalers to banks and large corporations.

Having come under corporate ownership, this land concentrated in the hands of leading local companies, transforming the area into one dominated by tourism-related businesses. One prominent local company, Kitaichi Glass, owns eleven of the twenty-two parcels of land in our target area. In addition to its original mainstay of glass manufacturing and sales, Kitaichi has opened restaurants and souvenir shops aimed at tourists, as well as glass-making workshops. In short, this area has changed from the business space of wholesalers to a commercial space aimed at tourists. The

---

[10]This is based upon a blueprint produced in October 1924 and reproduced by this author at the Otaru Branch Office of the Sapporo Legal Affairs Bureau. However, supplementary maps are used to compensate for the fact that this blueprint does not account for the plot at Sakaimachi No. 272-2, which was newly registered in April 1976.

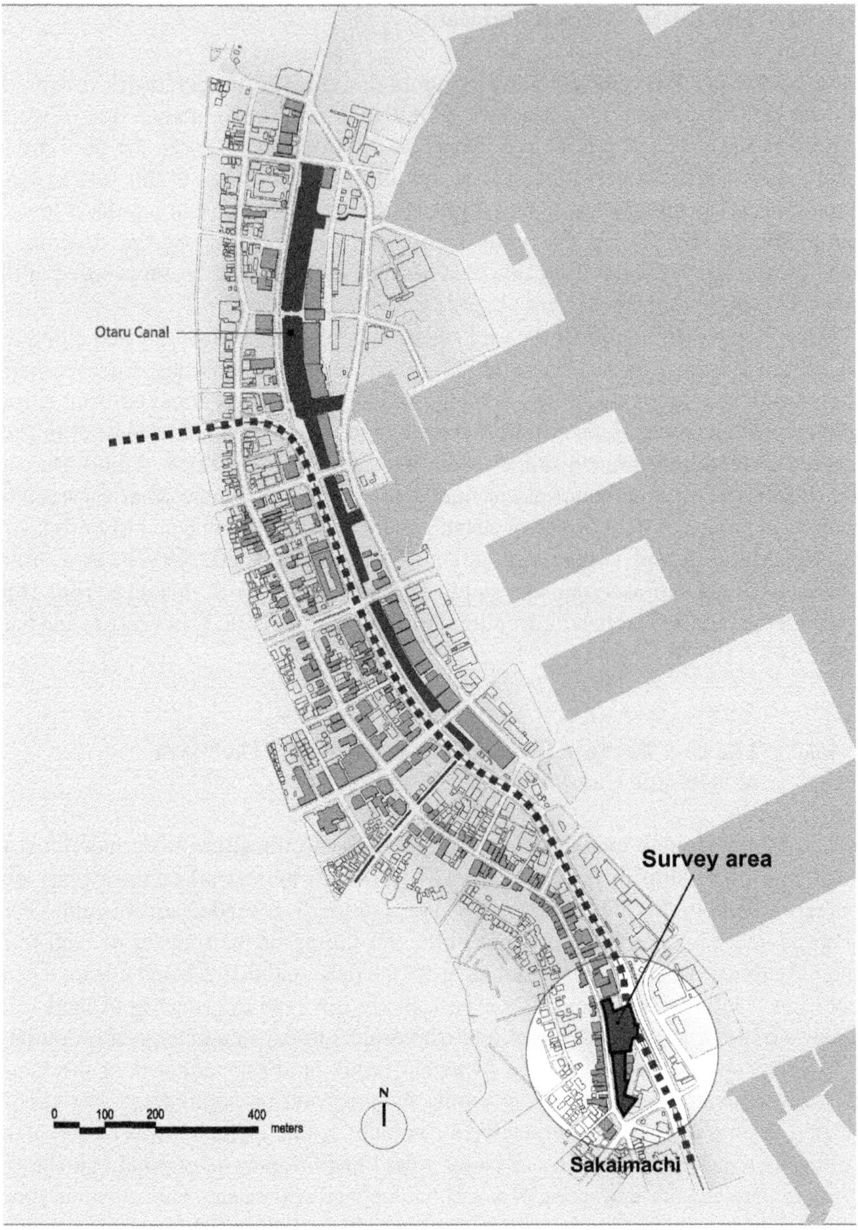

Fig. 6.6 Changes to land ownership in Otaru: an overview of the survey area (Author's illustration)

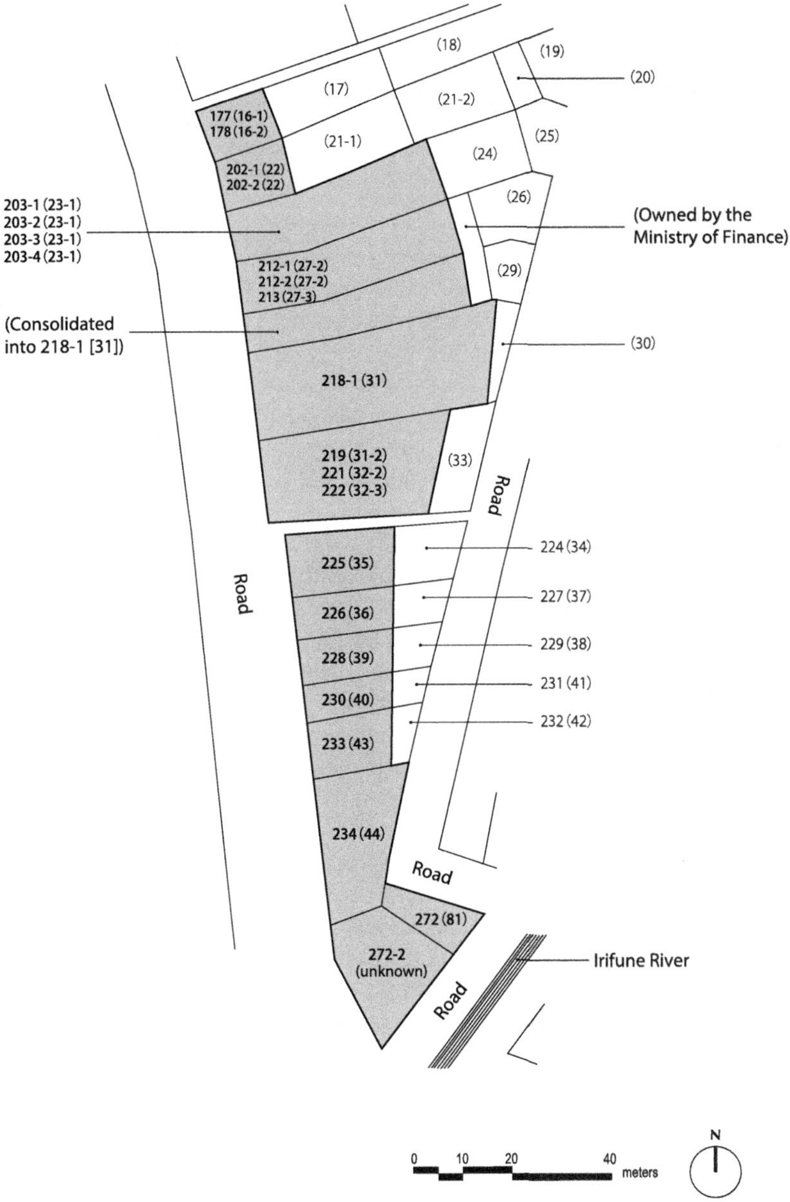
(18)
(19)
(20)
(17)
(21-2)
177 (16-1)
178 (16-2)
(21-1)
(25)
(24)
202-1 (22)
202-2 (22)
(26)
203-1 (23-1)
203-2 (23-1)
203-3 (23-1)
203-4 (23-1)
(Owned by the Ministry of Finance)
212-1 (27-2)
212-2 (27-2)
213 (27-3)
(29)
(Consolidated into 218-1 [31])
(30)
218-1 (31)
219 (31-2)
221 (32-2)
222 (32-3)
(33)
Road
224 (34)
225 (35)
Road
227 (37)
226 (36)
229 (38)
228 (39)
231 (41)
230 (40)
232 (42)
233 (43)
234 (44)
Road
272 (81)
272-2 (unknown)
Irifune River
Road
0 10 20 40 meters
N

◂**Fig. 6.7** Otaru land ownership survey: a close-up view of the target area. *Source* Created by the author based on official maps and similar drawings provided by the Otaru branch of the Sapporo Legal Affairs Bureau. *Notes* (1) The public map is rendered using the traditional Japanese system of measurement (shakkanhō), former location, and former address. (2) Current lot numbers are displayed first, followed by former lot numbers in parentheses. For example, "Sakaimachi 225" was formerly "Minatomachi 35."

period of land acquisition peaked between 1984 and 1986. This timing coincides with the conclusion of the protracted "canal war," and Otaru's emergence as a tourist destination. Since then, tourism-related industries have flourished and the number of visitors to Otaru has increased sharply. The origin of these trends is corroborated by the transfers in land ownership.

Next, let us compare rates of local versus outside ownership. Again, 80% of the land is held by local corporations (eighteen parcels are held by local companies, two parcels are held by outside corporations, and two parcels are held by the city of Otaru). The concentration of local capital is the reality behind the trend toward corporate ownership.

This data contradicts a lament heard frequently in the course of our fixed point observation survey and other interviews with local residents that: "Sakaimachi no longer belongs to us, the local residents. Outside investors from Sapporo or Tokyo are the ones who are making money there." The fact that a full 80% of the land in Sakaimachi is locally owned—and most of it held by Kitaichi Glass, a leading company founded in Otaru—is a direct refutation of the often-heard assertion by locals that profits have accrued solely to businesses operating with "outside capital."[11] Nevertheless, these assertions are widely believed to be true. How can we make sense of this contradiction?

False stories that are nevertheless held sacred are indeed "myths."[12] We can thus refer to this contradiction as "the myth of outside capital" (Horikawa 2013a). But how did this "myth of outside capital" emerge? And might it help us to understand the consciousness of local residents in the post-canal war period?

It is worth repeating that the disparity between land ownership data and residents' consciousness is not something that needs correction. Nor is it the result of a calculation error. On the contrary, this disparity has something vital to tell us. Rather

[11]One can detect yet another sentiment in the laments of local residents—namely, a sense of dismay that the people who worked the hardest for the cause of community development have received no reward for their considerable efforts, while those who took no part in such initiatives are now reaping all the benefits of the local tourism boom. This reveals a strong normative consciousness in which legitimacy is accorded to those who supported the preservation movement, and withheld from those who did not.

[12]There are two characteristics to myths: (1) regardless of their factual veracity, they are believed to be true and (2) they are deemed sacred and cannot be contested. See Horikawa (2013a).

Table 6.13 Changes to land ownership in the Sakaimachi District of Otaru

Location		Land classification	Owner according to the historical literature		
Former lot number	New lot number		1916	1920	1927
Minatomachi 16-1	Sakaimachi 177	Residential land	Yamada Kichibei (Otaru)	Yamada Kichibei (Otaru)	Hokkaido Takushoku Bank (Otaru)
Minatomachi 16-2	Sakaimachi 178				
Minatomachi 22-1	Sakaimachi 202-1	Residential land	Wada Tōkichi (Otaru)	Kakimoto Sakunosuke (Hakodate)	Otaru Shinbun Co., Ltd. (Otaru)
	Sakaimachi 202-2	Residential land			
Minatomachi 23	Sakaimachi 203-1	Residential land	Suita Tomizō (Otaru)	Ōno Yosamatsu (Otaru)	Hokkaido Bank (Otaru)
	Sakaimachi 203-2				
	Sakaimachi 203-3				
	Sakaimachi 203-4				
Minatomachi 27-2	Sakaimachi 212-2	Residential land	Sōma Teppei (Hakodate)	Sōma Teppei (Hakodate)	Sōma Teppei (Hakodate)
Minatomachi 27-3	Sakaimachi 213	Residential land			
Minatomachi 31-1	Sakaimachi 218-1	Residential land	Kimura Enkichi (Otaru)	Kimura Enkichi (Otaru)	Kimura Enkichi (Otaru)
Minatomachi 31-2	Sakaimachi 219				
Minatomachi 32-2	Sakaimachi 221	Residential land	Fujiyama Yōkichi (Otaru)	Fujiyama Yōkichi (Otaru)	Fujiyama Yōkichi (Otaru)
Minatomachi 32-3	Sakaimachi 222	Residential land			
Minatomachi 35	Sakaimachi 225	Residential land	Daiichi Bank (Otaru)	Shibusawa Sōko (Tokyo)	Shibusawa Sōko (Tokyo)
Minatomachi 36	Sakaimachi 226	Residential land	Adachi Keijirō (Otaru)	Adachi Keijirō (Otaru)	Adachi Keijirō (Otaru)
Minatomachi 38	Sakaimachi 228	Residential land	Adachi Keijirō (Otaru)	Adachi Keijirō (Otaru)	Adachi Keijirō (Otaru)
Minatomachi 40	Sakaimachi 230	Residential land	Satō Matsutarō (Atsuta-gun)	Satō Matsutarō (Atsuta-gun)	Satō Masao (Atsuta-gun)
Minatomachi 43	Sakaimachi 233	Residential land	Satō Matsutarō (Atsuta-gun)	Satō Matsutarō (Atsuta-gun)	Satō Masao (Atsuta-gun)
Minatomachi 44	Sakaimachi 234	Residential land	Honma Kikujūrō (unknown)	Kōmura Eitarō (Otaru)	Kōmura Eitarō (Otaru)
Minatomachi 81	Sakaimachi 272	mixed use	Otaru-ku (Otaru)	Otaru-ku (Otaru)	Otaru-shi (Otaru)
	Sakaimachi 272-2				

(continued)

Table 6.13 (continued)

Location		Land classification	Owner according to register or land registries									
Former lot number	New lot number											
Minatomachi 16-1	Sakaimachi 177	Residential land	Yamada Kichibei (Otaru, 1899 registration)	Yamada Ishi (Otaru, inherited in 1909)	Yamada Kichibei (Otaru, 1914, transfer of ownership?)	Hokkaido Takushoku Bank (Otaru, 1922, sale contract?)	Hokkaido Takushoku Bank Otaru branch (Otaru, 1925, bid acquisition)	Ogura Co., Ltd. (Otaru, transfer of ownership, divided and reregistered as Minatomachi 16-1, 16-2	..	..	..	..
Minatomachi 16-2	Sakaimachi 178							Ogura Co., Ltd. (Otaru, 1966, divided and reregistered as 16-2)	Imai Sangyō Co., Ltd. (Otaru, 1974, sale)	Kitaichi Co., Ltd. (Otaru, 1984, sale)	..	..
Minatomachi 22-1	Sakaimachi 202-1	Residential land	Wada Tōkichi (Otaru, 1889, sale)	Wada Tōkichi (Otaru, 1908, inheritance)	Kakimoto Sakunosuke (Hakodate, 1919, sale)	Otaru Zōsen Co., Ltd. (Unknown, 1922, bid acquisition)	Otaru Shinbun Co., Ltd. (Otaru, 1922, transfer of ownership)	Hokkaido Bank (Otaru, 1939, transfer of ownership due to confusion surrounding ownership rights)	Hokkaido Takushoku Bank (Sapporo, 1945, transfer of rights through merger)	Imai Sangyō Co., Ltd. (Otaru, 1974, sale)	Kitaichi Co., Ltd. (Otaru, 1984, sale)	Kitaichi Co., Ltd. (Otaru, 1985, Sakaimachi 202-1)
	Sakaimachi 202-2	Residential land										Kitaichi Co., Ltd. (Otaru, 1985, divided and reregistered as 202-2)

(continued)

Table 6.13 (continued)

Location		Land classification	Owner according to register or land registries									
Former lot number	New lot number											
Minatomachi 23	Sakaimachi 203-1	Residential land	Suita Rōkichi (Otaru, 1895, inheritance)	Suita Tomizō (Otaru, 1911, inheritance)	Ōno Yosamatsu (Otaru, 1918, sale)	Hokkaido Bank (Otaru, 1927, bid acquisition)	Hokkaido Takushoku Bank (Sapporo, 1945, transfer of rights through merger)	Hokkaido Takushoku Bank (Sapporo, 1972, registration of ownership through merger; divided and reregistered as 203-1)	..	..	..	..
	Sakaimachi 203-2							Hokkaido Takushoku Bank (Sapporo, 1972, registration of ownership by merger; divided and reregistered as 203-2)	Resolution and Collection Bank (Tokyo, 1999, sale)	Resolution and Collection Bank (Tokyo, 1999, merger)	Kitaichi Co., Ltd. (Otaru, 2003, sale)	..
	Sakaimachi 203-3							Hokkaido Takushoku Bank (Sapporo, 1972, registration of ownership by merger; divided and reregistered as 203-3)	Sakai Riyo (Otaru, 1974, sale)	Kitaichi Co., Ltd. (Otaru, 1986, transfer of ownership through exchange)	..	..

(continued)

Table 6.13 (continued)

Location		Land classification	Owner according to register or land registries									
Former lot number	New lot number											
	Sakaimachi 203-4							Hokkaido Takushoku Bank (Sapporo, 1972, registration of ownership by merger; divided and reregistered as 203-4)	Kitaichi Co., Ltd. (Otaru, 1986, transfer of ownership through exchange)	..	..	..
Minatomachi 27-2	Sakaimachi 212-2	Residential land	Sōma Honten (Hakodate, 1948, registration upon corporate merger)	Shimada Genzō (Otaru, 1949, sale)	Itōchu Corp. (Osaka, 1951, sale)	Taniguro Shōten (Otaru, 1957, sale)	Ōhashi Bussan Co., Ltd. (Gifu, 1969, sale)	Otaru City (Otaru, 1969, sale)	Ōhashi Bussan Co., Ltd. (Gifu, 1974, divided and reregistered as 212-2)	..	..	..
Minatomachi 27-3	Sakaimachi 213	Residential land	Sōma Honten (Hakodate, 1948, registration upon corporate merger and subdivision)	Kano Shigeharu (Otaru, 1948, transfer of ownership)	..	..	..	..	..	..	..	..
Minatomachi 31-1	Sakaimachi 218-1	Residential land	Kimura Enkichi (Otaru, 1902, unclear)	Kimura Enkichi (Otaru, 1964, consolidation with Sakaimachi 28-ban)	Kimura Enkichi (Otaru, 1964, divided and reregistered as Minatomachi 31-1)	Kitaichi Co., Ltd. (Otaru, 1983, sale)	..	..	..	..	..	..

(continued)

Table 6.13 (continued)

Location		Land classification	Owner according to register or land registries									
Former lot number	New lot number											
Minatomachi 31-2	Sakaimachi 219				Kimura Enkichi (Otaru, 1964, divided and reregistered as Minatomachi 31-2	Nippon Kompō Shizai Co., Ltd. (Tokyo, 1964, sale)	Asahara Kenzō (Otaru, 1977, sale)	..	..	..	..	..
Minatomachi 32-2	Sakaimachi 221	Residential land	Fujiyama Yōichi (Otaru, 1941, inheritance)	Kosaka Ryūnosuke (Otaru, 1951, transfer)	Kosaka Ayako, Kosaka Junko (Tokyo, 1955?, inheritance)	Koshizaki Shōten Co., Ltd. (Otaru, 1957?, sale)	Shōsan Shōji Co., Ltd. (Tokyo, 1970, company merger)	Kitaichi Co., Ltd. (Otaru, 1999, sale)	..	..	..	..
Minatomachi 32-3	Sakaimachi 222	Residential land	Fujiyama Yōichi (Otaru, 1941, inheritance)	Muramoto Seitarō (Otaru, 1949, sale)	Shōwa Sekiyu Co., Ltd. (Tokyo, 1949, sale)	Miyuki Shōji Co., Ltd. (Otaru, 1958, sale)	Provisional measure and temporary seizure (1981, creditor: Masaki Shōten, Kyoto)	Yamakō Okuno Co., Ltd. (Otaru, 1981, sale)	..	..	..	..
Minatomachi 35	Sakaimachi 225	Residential land	Hayashi Ryōtarō (Otaru, unknown, unknown)	Ōi Terumae? (Sapporo, 1903, sale)	Endō Matabei (Otaru, 1903, sale)	Daiichi Bank, Otaru branch (Otaru, 1915, transfer of ownership)	Shibusawa Sōko Co., Ltd (Tokyo, 1916, transfer of ownership)	Takai Nobushige (Sapporo, 1939, transfer of ownership)	Hokkai Seishi Co., Ltd. (Otaru, 1946, sale)	Nittō Co., Ltd. (Sapporo, 1980, sale)	..	..
Minatomachi 36	Sakaimachi 226	Residential land	Hayashi Ryōtarō (Otaru, 1893, registration)	Komachiya Jun (Otaru, 1903, sale)	Adachi Keijirō (Otaru, 1906, sale)	Adachi Ken'ichirō (Otaru, 1938, inheritance)	Fujii Jun'ichi (Sapporo, 1939, sale)	Nittō Co., Ltd. (Sapporo, 1957, sale)	..	..	..	..

(continued)

Table 6.13 (continued)

Location		Land classification	Owner according to register or land registries									
Former lot number	New lot number											
Minatomachi 38	Sakaimachi 228	Residential land	Fujii Jun'ichi (Sapporo, 1939, sale)	Nittō Co., Ltd. (Sapporo, 1957, sale)	..	..	..	..	..	..	..	..
Minatomachi 40	Sakaimachi 230	Residential land	Daimaru Fujii Co., Ltd. (Sapporo, 1947, sale)	Kitaichi Co., Ltd. (Otaru, 1984, sale)	..	..	..	..	..	..	..	..
Minatomachi 43	Sakaimachi 233	Residential land	Daimaru Fujii Co., Ltd. (Sapporo, 1947, sale)	Kitaichi Co., Ltd. (Otaru, 1984, sale)	..	..	..	..	..	..	..	..
Minatomachi 44	Sakaimachi 234	Residential land	Arata Shōkai Co., Ltd. (Otaru, 1966, sale)	Mitsubishi Oil Co., Ltd. (Tokyo, 1969, provisional registration of ownership transfer request)	Arata Shōkai Co., Ltd. (Otaru, 1998, deletion of provisional registration)	..	..	..	..	..	..	..
Minatomachi 81	Sakaimachi 272	Mixed use	Otaru City (unknown)	..	..	..	..	..	..	..	..	..
	Sakaimachi 272-2		Otaru City (Unknown, 1976, preservation of ownership)	..	..	..	..	..	..	..	..	..

Sources Register and Land register data: Compiled by the author using the register, closed register, land register, and public maps held within the Otaru branch collection of the Sapporo Legal Bureau (during 2002 and 2003 field surveys)
Historical literature
1. Nagano and Marufuji, eds. (1916) *Otaruku Kaisei Chiban Tochi Daichō, Otaruku Tochi Renraku Zenzu* (Sapporo: Marufuji Sokuryō Jimusho, pp. 157–161)
2. Marufuji and Wakamatsu, eds. (1920) *Saishin Otaruku Tochi Daichō (Including Kumausumura, Takashimamura, and Shioyamura), Otaruku Tochi Renraku Zenzu* (Otaru: Watanabe, Hisatarō, pp. 155–158)
3. Satō, ed. (1927) *Saishin Otarushi Oyobi Kōgai Tochi Daichō, Saishin Otarushi Oyobi Kōgai Tochi Jissoku Renraku Zenzu* (three volumes and document box) (Otaru: Sangyōshōreikai, Volume 1, pp. 213–217)
Note Parenthetical data indicates the owner's address, year of transfer, and nature of transfer

Table 6.14 Changes to land ownership in the Sakaimachi district of Otaru: dates of ownership transfer (Part 1)

Location		land use	1889	1890	1891	1892	1893	1894	1895	1896	1897	1898	1899	1900	1901	1902	1903	1904
Former lot number	New lot number																	
Minatomachi 16-1	Sakaimachi 177	residential land											□					
Minatomachi 16-2	Sakaimachi 178																	
Minatomachi 22-1	Sakaimachi 202-1	residential land	●															
	Sakaimachi 202-2	residential land																
Minatomachi 23	Sakaimachi 203-1	residential land							○									
	Sakaimachi 203-2																	
	Sakaimachi 203-3																	
	Sakaimachi 203-4																	
Minatomachi 27-2	Sakaimachi 212-2	residential land																
Minatomachi 27-3	Sakaimachi 213	residential land																
Minatomachi 31-1	Sakaimachi 218-1	residential land														□		
Minatomachi 31-2	Sakaimachi 219																	
Minatomachi 32-2	Sakaimachi 221	residential land																
Minatomachi 32-3	Sakaimachi 222	residential land																
Minatomachi 35	*Sakaimachi 225*	*residential land*															●●	
Minatomachi 36	Sakaimachi 226	residential land					□										●	
Minatomachi 38	Sakaimachi 228	residential land																
Minatomachi 40	Sakaimachi 230	residential land																
Minatomachi 43	Sakaimachi 233	residential land																
Minatomachi 44	Sakaimachi 234	residential land																
Minatomachi 81	Sakaimachi 272	mixed use																
	Sakaimachi 272-2																	
		subtotal	1				1		1				1			1	3	

(continued)

Table 6.14 (continued)

Location		land use	1905	1906	1907	1908	1909	1910	1911	1912	1913	1914	1915	1916	1917	1918	1919	1920
Former lot number	New lot number																	
Minatomachi 16-1	Sakaimachi 177	residential land					○					△						
Minatomachi 16-2	Sakaimachi 178																	
Minatomachi 22-1	Sakaimachi 202-1	residential land				○											●	
	Sakaimachi 202-2	residential land																
Minatomachi 23	Sakaimachi 203-1	residential land							○							●		
	Sakaimachi 203-2																	
	Sakaimachi 203-3																	
	Sakaimachi 203-4																	
Minatomachi 27-2	Sakaimachi 212-2	residential land																
Minatomachi 27-3	Sakaimachi 213	residential land																
Minatomachi 31-1	Sakaimachi 218-1	residential land																
Minatomachi 31-2	Sakaimachi 219																	
Minatomachi 32-2	Sakaimachi 221	residential land																
Minatomachi 32-3	Sakaimachi 222	residential land																
Minatomachi 35	Sakaimachi 225	residential land											△	△				
Minatomachi 36	Sakaimachi 226	residential land		●														
Minatomachi 38	Sakaimachi 228	residential land																
Minatomachi 40	Sakaimachi 230	residential land																
Minatomachi 43	Sakaimachi 233	residential land																
Minatomachi 44	Sakaimachi 234	residential land																
Minatomachi 81	Sakaimachi 272	mixed use																
	Sakaimachi 272-2																	
		subtotal		1		1	1		1			1	1	1		1	1	

(continued)

Table 6.14 (continued)

Location		land use															
Former lot number	New lot number		1921	1922	1923	1924	1925	1926	1927	1928	1929	1930	1931	1932	1933	1934	1935
Minatomachi 16-1	Sakaimachi 177	residential land		●			●										
Minatomachi 16-2	Sakaimachi 178																
Minatomachi 22-1	Sakaimachi 202-1	residential land		●△													
	Sakaimachi 202-2	residential land															
Minatomachi 23	Sakaimachi 203-1	residential land							●								
	Sakaimachi 203-2																
	Sakaimachi 203-3																
	Sakaimachi 203-4																
Minatomachi 27-2	Sakaimachi 212-2	residential land															
Minatomachi 27-3	Sakaimachi 213	residential land															
Minatomachi 31-1	Sakaimachi 218-1	residential land															
Minatomachi 31-2	Sakaimachi 219																
Minatomachi 32-2	Sakaimachi 221	residential land															
Minatomachi 32-3	Sakaimachi 222	residential land															
Minatomachi 35	Sakaimachi 225	residential land															
Minatomachi 36	Sakaimachi 226	residential land															
Minatomachi 38	Sakaimachi 228	residential land															
Minatomachi 40	Sakaimachi 230	residential land															
Minatomachi 43	Sakaimachi 233	residential land															
Minatomachi 44	Sakaimachi 234	residential land															
Minatomachi 81	Sakaimachi 272	mixed use															
	Sakaimachi 272-2																
		subtotal		3			1		1								

(continued)

Table 6.14 (continued)

Location		land use	1936	1937	1938	1939	1940	1941	1942	1943	1944	1945
Former lot number	New lot number											
Minatomachi 16-1	Sakaimachi 177	residential land										
Minatomachi 16-2	Sakaimachi 178											
Minatomachi 22-1	Sakaimachi 202-1	residential land				△						△
	Sakaimachi 202-2	residential land										
Minatomachi 23	Sakaimachi 203-1	residential land										△
	Sakaimachi 203-2											
	Sakaimachi 203-3											
	Sakaimachi 203-4											
Minatomachi 27-2	Sakaimachi 212-2	residential land										
Minatomachi 27-3	Sakaimachi 213	residential land										
Minatomachi 31-1	Sakaimachi 218-1	residential land										
Minatomachi 31-2	Sakaimachi 219											
Minatomachi 32-2	Sakaimachi 221	residential land						○				
Minatomachi 32-3	Sakaimachi 222	residential land						○				
Minatomachi 35	Sakaimachi 225	residential land				△						
Minatomachi 36	Sakaimachi 226	residential land			○	●						
Minatomachi 38	Sakaimachi 228	residential land				●						
Minatomachi 40	Sakaimachi 230	residential land										
Minatomachi 43	Sakaimachi 233	residential land										
Minatomachi 44	Sakaimachi 234	residential land										
Minatomachi 81	Sakaimachi 272	mixed use										
	Sakaimachi 272-2											
		subtotal			1	4		2				2

Table 6.14 Changes to land ownership in the Sakaimachi district of Otaru: dates of ownership transfer (Part 2)

Location		Land use	1946	1947	1948	1949	1950	1951	1952	1953	1954	1955	1956	1957	1958	1959	1960	1961
Former lot number	New lot number																	
Minatomachi 16-1	Sakaimachi 177	residential land																
Minatomachi 16-2	Sakaimachi 178																	
Minatomachi 22-1	Sakaimachi 202-1	residential land																
	Sakaimachi 202-2	residential land																
Minatomachi 23	Sakaimachi 203-1	residential land																
	Sakaimachi 203-2																	
	Sakaimachi 203-3																	
	Sakaimachi 203-4																	
Minatomachi 27-2	Sakaimachi 212-2	residential land			□	●		●						●				
Minatomachi 27-3	Sakaimachi 213	residential land			□△													
Minatomachi 31-1	Sakaimachi 218-1	residential land																
Minatomachi 31-2	Sakaimachi 219																	
Minatomachi 32-2	Sakaimachi 221	residential land						△				○		●				
Minatomachi 32-3	Sakaimachi 222	residential land				●●									●			
Minatomachi 35	Sakaimachi 225	residential land	●															
Minatomachi 36	Sakaimachi 226	residential land												●				
Minatomachi 38	Sakaimachi 228	residential land												●				
Minatomachi 40	Sakaimachi 230	residential land		●														
Minatomachi 43	Sakaimachi 233	residential land		●														
Minatomachi 44	Sakaimachi 234	residential land																
Minatomachi 81	Sakaimachi 272	mixed use																
	Sakaimachi 272-2																	
		subtotal	1	2	3	3		2				1		4	1			

(continued)

Table 6.14 (continued)

Location		Land use	1962	1963	1964	1965	1966	1967	1968	1969	1970	1971	1972	1973	1974	1975	1976	1977
Former lot number	New lot number																	
Minatomachi 16-1	Sakaimachi 177	residential land					△											
Minatomachi 16-2	Sakaimachi 178						△								●			
Minatomachi 22-1	Sakaimachi 202-1	residential land													●			
	Sakaimachi 202-2	residential land																
Minatomachi 23	Sakaimachi 203-1	residential land											△					
	Sakaimachi 203-2												△					
	Sakaimachi 203-3												△		●			
	Sakaimachi 203-4												△					
Minatomachi 27-2	Sakaimachi 212-2	residential land								●●					△			
Minatomachi 27-3	Sakaimachi 213	residential land																
Minatomachi 31-1	Sakaimachi 218-1	residential land			※													
Minatomachi 31-2	Sakaimachi 219				※●													●
Minatomachi 32-2	Sakaimachi 221	residential land									△							
Minatomachi 32-3	Sakaimachi 222	residential land																
Minatomachi 35	Sakaimachi 225	residential land																
Minatomachi 36	Sakaimachi 226	residential land																
Minatomachi 38	Sakaimachi 228	residential land																
Minatomachi 40	Sakaimachi 230	residential land																
Minatomachi 43	Sakaimachi 233	residential land																
Minatomachi 44	Sakaimachi 234	residential land					●											
Minatomachi 81	Sakaimachi 272	mixed use																
	Sakaimachi 272-2																△	
		subtotal			1		3			2	1		4		4		1	1

(continued)

Table 6.14 (continued)

Location		Land use	1978	1979	1980	1981	1982	1983	1984	1985	1986	1987	1988	1989	1990	1991	1992	1993
Former lot number	New lot number																	
Minatomachi 16-1	Sakaimachi 177	residential land																
Minatomachi 16-2	Sakaimachi 178								●									
Minatomachi 22-1	Sakaimachi 202-1	residential land							●									
	Sakaimachi 202-2	residential land																
Minatomachi 23	Sakaimachi 203-1	residential land																
	Sakaimachi 203-2																	
	Sakaimachi 203-3										△							
	Sakaimachi 203-4										△							
Minatomachi 27-2	Sakaimachi 212-2	residential land																
Minatomachi 27-3	Sakaimachi 213	residential land																
Minatomachi 31-1	Sakaimachi 218-1	residential land						●										
Minatomachi 31-2	Sakaimachi 219																	
Minatomachi 32-2	Sakaimachi 221	residential land																
Minatomachi 32-3	Sakaimachi 222	residential land				△●												
Minatomachi 35	Sakaimachi 225	residential land			●													
Minatomachi 36	Sakaimachi 226	residential land																
Minatomachi 38	Sakaimachi 228	residential land																
Minatomachi 40	Sakaimachi 230	residential land							●									
Minatomachi 43	Sakaimachi 233	residential land							●									
Minatomachi 44	Sakaimachi 234	residential land																
Minatomachi 81	Sakaimachi 272	mixed use																
	Sakaimachi 272-2																	
		subtotal			1	2		1	4		2							

(continued)

Table 6.14 (continued)

Location		Land use	1994	1995	1996	1997	1998	1999	2000	2001	2002	2003	subtotal
Former lot number	New lot number												
Minatomachi 16-1	Sakaimachi 177	residential land											6
Minatomachi 16-2	Sakaimachi 178												8
Minatomachi 22-1	Sakaimachi 202-1	residential land											9
	Sakaimachi 202-2	residential land											
Minatomachi 23	Sakaimachi 203-1	residential land											6
	Sakaimachi 203-2							●△				●	9
	Sakaimachi 203-3												8
	Sakaimachi 203-4												7
Minatomachi 27-2	Sakaimachi 212-2	residential land											7
Minatomachi 27-3	Sakaimachi 213	residential land											2
Minatomachi 31-1	Sakaimachi 218-1	residential land											2
Minatomachi 31-2	Sakaimachi 219												3
Minatomachi 32-2	Sakaimachi 221	residential land						●					6
Minatomachi 32-3	Sakaimachi 222	residential land											6
Minatomachi 35	Sakaimachi 225	residential land											8
Minatomachi 36	Sakaimachi 226	residential land											6
Minatomachi 38	Sakaimachi 228	residential land											2
Minatomachi 40	Sakaimachi 230	residential land											2
Minatomachi 43	Sakaimachi 233	residential land											2
Minatomachi 44	Sakaimachi 234	residential land											1
Minatomachi 81	Sakaimachi 272	mixed use											0
	Sakaimachi 272-2												1
		subtotal						3				1	48

(continued)

Table 6.14 (continued)

Sources Register and land register data: compiled by the author using the register, closed register, land register, and public maps held within the Otaru branch collection of the Sapporo Legal Bureau (during the author's 2002 and 2003 field surveys)

Note (1) ○=inheritance, ● = sale,△ = ownership transfer,□ = registration,※ = lot consolidation,shaded area = corporate ownership

(2) The subtotal column counts all transfers of ownership rights. Strictly speaking, there are cases in which the registration does not match the transfer of ownership, but these are nevertheless counted uniformly as cases of "transfer." Only cases of lot consolidation or division were excluded

(3) Different methods of counting explain the discrepancies between the vertical and horizontal subtotals. When counting by lot, pre-subdivision ownership transfers are double-counted twice. But there is no such overlap in the annual totals

than simply dismissing this local myth, we should attempt to decipher the social perspectives it reflects.[13]

This myth has been continuously reproduced as it is shared widely throughout Otaru society. Even the *Hokkaido Shinbun* has reported that the majority of Sakaimachi vendors are from outside both the city and Hokkaido and conduct their business in Sakaimachi with no sense of responsibility to the local region or contributions to local tax revenue. One article clearly states that "90 percent" of these businesses are funded by "outside capital" (*Hokkaido Shinbun*, Otaru edition, March 17, 2007; March 29, 2007). These new reports confirm the existence of a powerful myth.

As we have seen, my own research into local land ownership tells a different story: 80% of land in Sakaimachi is held by local businesses, while less than 20 percent belongs to outsiders. Furthermore, if we aggregate the data acquired over the course of the fixed point observation surveys pertaining to local versus outside businesses, the rate of "local Otaru capital" within Sakaimachi appears to be 49% (Horikawa and Fukaya, ed. 2012: 66–67). In short, 80 percent of the land is owned by local capital, and half .of the companies operating businesses in the area are also local concerns. The idea that "most Sakaimachi businesses belong to outside capital" is a myth.

How can we explain the persistence of a belief that has no basis in fact?

One answer has to do with the idea of a "dominant story." This is a hypothesis developed by Fukudome Chika, a 2011 participant in my social research practicum (Horikawa and Fukaya, eds. 1992). Fukudome suggested that Otaru's canal war gave rise to a certain "town legend" (*machi no monogatari*) (Yasumura 2006) or "dominant story." According to this dominant story, the "residents who fought to protect the local landscape" are responsible for Otaru's subsequent prosperity (Horikawa and Fukaya, eds. 2012: 68). Consequently, shop designs, products, and business practices that disregard the city's history are part of a "minor story" that contravenes the "town legend." The gravest offense of these businesses is their refusal to abide by the dominant story. These business owners are thus branded "outsiders" and excluded from the category of local citizenship. The use of the term "outsider" forces the merchants of Sakaimachi to make a decision: abide by the dominant story or live as non-citizens. This is the gist of Fukudome's hypothesis (Horikawa and Fukaya, ed. 2012: 59–78).

Fukudome's argument is persuasive but may not explain everything. The concept of a "town legend" is most effective within discussions of local history. But is a "dominant story" really powerful enough to distinguish between insiders and outsiders? I believe that if we supplement Fukudome's argument with another hypothesis regarding "burden and profit," we can finally explain this myth. According to the latter hypothesis, which could also be called "the free rider hypothesis," people are classified as "insiders" or "outsiders" depending upon their perceived contribution to Otaru's emergence as a "tourist city."

[13]The ideas presented here are my own, but they have been honed by my debates with participants in my social research practicum at Hosei University. I would like to offer particular thanks to my 2012 teaching assistant Fukaya Naohiro and my students Sasaki Kenta and Mukainakano Kazuki.

According to this hypothesis, people who never involved themselves in the effort to save the Otaru canal or subsequent community and tourist development efforts, but nevertheless set up shop in Sakaimachi to profit from the booming tourist trade, have been deemed "outsiders." The category of "outsider" is not congruent with administrative, geographical, or taxation divisions of insiders and outsiders. Rather, it is *a social demarcation* made according to whether or not a given person has worked hard to contribute to Otaru. Those who did make these efforts and went on to open stores in the area are seen as legitimate. The others are considered "free riders" who have exploited the fruits of the canal preservation campaign and Otaru's newfound fame as a tourist destination. This is what makes "outsider" a derogatory term.

In a schematic of this phenomenon, "A" represents those who acted as responsible local citizens, either by joining the canal preservation movement or participating in subsequent community development efforts, while "B" represents those who shirked this responsibility. There is a further division between those who went on to do nothing (B1) and those who subsequently sought to profit in Sakaimachi (B2). Naturally enough, it is those in the "B2" category who are viewed as outsiders. These "outsiders," who did not stand up and fight for Otaru but nevertheless went on to make a killing in the tourist trade are seen as "outside capital" for illegitimately exploiting local resources.

6.4.4 The Paradox of Tourism Development: The Logic of Landscape Change

The Otaru townscape, which is the city's primary "resource for tourism," has been transformed as a consequence of corporate land ownership and the growing conversion of buildings into souvenir shops and other forms of retail space. This conclusion leaves us with the sense that something strange or paradoxical is at work.

The paradox is that while Otaru has become a major tourist attraction, its traditional townscape has been transformed—and is in fact now disappearing. What happens when a city that depends upon tourism as its "selling point" loses the very resource that made it a tourist destination in the first place—its traditional townscape? Indeed, this odd state of affairs is better described in stronger terms as the "crisis of a tourist city." Why is Otaru putting the noose around its own neck? According to a common version of the local history, "the 'canal war' was a major conflict, but Otaru experienced a magnificent reversal in fortune as a result, and became a city of tourism." Why is the landscape that has been so crucial to Otaru's good fortune disappearing?

Attracting customers is the overriding concern of anyone working in the tourism industry. Business owners attempt to attract as many visitors as possible by envisioning the most typical tourist course through Otaru and the most standard attributes of a tourist. They outfit their businesses to make them as appealing as possible to this idealized notion of the "tourist" (whether or not this tourist actually exists) and to meet his or her anticipated needs. This creates the paradox of tourism development. In order for each individual store to appeal to tourists, it must embody the type of "tourist shop" generally considered to be the most attractive to tourists. But given the common understanding of what this means, these shops end up becoming uniform in nature. In other words, in order to attract the greatest number of customers, stores feel they must abandon any individuality that may not meet the "needs of tourists" and refashion themselves to more accurately accommodate these needs (Coontz 1992: 61–62).[14] Individual shopkeepers internalize the general assumptions regarding tourist needs and abandon the individual character of their shops in order to better represent the stereotypical image of the town (Fig. 6.8 and 6.9)

This social mechanism can be explained with reference to "The Gift of the Magi," the well-known 1906 short story by O. Henry. "The Gift of the Magi" is a story of self-sacrifice, in which a husband and wife each sell their most precious belonging in order to buy a gift for the other. Jim sells his precious gold watch in order to purchase a comb to adorn the beautiful hair of his wife, Della. Yet unbeknownst to Jim, Della has cut and sold her prized locks in order to buy him a platinum watch chain. The difference between the case of Otaru and the O. Henry story lies in the anticipation of the actions of others. In the case of Otaru, store proprietors will change the style of their store according to their expectations of others. In "The Gift of the Magi," there is no such expectation—only an unexpected climax to the story. Della did not expect Jim to sell his gold watch, and Jim did not expect Della to cut her long, beautiful hair. This gives the story its moving conclusion. If we are to summarize the mechanism at work in Otaru, however, it is this: the internalization of expectations regarding the behavior of others (tourists) led to a loss in individuality and the standardization of the Otaru townscape.

Thus, tourists who visit Otaru to view *a unique landscape* will actually gaze upon a *highly uniform, standardized townscape.* Therein lie the paradox and the source of our sense of strangeness. We are witnessing a cyclical process in which "genuine" stores attempt to respond to those who embody the purest type of "tourist behavior." How does this cyclical process unfold and perpetuate itself? To begin, a certain shop will undergo a renovation aimed at attracting customers and experience great success. Even if this shop sells crab from another part of Hokkaido rather than local fish, the simple fact of its "success" triggers this cycle. Other store proprietors may privately disapprove, but the first store's "success" at anticipating "the needs of tourists" pressures them to outfit their store in a similar way and adopt the same sales strategy.

Why do other store proprietors largely follow in the footsteps of a "success story?" Why do they become enfolded into this cyclical process, instead of standing their ground and maintaining that "I won't operate that kind of business?" I have no intention of attributing this to the lack of awareness among local business owners. Our objective is not to condemn the choices of individual shop proprietors, but rather to clarify why a variety of different businesses end up adopting similar strategies.

[14]Coontz (1992) makes this argument with regard to feminism, a completely different topic. Nevertheless, my analysis here owes a great deal to the logical structure of her argument.

Fig. 6.8 One example of landscape change in Sakaimachi, Otaru (H District, Building 94). Photographs date from (**a**) September 1998, (**b**) September 1999, (**c**) September 2008, and (**d**) September 2009. All photographed by the author

Fig. 6.8 (continued)

Fig. 6.9 Building 94 of H District in 2016. Photographed by the author (September 2016)

Given that tourist groups spend just a short period of time in their destination and make only one trip, shop owners seek to benefit immediately from the crowds of tourists who pass in front of their stores. Without long-term strategies or community trust, they deem it better to attract customers immediately and convince them to stop and buy even just one item.

For these stores, which research confirms are concentrated in the rapidly changing Sakaimachi area, the most crucial challenge is to convince tourists that they are "tourist shops." This is more important than demonstrating their historic character or the authenticity of their building or inventory. They must ensure that their stores are not mistaken for ordinary houses. Nor should they be mistaken for stores selling everyday goods. This is because any association with the every day must be avoided, for tourism trips are supposed to be comprised of the unusual—the extraordinary. There is little meaning in purchasing an item at a sightseeing spot that could just as easily have been bought at home. The seeking out of the extraordinary and non-routine is one essential aspect of tourist behavior (Urry 1990 = 1995). The idea of going somewhere new and acquiring something unusual, something that can only be found in that place, is the original form of tourism. Tourism is supposed to involve the extraordinary—the unusual and the unexpected. That is precisely why the stores lining the streets of Sakaimachi must demonstrate to tourists as clearly as possible that they are, in some way, extraordinary and not-to-be-missed tourist destinations. In a region of heavy tourism, the pressure posed by nine million visitors and the success story of a neighboring shop suggests that shop managers have adopted this as a rational business strategy. In this way, selling a unique historical environment to tourists creates the paradox described earlier: the creation and replication of the most uniform kind of tourist shop—the type that could be found in any popular tourist destination.

The creation of the "tourist city, Otaru" came at the high cost of a protracted "canal war" and the lingering divisions created by this conflict. And what did the "tourist city, Otaru" end up losing? Ironically, Otaru lost the very landscape that its own residents fought to defend.

6.5 The Beginning of the End: The "Publicness" of the Lost Landscape

We know what Otaru lost as a result of the paradoxical effects of tourism. Yet readers may well ask why Otaru tourism has not yet reached its saturation point as a result of this growing uniformity. In other words, why isn't the number of visitors to Otaru decreasing? This is our final point of discussion.

The Otaru Tourism Promotion Council identifies the characteristics of Otaru tourism as follows:

> Seventy percent of the five million visitors to Otaru are from Hokkaido, and primarily from the central area of the prefecture. Repeat visitors account for 90 percent of this total. Surprisingly, of the remaining 30 percent of visitors from outside Hokkaido, 40 percent are repeat visitors. Thus, in total, 65 percent of visitors to Otaru are repeat visitors. This suggests a number of things.
>
> First, a repeat visitor will ask himself, "what should I go and do in Otaru this time?" This suggests that a single hub or "set menu" for tourism, such as the canal, the music box museum, and sushi will not sustain the interest of a repeat visitor (although there is no denying their importance as representative aspects of Otaru tourism). Tour planners at travel agencies have also told us: "Keep touting new products or new approaches to Otaru. Otaru has so many resources for tourism, so wouldn't it be possible to create a tourism network in the city, and design new products across a broader-area network? Time is running out for today's kind of tourism."
>
> The tourism industry is currently experiencing a major upheaval due to the "personalization" of travel. In one hotel, the number of group tours fell by 60 percent from the previous year, but the difference was made up through an increase in the number of individual guests. Travel agencies have also told us that "we're certain that personal travel is the future mainstream." Agencies are offering customers the ability to choose their own air travel and hotel, and freely decide upon their course.
>
> Otaru Kankō Yūchi Sokushin Kyōgikai Chōsa Kenkyūbu, ed. 1998: 1L

This report makes two key points. The first is that the majority of visitors to Otaru are making repeat visits. The second is that "personal" tourism—as opposed to group tours—has become the norm. The typical visitor to Otaru hails from elsewhere in Hokkaido, traveling alone or in a small group, chooses his or her activities freely and is making a repeat visit to the city.

Are these characteristics consistent with the "paradox of tourism" discussed in the preceding section? Why would "repeat visitors" on a "personal" trip choose to return to Otaru if the uniformity of the city's offerings is increasing? Why aren't tourist numbers decreasing as tourists grow tired of Otaru? Isn't there a contradiction between the paradox of tourism development and the central claims of the Otaru Tourism Promotion Council report?

In fact, there is no contradiction between the two. As observed by Myōen et al. (1993), continued capital investment from the late 1980s onward allowed Otaru to retain its novelty as a tourist destination. The successive inauguration of the Otaru glass boats, the Ishihara Yūjirō Memorial Hall, the Hokuren Fūdokan, the Otaru Beer brewery, and Kitaichi Glass, together with tourist facilities such as the Orugōrudō and Kaimeirō music box store, and the recent proliferation of confectionery stores, continue to attract repeat visitors. All of this, together with a recent increase in overseas visitors, has helped to obscure the paradox of tourism development. Nevertheless, Otaru's unique townscape continues to disappear. The fact that a series of new investments in tourist facilities has inspired many Hokkaido-area visitors to make repeat visits, and the growth of inbound tourism (primarily from China), has only prevented the paradox of tourism development from becoming more apparent.

Next, if "personal" travel has become the primary form of tourism for the future, it will have a major impact on the development of Otaru tourism. According to Akama Hajime, a key member of the Otaru Tourism Promotion Council and a former executive at one of Otaru's top hotels:

> {...} As I wrote here [in the 1998 report of the Otaru Tourism Promotion Council], ultimately, when you build a town that its citizens can be proud of, that can become a resource for tourism. This would transform tourism, [so that tourists would] want to peek at how local residents breathe. Up until now [tourists] have just wanted to go look at a beautiful landscape, but that's over now. And I think they would prefer to get closer to local citizens' everyday lives than peer into a bunch of souvenir stores. That's what I would want. When I go overseas I like to see how ordinary people live—places like the markets, and where children go to play. Don't you agree? {...} And I want you to understand that I'm not saying just looking at people's lives is interesting. [My point is] that seeing a way of life that is different from your own makes you think. And some tourists will enjoy discovering things on their own legs and with their own heads, and others won't. And the ones who don't enjoy it can get on a bus and [carry on] with the kind of tourism that's the same as staring at a goldfish in a fishbowl. But I think that kind of tourism is dying out. So tourists will also need to be competent. That's why I'm (always) saying that people's values change with the times, and what you can sell to [people] changes, too. You know (that society) is maturing when the idea of 'brands' (whether for clothing or cars or anything else) starts to feel unfashionable.
>
> Interview with Akama Hajime in Otaru, September 4, 1998. Information in parentheses has been added by the author, information in brackets by the translator.

Akama makes his argument in a roundabout way, but his point mirrors the 1998 report by the Otaru Tourism Promotion Council. For example, Akama says: "Up until now [tourists] have just wanted to go look at a beautiful landscape, but that kind of [tourism] is over now," and "[they] can get on a bus and [carry on] with the kind of tourism that's the same as staring at a goldfish in a fishbowl. But I think that kind of tourism is dying out." Akama's point is that the kind of tourism in which people simply stare out at the scenery from their bus window is "over now." Today's tourists want to walk the town and "discover things on their own legs and with their own heads"—even how the residents of their destination live and breathe.

These changes to travel patterns have not necessarily been positive for Otaru. Throughout the 1990s, new investment concealed the gradual loss and standardization of Otaru's townscape. In the 2000s, however, these losses have become more visible—and tourism began to decline. This decline signals a growing disinterest on the part of potential tourists, but as we have seen, for Otaru's shop keepers intent upon attracting customers, the logic of renovation continues to promote the standardization and homogenization of storefronts even further. This may have accelerated the pace of alienation and estrangement among visitors to Otaru and will likely spell doom for all but the few stores that have successfully established their own niche. Sakaimachi, once a street of wholesalers, has been stripped of this history and now lacks the physical resources (the historic buildings) that might allow it to respond to changing tourism trends. The "trend toward urban monoculture" is clearly at work when capital is only invested in a specific direction and physical facilities and buildings are constructed toward a single purpose. Capital can be withdrawn quickly when the area's fortunes wane, leaving the remaining companies with little resilience. Eventually, the area is neglected and abandoned. One need only summon up the image of the Otaru canal in the early 1970s: filled with sludge and rotting, half-submerged boats.

Since the latter half of the 1980s, new investments expanded Otaru's tourist areas, keeping this scenario a distant threat. The threat is much more real now. We can detect the effects of the process described earlier in this chapter, in which tourist-oriented businesses over-adapt to the tourist "gaze," destroying the unique characteristics of the town in the process. Falling land prices have prompted the construction of high-rise apartments in the canal and port area of Otaru. Today high-rise apartment buildings stand on land immediately adjacent to designated historic preservation areas, in clear view of the structures that serve as the very "face" of the city—the Otaru Warehouse and the Ōie Warehouse (Figs. 6.10, 6.11 and 6.12). Since 2012, the uptick in inbound tourism has boosted Otaru's tourist numbers, but international visitors are simply compensating for the growing disaffection of Japanese tourists.

Let us conclude with a summary of the key findings in this chapter.

The goal of the canal preservation movement was, obviously enough, saving the canal. This canal was the origin of "the port city of Otaru." We could thus say that the canal preservation movement was dedicated to saving "the port city of Otaru." The canal and the port had acted as the fundamental engine of Otaru's economy, and the consciousness of this fact informed the common description of the canal as "the origin" of Otaru. It also informed Mineyama Fumi's manner of describing the canal as "dead," and later on, the strategic identification of "the canal as a resource for tourism." The canal was spoken of as the "origin" of Otaru—not in the present tense, but in the *past* tense, revealing a shared recognition that present-day Otaru existed in its current form because of the canal. The canal would not have been targeted for preservation if it were not understood as a symbol of the past. A sense of crisis that the canal had become a thing of the past prompted the cries for its preservation, and

Fig. 6.10 A view of the Otaru warehouses and neighboring apartment buildings. Photographed by the author (September 2013)

Fig. 6.11 The symmetrical Façade of the "Ōie Warehouse," with an apartment building visible behind. Photographed by the author (September 2013)

Fig. 6.12 A view of the Ōie warehouse and apartment buildings from Kitahama. Photographed by the author (September 2009)

yet the slogan "save the canal" was another way of saying "farewell" to a bygone age. Thus, while the first chapter of the canal war was defined by the effort to save the canal and mount a canal-based form of community development, it was also a declaration that the "*port city* of Otaru" was no more.[15]

Following the conclusion of the "canal war" in 1984, Otaru became a city of tourism. The two main actors in the canal controversy—the preservation movement and the city government—retired from the center stage; the dispute was deemed "settled" and memories grew dimmer. Local government officials carried on with administrative tasks as before. Their personal relationships torn to shreds by the break-up of the preservation movement; former activists had little choice but to rededicate themselves to their homes and workplaces in an attempt to heal their wounds. In the absence of the principal actors in the "canal war," private tourist businesses led the growth in local tourism. One could say that tourism operators took to the stage vacated by the city government and the preservation movement and developed Otaru tourism without reference to a local tourism policy (Horikawa 2009a). This was a case of "unintended tourism" or "tourism as accomplished fact."

After nearly a decade of silence, former preservation activists began to resume their involvement in local affairs, and a new debate about the need for a local tourism authority arose. This was the strong assertion of the Otaru Tourism Promotion Council, staffed by former members of the canal preservation movement. Otaru

[15]In this sense, the essence of the "second chapter" could be described as the social implementation and conflict surrounding the reality and quality of Otaru as a *city of tourism*.

had been a port city for so long that it was difficult for people to acknowledge or accept the fact that tourism had become the city's primary industry. This is why the Otaru Tourism Promotion Council used data to demonstrate that the Otaru economy was, in fact, already that of a tourist city. Otaru was no longer a port city (Otaru Kankō Yūchi Sokushin Kyōgikai Chōsa Kenkyūbu, ed. 1998).

The Otaru city government responded to this reality by adjusting its port-related policies and developing new measures for tourism and community development. These were rooted in an implicit understanding that tourism development would make use of Otaru's unique history. Whether or not the point was made explicitly, there was a general recognition that Otaru's "selling point" was no longer the cargo-handling functions of the port, but the historic townscape of the city itself. Residents observed the design code established by the city government because they understood this as the creation and maintenance of a collective good that would assure an income for the citizens of Otaru. In some cases, local citizens chose to build new homes in the traditional architectural style of Otaru. The buildings that comprised the Otaru townscape were constructed, remodeled, or repaired according to a "nostalgic" design that evoked "the stone warehouses" of the canal district. The landscape survey data presented earlier in this chapter shows that while the exteriors of historic buildings were utilized as a tourism resource, their interiors were often updated to serve new and different purposes—at the very least, this was the basic design trend from the 1990s onward. In other words, this is how both the city government and the citizens of Otaru made sense of the protracted "canal war."

As a result, Otaru experienced an unprecedented boom in tourism. At its peak, Otaru welcomed more than nine million visitors in a single year. In a city with a population of less than 140,000, this is a truly staggering number. As the living space of local residents was rapidly coated over with a tourism-oriented veneer (in the trend toward urban monoculture), local tourism was increasingly ridiculed by residents as "Dejima tourism" or "cheap, souvenir-buying tourism." Nevertheless, a number like nine million forced the collective realization (even among the city's most established interests, who took pride in Otaru's status as a port city) that Otaru was now making its living as a tourist destination. Meanwhile, the mass media and researchers viewed Otaru as a "prime exemplar of tourism development." The "declining town of Otaru" underwent a protracted "canal war" to rise again, phoenix-like, as "the tourist city of Otaru:" this was how the post-1960s history of Otaru was typically discussed. The pain and calumny that characterized the final stages of the canal preservation campaign were consigned to oblivion and never spoken of. Also forgotten were the many layers to the debate surrounding the canal's fate. How does one live in a city? Why preserve? Why construct a highway? How can citizens review or reexamine public works projects? These questions, often repeated as the preservation movement began to cleave in two, were also forgotten. To the extent that the canal war was mentioned at all, it was always referred to as a necessary evil—an essential prerequisite to the grand urban regeneration that followed.

The data generated through fixed point observations of the changes to the urban landscape of Otaru indicates another trend, which hits at the very heart of this often-repeated "success story." Survey data suggests that Otaru's prosperity as a city of tourism may be finite (Horikawa and Fukaya, eds. 2013: 14–16). From around 2010, the data suggests that Otaru's tourism development efforts may have reached the "beginning of the end." The reason for this is the steady increase in new buildings that do not convey the "nostalgic" and "historic" atmosphere of Otaru—much less the geographic or historical context of the city. The city's design policy has been the subject of repeated debate, as mass-market clothing retail chains and pachinko parlors have set up shop in Otaru. By 2010, however, a new and qualitatively different trend was well underway.

Before 2010, the advent of new businesses with no relation to the tourist trade had been the focus of debates surrounding the city's design policy. Since then, however, stores targeting tourists in the very heart of Otaru's tourist zone have become the primary subject of debate. It is essential to note that these changes occurred not in a different industry but within the very heart of the local tourist trade.

In a second major departure from the typical trend in landscape change, after 2010 the construction of new buildings did not reference the city's design policy or the traditional "grammar" of Otaru construction. Rather, builders have referenced the "stylish" and "cutting edge" architecture of metropolises like Tokyo or New York. The fundamental logic of design has changed from the creation of "buildings that convey a feeling of nostalgia and the unique history of Otaru" to "stylish, modern buildings like the ones in Tokyo" (Figs. 6.13 and 6.14). Previously, when a new building was constructed, its quality was discussed in terms of the extent to which the design conveyed a "distinctly Otaru feel." In other words, the focus of the debate

Fig. 6.13 Changes to the "Grammar" of the Sakaimachi townscape (1) Photographed by the author (September 2015)

Fig. 6.14 Changes to the "Grammar" of the Sakaimachi townscape (2) Photographed by the author (September 2015)

was the question of how well new buildings measured up to the standards of "unique" Otaru. For city authorities, the goal was to keep this standard as high as possible. For building owners, of course, the challenge was to meet the requisite design standard with a minimum of expenditure. But despite this fierce conflict of interests, both sides understood the accepted measure of design.

The recent construction of "stylish, modern buildings like the ones in Tokyo" suggests this measure is no longer shared. The buildings shown in Figs. 6.13 and 6.14 are not built with the signature soft stone traditionally used in the construction of Otaru warehouses and have flat roofs, not the sloping roofs that are so characteristic of Otaru architecture. The overall appearance is entirely modern, with side walls of frosted glass, stainless steel, and concrete. This modern architectural style is entirely unrelated to the "architectural grammar" adopted and disseminated at the conclusion of the canal war.[16]

Another development supports the conclusion that Otaru's tourism has reached "the beginning of the end." The local myth that "sushi restaurants don't go out of business" has been destroyed as local sushi restaurants have begun to do just that. The transformation of Otaru from a "city of sushi" to a "city of sweets," and the supplanting of Kitaichi Glass by the confectioner LeTAO as the dominant business at the heart of the local tourist trade areother harbingers of Otaru's waning fortunes. In particular, the fact that the core Otaru tourism industry is no longer glass manufacturing (a traditional local industry), but rather a confectionery maker without roots in

[16]For more information, readers may refer to Chap. 2 of this volume and Fig. 2.5.

Otaru, runs counter to Otaru's community development initiatives centered upon the unique history and character of the region.[17] An unintended consequence of Otaru's canal war was the subsequent boom in "accidental" tourism, which evolved into a form of "Dejima tourism" entirely separate from the everyday lives of Otaru's residents. However imperfectly, this process had developed a shared "grammar" from which to draw—a grammar that has now been discarded.[18] The publicness embodied in the landscape of Otaru is now being lost.

This carries implications for Otaru's standing as a city of tourism. We must consider Otaru's very survival, given the trends toward extreme population aging, the decreasing number of visitors to Otaru, and a dwindling local population. Put differently, it poses the question of whether Otaru is already a "depopulating" city, as suggested by many local informants. Readers might immediately object to such a suggestion: "you say that the number of visitors has dropped to five million per year. But that is still a significant number of tourists, and 'depopulated' is a descriptor for deserted rural villages, not a city like Otaru." This type of objection is rooted in the common understanding of Otaru's "great urban revival."

There are at least three implications to the suggestion that Otaru is already a "depopulating" city: (1) this is occurring in the demographic sense (i.e., there has been a measurable decline in the resident population); (2) the resources for tourism are decreasing; and (3) former preservation activists are dying. Given the measurable decline in the local population, in the broadest sense of the word Otaru is indeed a "depopulating" city. With regard to the second point, the results of my fixed point observation surveys demonstrate that, contrary to appearances, the number of historic buildings is decreasing year by year. Furthermore, now that the city's shared architectural and design "grammar" have been abandoned, the traditional townscape of Otaru—one of its primary resources for tourism—will rapidly disappear. And finally, as memories and lessons of Otaru's "canal war" are consigned to oblivion or reformulated as a grand success story, the significance attributed to the preservation of the canal and the accumulation of deeper arguments developed over the course of

[17]In the course of fieldwork conducted in 2014, I heard one staff member at PATHOS LeTAO, the largest storefront of the confectionery chain, attempt to lure in potential customers by calling out to them that PATHOS was the largest store in the chain, and that the same products could be found at Chitose airport in Sapporo (Author's observation survey in Sakaimachi, Otaru, September 5, 2014). Making appeals based on the extent of the product line-up is the sales strategy of a mass-market retailer, and is entirely different from a sales strategy that emphasizes the unique and rare quality of products to potential customers. This episode supports my conclusion that the post-canal war "grammar" of community development, which was developed around the concept of Otaru's unique attributes, has been discarded.

[18]The effort in recent years to bring casinos to Otaru is yet another example of the growing deviation from the traditional "grammar" of community development in Otaru. This is because casinos are entirely self-contained entertainment facilities, with absolutely no connection to the outside world. They exist with no reference to local environmental conditions or history, or to regional social conditions. As demonstrated by repeated competitions among cities across Japan to establish casinos, facilities that could just as easily be established anywhere in the country are the polar opposite of the community development efforts in Otaru, which focused on the intrinsic characteristics of the city.

the long preservation campaign are being lost—and can no longer be handed down to the next generation.

Today's Otaru must confront a number of questions. How can the lessons learned in the course of the "canal war" be passed down to the next generation? How can the city address the shrinking local population and the disappearance of the traditional townscape? The "beginning of the end" is not a mere warning for the future—it is already well underway. Otaru must grapple with the reality of its status as a depopulating city.

What did Otaru gain? And what has Otaru lost? Drawing upon the data generated by the fixed point observation surveys, I can conclude that although the very scenery that Otaru residents sought to protect enabled the city to become "a shining example of the tourist city," the city is now losing the very landscape that propelled its rise as a tourist destination. The data presented in this chapter reveals a paradox: by becoming a tourist city, Otaru lost the very environment that originally represented such a vital resource for tourism. Having once been revived by tourism, Otaru must now attempt a new form of community development for the post-tourism era that seems imminent. Will Otaru survive as a tourist city? Or will it become a depressed provincial city once again? The conditions imposed by the scenery of Otaru, even as it loses its publicness, already represent a reality from which the city cannot escape.

References

Coontz, S. (1992). *The way we never were: American families and the nostalgia trap.* New York, NY: Basic Books.

Halbwachs, M. ([1950] 1968). *Le mémoire collective* (2e edition, Bibliothèque de sociologie contemporaine) Paris: Presses Universitaires [Koseki, Tōichirō, trans. (1989) *Shūgōteki Kioku.* Kyoto: Kōrosha].

Hama, H. (2010). Kioku to Basho: Kindaiteki Jikan, Kūkan no Henyō. *Shakaigaku Hyōron, 60(4)*, 465–480.

Hamada, T. (1981). Rekishiteki Kankyō no Seibi ni Kan Suru Kenkyū: Otarushi Ironai Chiku o Kēsu Sutadi Toshite. Graduation thesis, Department of Engineering, Hokkaido University.

Horikawa, S. (1998a). Toshi Kūkan to Seikatsusha no Manazashi. In A. Ishikawa, K. Satō, & K. Yamada (Eds.), *Mienai Mono o Miru Chikara: Shakai Chōsa to Iu Ninshiki* (pp. 133–149). Yachiyo Shuppan.

Horikawa, S. (1998b). Rekishiteki Kankyō to Chiiki Saisei: Machinami Hozon ni Okeru 'Bashosei' no Sōtenka. In H. Funabashi & N. Iijima (Eds.), *Kankyō* (Kōza Shakaigaku 12) (pp. 103–132). University of Tokyo Press.

Horikawa, S. (2000a). Rekishiteki Kankyō Hozon to Kankō Kaihatsu: Otarushi ni Okeru Shakaigakuteki Jisshō Kenkyū (1997–1999). *Kenkyū Nenpō, 4*, 21–36. Hōsei Daigaku Tama Chiiki Shakai Kenkyū Sentā.

Horikawa, S. (2000c). Unga Hozon to Kankō Kaihatsu: Otaru ni Okeru Toshi no Shisō. In S. Katagiri (Ed.), *Rekishiteki Kankyō no Shakaigaku* (Shirīzu Kankyō Shakaigaku 3) (pp. 107–129). Shinyōsha.

Horikawa, S. (2001). Keikan to Nashonaru Torasuto: Keikan wa Shoyū Dekiru ka. In H. Torigoe (Ed.), *Shizen Kankyō to Kankyō Bunka* (Kōza Kankyō Shakaigaku 3) (pp. 159–189). Yūhikaku.
Horikawa, S. (2003). Toshi to Hozon no Fīrudowāku: 'Komyunikēshon Toshite no Chōsa' e no Sukecchi. In I. Watado, Y. Hirota, J. Tajima (Eds.), *Toshiteki Sekkai/Komyuniti/Esunishiti: Posuto Metoroporisu no Toshi Esunogurafi Shūsei* (pp. 189–211). Akashi Shoten.
Horikawa, S. (2008a). Kankō Kaihatsu to Tochi Shoyū no Hensen Katei ni Kan Suru Kenkyū Nōto. In Hōsei Daigaku Daigakuin Eko Chiiki Dezain Kenkyūsho (Ed.), *Monbukagakushō Gakujutsu Furontia Suishin Jigyō, Hōsei Daigaku Daigakuin Eko Chiiki Dezain Kenkyūsho 2007-nendo Hōkokusho* (pp. 379–387). Hōsei Daigaku Daigakuin Eko Chiiki Dezain Kenkyūsho.
Horikawa, S. (2009a). *Kichō Hōkoku 'Hozon Undō no Keika to Seika,' Otaru Shinpojiumu Jikkōiinkai, ed. Otaru Unga to Sekizō Sōkogun no Hozon Undō Kara Nani o Uketsugu ka: Chiiki ni Iki, Chiiki o Mamoru…Machizukuri Undō no Senkusha Mineyama Fumi-shi ga Tsutaeru Koto Shinpojiumu Kaisai Hōkoku* (pp. 9–30). Sapporo: Nihon Kenchikugakkai Hokkaido Shibu (A4, 76 pp.).
Horikawa, S. (2010). Basho to Kūkan no Shakaigaku: Toshi Kūkan no Hozon Undō wa Nani o Imi Suru no ka. *Shakaigaku Hyōron, 60(4)*, 517–534.
Horikawa, S. (2011). Kindai Toshi no Mizubeto Kōkyōken: Toshin no Mizube wa Dare no Mono ka. *Kantō Toshi Gakkai Nenpō, 13*, 50–59.
Horikawa, S. (2013a). Keikan ni Umekomareta Kōshi Kannen: Aru Toshi Shinwa no Kaidoku, Shiron. In S. Tanaka (Ed.), *Toshi Kankyō ni Okeru Seikatsu Kōkyōsei ni Kan Suru Hikaku Shakaigakuteki Kenkyū* (2007–2012-nendo Kakenhi Kiban Kenkyū A [Kenkyū Kadai Bangō 20243030] Kenkyū Seika Hōkokusho) (pp. 138–157). Nagoya Daigaku Daigakuin Kankyōgaku Kenkyūka Tanaka Shigeyoshi Kenkyūshitsu (429 pp.).
Horikawa, S. (Ed.). (1998). *Otarushi ni Okeru Rekishiteki Kankyō Hozon to Kankō Kaihatsu: 1997-nendo Hōsei Daigaku Shakaigakubu Shakai Chōsa Jisshū Hōkokusho*. Hōsei Daigaku Shakaigakubu Shakai Chōsa Jisshūshitsu (239 pp.).
Horikawa, S. (Ed.). (1999). *Otarushi ni Okeru Rekishiteki Kankyō Hozon to Kankō Kaihatsu (2): 1998-nendo Hōsei Daigaku Shakaigakubu Shakai Chōsa Jisshū Hōkokusho*. Hōsei Daigaku Shakaigakubu Shakai Chōsa Jisshūshitsu (285 pp.).
Horikawa, S. (Ed.). (2000) *Otarushi ni Okeru Rekishiteki Kankyō Hozon to Kankō Kaihatsu (3): 1999-nendo Hōsei Daigaku Shakaigakubu Shakai Chōsa Jisshū Hōkokusho*. Hōsei Daigaku Shakaigakubu Shakai Chōsa Jisshūshitsu (367 pp.).
Horikawa, S., & Egami, W. (2002). Kankyō to Komyuniti: Nihon. In S. Kurasawa (Ed.), *Kaiteiban Komyunitiron* (Hōsō Daigaku Iinsatsu Kyōzai 83081-1-0211) (pp. 102–115). Hōsō Daigaku Kyōiku Shinkōkai.
Horikawa, S., & Fukaya, N. (Eds.). (2012). *Toshi Gabanansu no Shakaigakuteki Jisshō Kenkyū (3): 2011-nendo Hōsei Daigaku Shakaigakubu Shakai Chōsa Jisshū Hōkokusho*. Hōsei Daigaku Shakaigakubu Shakai Chōsa Jisshūshitsu (192 pp.).
Horikawa, S., & Fukaya, N. (Eds.). (2013). *Toshi Gabanansu no Shakaigakuteki Jissho Kenkyū (4): 2012-nendo Hōsei Daigaku Shakaigakubu Shakai Chōsa Jisshū Hōkokusho*. Hōsei Daigaku Shakaigakubu Shakai Chōsha Jisshūshitsu (194 pp.).
Horikawa, S., & Matsuyama, Y. (Eds.). (2016). *Toshi Gabanansu no Shakaigakuteki Jissho Kenkyū (5): 2015-nendo Hōsei Daigaku Shakaigakubu Shakai Chōsa Jisshū Hōkokusho*. Hōsei Daigaku Shakaigakubu Shakai Chōsha Jisshūshitsu (168 pp.).
Horikawa, S., & Matsuyama, Y. (Eds.). (2017). *Toshi Gabanansu no Shakaigakuteki Jissho Kenkyū (6): 2016-nendo Hōsei Daigaku Shakaigakubu Shakai Chōsa Jisshū Hōkokusho*. Hōsei Daigaku Shakaigakubu Shakai Chōsha Jisshūshitsu (186 pp.).
Horikawa, S., & Morihisa, S. (Eds.). (2008a). *Toshi Gabanansu no Shakaigakuteki Jissho Kenkyū: 2007-nendo Hōsei Daigaku Shakaigakubu Shakai Chōsa Jisshū Hōkokusho*. Hōsei Daigaku Shakaigakubu Shakai Chōsha Jisshūshitsu (202 pp.).

Horikawa, S., & Morihisa, S. (Eds.). (2009). *Toshi Gabanansu no Shakaigakuteki Jissho Kenkyū: 2008-nendo Hōsei Daigaku Shakaigakubu Shakai Chōsa Jisshū Hōkokusho.* Hōsei Daigaku Shakaigakubu Shakai Chōsha Jisshūshitsu (220 pp.).

Hōsei Daigaku Daigakuin Eko Dezain Kenkyūsho, Nippon no Minatomachi Kenkyūkai. (Eds.). (2005). Nihon no Kindai Minatomachi: Sono Kiso to Kūkan Keisei Genri no Hakken. Hōsei Daigaku Daigakuin Eko Dezain Kenkyūsho, Nippon no Minatomachi Kenkyūkai (177 pp.).

Lynch, K. (1972). *What time is this place?*. Cambridge, MA: MIT Press.

Marufuji, U., & Wakamatsu, S. (Eds.). (1920). Saishin Otaruku Tochi Daichō (including Kumausumura, Takashimamura, and Shioyamura), Otaruku Tochi Renraku Zenzu. Otaru: Watanabe, Hisatarō (496 pp. register, 8 diagrams).

Maruya, S., & Yamazaki, M. (1987). *Nihon no Machi.* Bungeishunjū.

Matsushita, S. (1986). Rekishiteki Machinami ni Okeru Keikan Seibi Shuhō ni Kan Suru Kenkyū: Otarushi Ironai-dōri, Midoriyamate-dōri no Bāi. Graduation thesis, Hokkaido University, Department of Engineering.

Muramatsu, T. (1986). *Uminekoya no Kyaku.* Asahishinbunsha.

Myōen, K. (1992). Otaru ni Okeru Keikan no Henyō Kōzō ni Kan Suru Kenkyū: Unga Shūhen Chiku o Megutte. Hokkaido Daigaku Daigakuin Kōgaku Kenkyuka Kyojūchikeikakugaku Kōza Shūshi Ronbun (A4, 96 pp.).

Myōen, K., Adachi, F., Morishita, M., Satō, K., Saeki, S., Yanagida, R., et al. (1993). Otaru no Rekishiteki Kankyō no Henyō Kōzō ni Kan Suru Kenkyū: Unga, Ironai, Sakaimachi Chiku no Bāi. *Nihon Kenchikugakkai Hokkaido Shibu Kenkyū Hōkokushū, 66,* 529–532.

Nagano, K., & Marufuji, U. (Eds.). (1916). Otaruku Kaisei Chiban Tochi Daichō, Otaruku Tochi Renraku Zenzu. Sapporo: Marufuji Sokuryō Jimusho (371 pp. register, 8 diagrams).

Natake, N. (2007). *Toshi no Tenkai to Tochi Shoyū: Meiji Ishin Kara Kōdo Seichōki Made no Osaka Toshin.* Nihon Keizai Hyōronsha.

Natsubori, M. (1980). Otaru Unga o Tsubusu na. *Chūō Kōron, 95-3* (Tsūkan 1117), 345–360.

Natsubori, M. (1992). Yameru Kokyō: Unga Sensō no Machi. *Bessatsu Bungei Shunjū, 201,* 428–468.

Nishimura, Y., Ōnishi, T., Ōgaki, S., Kishii, T., & Koide, K. (Eds.). (2003). *(Toshi Kōgaku Kōza) Toshi o Hozen Suru.* Kajima Shuppankai.

Okamoto, S. (2003). *Ginza: Tochi to Tatemono ga Kataru Machi no Rekishi.* Hōsei Daigaku Shuppankyoku.

Otaru Kankō Yūchi Sokushin Kyōgikai Chōsa Kenkyūbu. (Ed.). (1998). Otaru o Kangaeru: Otaru Kankō no Genjō to Korekara. Otaru Kankō Yūchi Sokushin Kyōgikai (22 pp.).

Otaru Shinpojiumu Jikkō Iinkai. (Ed.) (2009). *Otaru Unga to Sekizō Sōkogun no Hozon Undō Kara Nani o Uketsugu ka: Chiiki ni Iki, Chiiki o Mamoru...Machizukuri Undō no Senkusha Mineyama Fumi-shi ga Tsutaeru Koto Shinpojiumu Kaisai Hōkoku.* Sapporo: Nihon Kenchiku Gakkai Hokkaido Shibu (A4, 76 pp.).

Otaru Unga Mondai o Kangaeru Kai. (Ed.) (1986a). *Otaru Unga Hozon no Undō, Rekishihen.* Sapporo: Otaru Unga Hozon Undō Kankōkai.

Satō, S. (Ed.) (1927). Saishin Otarushi Oyobi Kōgai Tochi Daicho, Saishin Otarushi Oyobi Kōgai Tochi Jissoku Renraku Zenzu (three volumes and document box). Otaru: Sangyōshōreikai (614 pp., 41 diagrams).

Shisei no Ayumi Henshū Iinkai. (1988). *12-nen: Sanki Shimura Shisei no Ashiato.* Otarushi.

Tokyo Daigaku Shakaikagaku Kenkyūjo. (Ed.). (1952). *Sengo Takuchijūtaku no Jittai: Takuchijūtaku Sōgō Kenkyū 1.* Tokyo Daigaku Shuppankai.

Tokyo Daigaku Shakaikagaku Kenkyūjo. (Ed.). (1968). Toshinchi ni Okeru Takuchijūtaku no Hensen: Tokyo Taitōku Takemachi no Bāi (Tokyo Daigaku Shakaikagaku Kenkyūjo Chōsa Hōkoku Dai 9-shū). Tokyo Daigaku Shakaikagaku Kenkyūjo (B5, 216 pp., not for sale).

Urry, J. (1990). *The tourist Gaze: Leisure and travel in contemporary societies*. London: Sage [Kabuto, Hirokuni, trans. (1995). *Kankō no Manazashi: Gendai Shakai ni Okeru Rejyā to Ryokō* (Riburaria Sensho). Hōsei Daigaku Shuppankyoku].
Urry, J. (1995). *Consuming places* (The International Library of Sociology). London: Routledge.
Yasumura, K. (2006). *Kankō Machizukuri no Rikigaku: Kankō to Chiiki no Shakaigakuteki Kenkyū*. Gakubunsha.

Chapter 7
To Preserve Is to Change

Abstract In this final chapter, I review the Otaru case study and its theoretical implications. The "rationale of preservation" advanced by the preservation movement reveals the problematic nature of the traditional question, "to develop or preserve?" and demonstrates that what the movement really wanted was to achieve community-centered, autonomous control of change. In short, the preservation movement advanced the paradoxical argument that "to preserve is to change." This analysis of the preservation movement suggests the first step toward a new vision for controlling urban spaces, and makes theoretical contributions to the fields of urban sociology and environmental sociology. The volume, as summarized in this final chapter, is an attempt to simultaneously address the "sociality of the material world" and the "the material nature of society."

Keywords Community · To preserve is to change · Sociality of the material world · The material nature of society

7.1 The Gaze from Otaru/on Otaru

Until now, our gaze has remained fixed on the Otaru canal preservation movement. What has been brought into view? Let us begin with a summary.

The preceding chapters have clarified the significance of Otaru's fight to preserve the canal. This can be summed up in a single phrase: "to preserve is to change." The Otaru canal preservation movement was a pioneering and radical attempt to explore how urban change—so commonly thought to be an everyday and *positive* phenomenon (Holleran 1998)—could be guided by civil society. Moreover, the Otaru preservation movement rejected the idea of "keeping the historic townscape frozen in time" as a kind of outdoor museum and base for tourism development (Barthel 1996; Barthel-Bouchier 2001).

Underlying the debate over the pros and cons of the Rinkōsen road construction plan was a deeper disagreement over the correct strategy and philosophy of urban redevelopment. It is this deeper disagreement that explains the gravity and persistence of the social confrontation over the canal's fate.

S. Horikawa, *Why Place Matters*,
https://doi.org/10.1007/978-3-030-71600-4_7

The Otaru municipal government pressed for the reclamation of the canal and the construction of the new Rinkōsen road in its place. Local authorities understood the road project as a way of replacing an outdated system of cargo transport (port-barge-canal-warehouses-*denuki kōji*) with a more modern one (truck and road). In the sense that the city was replacing an old function with a new one, this can be described as a "scrap and build" approach. Local officials thought it perfectly natural to destroy the useless canal in order to give the same physical site a new functionality. Indeed, anyone who understood the space of the canal as an entirely colorless, utterly replaceable "cube" (in other words, anyone who subscribed to the idea of land as "space") would agree with their decision. What initially appears as a non-ideological and impartial view of "space," however, is in fact one element of the philosophy that has guided modern urban planning in Japan. One could actually describe it as an extremely ideological understanding of space.

By contrast, at the heart of the preservation activists' anguished cry ("Otaru will no longer be Otaru if the canal is destroyed") was a different redevelopment strategy, which focused not upon destruction but rather upon the re-use of old structures as resources for "urban rehabilitation." Obviously, this strategy was informed by a very different perspective on land: not as space, but as "place."[1] "Urban rehabilitation" involves protecting the city's identity by preserving old structures, while gently updating their interiors to meet the needs of the present day. In contrast to the city's "scrap and build" approach, one could describe this as a philosophy of "urban regeneration." The evolution of the guiding ideal of the canal preservation movement, from "cryopreservation" to "community development," could also be described in terms of the competing views of the canal as "space" and "place."

The canal preservation movement was thus far more than a simple dispute over a road construction project. Beneath the first layer of debate regarding the pros and cons of the construction of the Rinkōsen road was another layer: a debate over the validity of the "developmental" ideology that prescribed road construction in the first place. And deeper down was yet another layer: the debate over what a city should be, what a city should look like, and how people should live in a city. Over time, these various layers—the technical, procedural, and ideological—came together to comprise "the canal issue" (Horikawa 2009, 2010, 2011). Beginning with the question of "what kind of city do we want to live in?" local residents turned to activism in an attempt to assert control over changes to their city. This was the true essence of the multilayered canal war.

This conclusion addresses the broader theme of the city and the public domain. Let us listen to the voices of former canal preservation activists as we consider this further.

Sasaki Kyōjirō, a born and bred resident of Otaru, recalls the canal from his childhood days as a kind of location with no address, a place where "anyone could

[1]The historical and sentimental associations of "place" may seem suspiciously ideological. However, what the research makes clear is that "placeness" is actually the most universal form of spatial understanding (Ashiwara 1979, 1983; Jinnai 1985, 1988, 1993; Hara 1987). This sense of "placeness" is seen as an "ideological" approach only because we have accepted as self-evident the implicit assumptions underlying modern urban planning.

go" (Horikawa 2005: 194–195). According to Sasaki's account, between the mid-1950s and 1960s, there was a strong distinction between the different neighborhoods of Otaru, and mischievous youngsters would defend their neighborhoods against invasions from groups of children from neighboring areas. According to Sasaki, the port and canal district was the only area exempted from these turf wars—the only place that anyone was free to visit:

> You see, that (people's territorial sense and competitiveness with adjacent neighborhoods) is really strong in Otaru. You couldn't cross a single street, for example, without being out of bounds. {...} So, [if you left the neighborhood], you would round up a whole gang of kids to go with you. We'd all go together. {...} And if we were caught, we were imprisoned [by the children of the neighborhood we had entered]. Unless our parents came with some sort of 'tip' to get us out, we were stuck. {...} And the only place without these kinds of territories (territorial demarcations) was the area around the canal—what today you'd call the port district. {...} You could ask anyone and get the same answer. Ask Ogawara-san (Ogawara Tadashi) and he'll tell you the same thing: the port district was the only place that didn't have these different territories. And that was because we'd all take our fishing rods and go fishing there. Anyone could go there. And barges weren't in use [anymore] then, and so we'd play *Yoshitsune no hassōtobi* or, because the water was still relatively clean back then, dive down for...things like *babagai* [Swift's scallops]...
>
> Interview with Sasaki Kyōjirō in Otaru on September 2, 1997. Information in parentheses has been added by the author, information in brackets by the translator.

Sasaki's recollection of his youthful antics paints a vivid and indelible picture. Yet the real significance of his story lies with the canal itself, and the fact that it was "the only place" excluded from the turf wars of his youth. The port district was the "only place" where "anyone" could go.

Was the area around the canal nothing more than a paradise for the rambunctious children of Otaru? In fact, Yamaguchi Tamotsu tells us that in 1976 the canal was equally a place where adults could spend time freely:

> In Otaru, it's a common sight to see old men put their grandchildren and a fishing rod on the back of their bicycles and head to the port to fish. *The canal is just a five-minute walk from the station, and it was such a wide expanse of open land*—overgrown with weeds, of course. *Somehow it felt logical that this was the case.* At that point [Japan was] in a period of (high) economic growth—ten percent growth—so it didn't seem possible that there could be any vacant land left. Old alleyways were disappearing, and vacant land was vanishing from the cities. It was a time when places where the ownership wasn't clear—places that, amazingly enough, anyone could still use—were disappearing quickly. I felt keenly how *right* this place was.
>
> Interview with Yamaguchi Tamotsu in Otaru, September 4, 1998. Emphasis and information in parentheses has been added by the author, information in brackets by the translator.

This is part of Yamaguchi's explanation of how he came to live in Otaru and involve himself in the effort to save the Otaru canal, which he relates in Chap. 5. Unlike other port cities, where ordinary people could not freely access the port area, the Otaru port district in 1976 functioned as a public "fishing pond" and city park, conveniently located just five minutes from the Otaru train station. Yamaguchi discerned a certain logic or rationality—what *civilians* would understand to be a rational use of land—in the scene of ordinary citizens enjoying fishing expeditions or walks along the canal, in an area that would typically fall under the strict control of port authorities. The "rationality" identified by Yamaguchi had no relation to local bureaucracy or the port laws. Rather, at a time when the impoverished city of Otaru had no funds to maintain a city park, the existence of the canal as a public space accessible to everyone *made sense to local residents.*

Yamaguchi also saw the canal as an interface between different districts of the city and their residents. He continues:

> The port is what originally brought this town into being. In that sense, the port is a kind of cornerstone of this town. (For example,) The grove of a village shrine is located right in the center, and if that grove is destroyed, the (village) community collapses entirely. In the same way, I felt the port was immensely important to Otaru.
>
> It became clear that the port would be cut off from the town by the six-lane road (in the city's plan). I sensed there would be some form of divine retribution if that happened. The topography of the town was so good—on the palm of my hand, here (*pointing to his palm*) is the ocean, and here (*pointing to the fingers extending from the palm*) were the different communities. On the north side of Otaru there's what is called the Echigo group—this is where a lot of herring fishermen from Niigata settled. {…} And people living in the town (center) are (originally from) Hokuriku—they're merchants. {…} That was where the merchants lived. If you look at it this way, there are differences (in the people and personalities of each district). They've *very cleverly segregated themselves* into different areas. Of course (in reality) there are finer distinctions as well. *Yet all of these areas were connected to the water*, and would be *cut off* (by the Rinkōsen road). That's an unhappy fate for the town, and *extremely shameful from a design perspective*, you know.
>
> Interview with Yamaguchi Tamotsu in Otaru, September 4, 1998. Emphasis and information in parentheses has been added by the author.

Yamaguchi understood the Otaru canal to be the heart of the local community, much like the grove of the village shrine, connecting the diverse groups of people living in the surrounding area. While the fishermen in their coastal neighborhood and the merchants living in the heart of the city may have "cleverly segregated themselves," every neighborhood remained "connected to the water." Here we can understand the Otaru canal as an interface that skillfully linked the "port system" along the coast of Otaru to the "city residential system" closer to the mountains. This represented a form of *city design* cultivated and refined by the residents of Otaru over many generations. It was not developed quietly by one person according to a blueprint, but rather by Otaru's residents, over the course of many intense arguments surrounding competing strategies for development that preceded the construction of the port's breakwater, port facilities, various land reclamation projects, and the canal itself, as recounted in Chap. 3. Moreover, as the life histories presented in Chap. 5 confirm, preservation activists clearly recognized this form of city design; indeed, it formed the basis for their efforts at preservation.

All of this suggests that "city design" should not be discussed in terms of state-of-the-art architecture or intellectual trends, but rather with regard to the living styles, know-how, and "placeness" accumulated by residents in the practice of their everyday lives. *It is our gaze upon Otaru* that reveals the essence of city design to be "social control over change"—not city planning laws that impose centralized control.

Significantly, preservation activists in Otaru did not discuss city design as a narrowly defined technical or artistic issue, but rather as an *ideological issue* that posed the question: "What kind of city do we want to live in?" The idea that "a city should be a place where people can live with peace of mind" demands a mechanism for avoiding the kinds of risk that force people from cities. This is precisely why the preservation campaign asserted the need for a resident-led form of community development that would prevent rapid changes to the urban environment and its underlying structure.

Yamaguchi Tamotsu discusses the importance of "vast swathes of open land" to a city, and condemns the state of a city in which an "extremely shameful" design would "entirely cut off" communities traditionally "connected to the sea." Ogawara Tadashi describes the importance of "space…buildings [that] feel as if it they grew up [naturally] from underground roots." Their words are a strong indictment of modern Japan's growth-focused city management and city planning legislation, which fails to arbitrate conflicts between property owners or matters pertaining to absolute land ownership. One of the most significant implications of the townscape preservation movement in Otaru (and elsewhere across Japan), is that activists did not merely address the potential transformation of particular buildings, but rather addressed preservation as an "urban ideology," with the goal of creating the city as a public domain. This was a direct critique of modern Japanese cities. Yamaguchi and Ogawara's words reflect their pragmatic and critical thought processes; both men avoid more saccharine expressions of nostalgia. Their words are not those of men indulging a passing enthusiasm.

What are the key findings of my research in Otaru?

First, buildings alone do not constitute a townscape. A townscape is an organic spatial system that residents develop in response to the environmental conditions of a specific local community. It is through human habitation that buildings take on meaning. My study of the Otaru preservation movement clarifies the "rootedness" of place.

Second, we have seen how a townscape is in fact an accumulation of residents' expertise at living in the city, and reflects how residents use and manage the city—their proficiency at living in a particular setting. The composition and layout of both Kawagoe and Otaru display residents' literal *mastery* of daunting geographical conditions, such as the narrow and often hilly terrain. Yet modern methods of city planning make no attempt to preserve the micro-level knowledge of residents regarding the utilization and management of urban space. On the contrary, their "micro-expertise" has been repudiated and destroyed. I believe this may serve as the third conclusion of this book. Indeed, Otaru confronted this very challenge.

Fourth, I have provided a clearer picture of the movement opposing modern city planning, and the logic by which it operated. In the course of detailed deliberations in Otaru, activists suggested different concepts with which to oppose modern city planning, including "preservation," "preservation with development," "tourism development," and the "educational power of the environment." Most importantly, we discovered that preservation activists' understanding of "preservation" did not negate change, but actually affirmed it, because "to preserve is to transform." Given the inevitability of change, activists aspired to achieve change in the form and at the pace of their choosing. While the movement initially used the term "preservation," this was simply because the overwhelming pressure for development forced the movement to adopt such a reactionary term. I would describe their real goal as "social control over change."

With this deeper understanding of the Otaru preservation movement, we can now identify two different levels of "preservation" sought by the movement's participants.

Unsurprisingly, individual historic buildings were the first focus of preservation efforts. As we have seen, the movement did not seek to keep these buildings frozen in time, but allowed a certain degree of transformation to accompany preservation. Here the crucial point was the preservation of groups of buildings as a concrete physical entity upon which new forms of *machizukuri*, or community building, could be based.

The second target of preservation was a "lively" or enlivened way of daily life that the canal and old buildings provided to Otaru residents. If the daily lives which interweave with historic buildings are not also protected from unchecked development, the preservation of the buildings as mere museum pieces has little meaning. As the *Yume no Machizukuri* executive committee manifesto proclaims, it was essential to fill "old vessels" with "new energy." If we add the words "a daily life with" before "new energy," we can understand why residents experienced the destruction of the old "vessels" (historic buildings, the canal) as the destruction of their way of life, and why such destruction meant that "Otaru would no longer be Otaru." When Fumi Mineyama discussed what it meant to "live in the region" (*chiiki ni ikiru*), she was talking about what it meant to live in Otaru *as* Otaru.

The movement to preserve the Otaru canal thus sought to preserve both specific buildings and the modes of daily life that supported and reproduced this particular space and landscape. Put simply, over time the goal of the preservation movement changed from "protecting the canal" to "protecting the canal, which represents our way of life." This is the real meaning behind the often invoked description of the movement's evolution "from static preservation to community development." We must remember that despite the evolution of the preservation movement, activists remained focused upon the intrinsic environment of the canal until the very end. Confounding the expectations of the modernization theories ascendant in postwar sociology, the Otaru movement was based upon the "immediacy of physical objects." If the form of the physical environment changed, the local society would change as well; this transformed society would then be "lived" in a new way. The crisis embedded in the reciprocal relationship between environment and society was expressed in the refrain of preservation activists: "Otaru will no longer be Otaru" without the canal.

How should we describe the townscape preservation movements that emerged in 1960s Japan? At the very least, the scope of this book has identified certain key elements:

The first has to do with the framing of the issue. Activists made clear that the issue of preservation was not related to the status of "single" architectural structures, but rather to "groups" of these structures (or to the entire "city" itself). This is well expressed by their choice of the word "townscape" (*machinami*) to describe the target of preservation.[2] After all, a single building cannot comprise a townscape. Only larger groupings of buildings, and their overall effect, can produce a townscape. *Disaggregated, each and every new building that replaces a historic structure may be legal*, and the overarching problem of preservation vanishes into thin air. Townscape preservation movements were thus was an effort to take on the environment as a whole in order to address problems that persisted despite the legality of development and change.

Second, the preservation movement presented a prototype for community development based on the "rehabilitation of historic stock," not "scrap and build." In Japan, old things are torn down and rebuilt, severing continuity and gradually obscuring memories of what came before. Nevertheless, townscape preservation movements did *not* urge a return to tradition, nor repudiate change. In reality, preservation activists advocated the reuse of the city's historic "stock" of buildings and tolerated a certain level of change, as long as overall continuity was guaranteed.

These first two points lead directly to a third: this volume has made it abundantly clear that Japan's townscape preservation movements actually used the word "preservation" to argue for the necessity of social control over change. In other words, the central concern of preservation activists has been the question of governance. Preservation is not necessarily the repudiation of change. In the sense that the preservation movement advocated a specific form of change ("change to prevent change") activists believed that "to preserve is to change."

Fourth, the townscape preservation movement set forth a foundational principle for urban planning: development needed to match local ecological and historical conditions. Faintly detectable in the words used by Ogawara Tadashi and Mineyama Fumi to discuss the Otaru townscape (*machinami*) is the implication that buildings are of corresponding (*narabitatsu*) or average (*hitonami*) size. We should take heed of the significance of this implication.

What implications does the praxis of this preservation movement carry for sociological theory—and especially for urban sociology and environmental sociology? Up until now, our gaze has been trained *upon* Otaru. It is time to reverse this gaze, and consider what we can discover by looking out *from* Otaru. Let us turn to this question in the following section.

7.2 Community Theory: The Implications and Limitations for Urban Sociology

As discussed in the introduction to this volume, an examination of post-1945 Japanese urban sociology reveals that the issue of residency has remained the main subject of discussion.

---

[2]The character for *nami* in the compound *machinami* (townscape) also appears in the word *narabitatsu* (to stand in a row), thereby invoking the image of a line of buildings.

Nitagai Kamon and Okuda Michihiro are leading advocates of this approach. Nitagai and Okuda occupy diametrically opposing theoretical positions, yet no one could dispute their status as leading representatives of Japan's postwar urban and regional sociology.

Nitagai argues that urban issues emerge from the conflict between the city residents' logic of "to live" (utility value) and the logic of capital's private ownership of the city (surplus value). Nitagai envisioned the collective recapture of urban space based upon the logic of "to live" (Matsubara and Nitagai, eds. 1976: 331–396). Nitagai's concept of "to live" (*sumu koto*) is different from the everyday expression of "living" (*sumu*). Specifically, Nitagai understands "living" as the simple act of dwelling in a particular place, while the concept "to live" also incorporates the idea of cooperation within the local society. By making this distinction, Nitagai depicts the state of modern cities, in which people are forced to fixate upon mere "living," in the biological sense. In order to revive residents' ability "to live" in the modern city, Nitagai problematizes the act of mere "living."

Okuda Michihiro identifies a similar crisis of residence within a very different theoretical and conceptual framework.

First, Okuda identifies two categories of city. The first is the "city" (*toshi*) as work space. The second is the "region" (*chiiki*) as the foundation of residents' ability "to live." Okuda sees the modern city as the multilayered product of these two lineages, and defines the problem as the destruction of the "region" by the "city." Okuda sought the specific dynamic by which resident-led movements from the "region" opposed to the pressures of the "city" could revive the multilayered city (what Okuda calls "community") (Okuda 1993: 33–61). Put differently, Okuda criticized the "monoculturization" of the work space of big cities, and argued that "community" depended on the coexistence of the layers of "city" and "region." This perspective remains consistent throughout Okuda's subsequent work (Okuda 2000).

The arguments made by Nitagai and Okuda are clearly relevant to my analysis of preservation movements. The preservation activists who wanted to live within the structures and landscape handed down through the generations were not seeking to exist merely in the biological sense; they aspired "to live" in Nitagai's sense of the phrase: cooperatively and embedded within the local society.

However, we need to focus upon how Nitagai and Okuda built their theories, as this illuminates the issues that demanded discussion among Japan's postwar sociologists, and clarifies what they posited as the "enemy."

Regardless of whether or not they used the word "capital," Nitagai and his colleagues critiqued the postwar pro-development approach to urban planning, and saw as its counterpoint the resident-led opposition movements that emerged to defend livelihoods and "living space" (*seikatsu kūkan*). The fact that the term "resident-led movement" defined the post-war period confirms their focus.

At the same time, the use of such abstract, non-place-specific terminology as "utility value" or "space" by Nitagai and Okuda suggests that their vision of the "region" in which people wish "to live" was, ultimately, a modern one—in other words, an entirely replaceable living space. Their subject, moreover, was the "universal citizen" posited by postwar modernization theory who, unfettered by (what were seen as) negative attachments to "hometown" or the "feudal land system," simply sought "to live" in any given place. Their image of space was also straightforward: universal and non-specific. The emerging residential class in the new housing developments along the outskirts of big cities—people who had left their ancestral homes in rural villages, and therefore broken free of the feudalistic social order to live and conduct themselves as "citizens" in their new home—were precisely what Nitagai and Okuda had in mind. When these "citizens" rose up against pollution or inadequate infrastructure, this took the form of a neighborhood movement based on "the power of resistance focused on 'to live' (*sumu koto*) (Sato 1993: 166). For our purposes, however, the most important point is this: the rationale of opposition to the major, modernizing restructuring of cities did not employ terms such as "love of one's hometown," or the "feudal land system," but rather a universal conception of "space" in which "to live."

This type of argument carried enormous significance during Japan's period of rapid economic growth, as the national landscape underwent dramatic upheaval. The language used to resist development could not evoke the pre-war past—it had to be tinged with positive and universal values.

However, a detailed examination of the Otaru case and other townscape preservation movements raises a crucial question: can the "power of resistance" in order "to live" stem from *any* generic space? Or does it stem from a specific place—a place with a name? The first scenario is consistent with modernization theory, but the second is more complicated: we must take into account unique environmental conditions, such as the region's natural features and collective historical memory. As has been stated so many times before, "living space" is not the homogenous, interchangeable "space" assumed by urban planning, but a "place." The power to resist development, and the very form of this resistance, inevitably depends upon the specific environmental conditions of the area in question. The Shiohama canal in Tokyo and the Otaru canal are both canals, but they differ not only in their construction and length, but in their significance to the local community.

If we return once more to the case of Otaru, residents' deep attachment to a specific piece of land and to a particular place, which prompted their defense of the Otaru canal, would be contemptuously dismissed as "feudalistic" by Japan's postwar sociological theory—and particularly by urban sociology. Postwar urban sociology in Japan has relegated the obsession with "placeness"—for example, the deep affection for one's own hometown—to the background, and argued that the more crucial issues are the creation, by the modern "citizens" of Japan's postwar democracy, of "publicness" and "fundamental principles of citizenship." Townscape preservation movements were an attempt to bring "placeness" back to the foreground. Going further, townscape preservation movements sought to dispel the association between such fixations with place—"the love of one's hometown," for example—and narrow-minded, illiberal nationalism, and reconsider them within a more existential register, in relation to such questions as "what kind of city do people want to live in?" or "what kind of existence do we want to have?" (Ueda 2016: 41–46). While postwar Japan's neighborhood movements have been discussed within the framework of resident-led "demand" and "obstruction" campaigns (Nishio 1975)—townscape preservation movements tried to return our attention to "the existential question" obscured by this framework.

This is why sociology must devise a theory of space that allows for "placeness." While some might prefer to describe this as a "spatial turn" (Yoshiwara 2002), there is a common desire to escape the monistic, Le Corbusier-like depiction of modern space.

Urban sociology, and community theory in particular, thus comprises a large part of my intellectual inheritance. Indeed, the extent of my intellectual debt to Nitagai Kamon and Okuda Michihiro is unfathomable. I have also drawn extensively from the work of the Chicago School and W.F. Whyte (Whyte 1955), the new urban sociology of Manuel Castells (Castells 1983), my own concept of "space versus place," works in the fields of architecture, urban planning, and architectural history, and the concept of "the organic spatial order." In the process, I have become keenly aware of urban sociology's neglect of "the historic environment." Given the extent of urban sociology's interest in city-based and city-focused social movements, how can we explain the field's utter disregard for townscape preservation campaigns? Before racing ahead with Okuda and Nitagai to an examination of the "universal" citizen in the "abstract" environment, isn't it necessary to stop and take a closer look at the specific environment of an individual community, and at what preservation campaigns in such places are actually trying to protect? Despite the extent of my debt to urban sociology, such concerns have drawn me closer to the field of environmental sociology.

7.3 The Implications and Limitations of Environmental Sociology

Environmental sociology in Japan has rapidly expanded its reach since the 1990s. Given the number of overviews of the field (Iijima 1998, 2001; Horikawa 1999, 2012; Umino 2001; Seki 2005; Horikawa 2017), I will limit this discussion to an explanation of how I have drawn from the field of environmental sociology.

As I once observed, "looking back, almost all of the environmental sociology research conducted in Japan has been framed by the question of suffering incurred by environmental destruction" (Horikawa 2012: 7). The question of damage, or more specifically, the question of the kinds of damage incurred by environmental destruction (Horikawa 2012: 7), has informed the development of many different concepts or theoretical positions in environmental sociology: for example, Iijima Nobuko's "structural theory of suffering," the "beneficiary sphere/suffering sphere" theory (*jyueki-ken/jyuku-ken ron*) developed by Funabashi Harutoshi and Hasegawa Kōichi, and the "social dilemma theory" developed by Funabashi and his colleagues (Horikawa 2017b). These theories are also part of my intellectual inheritance—indeed, they have become like flesh and blood relatives over the course of my research.

There is, however, an entire area that is immediately obscured by the use of the word "suffering." This includes matters that evade immediate identification as a problem—neighborhood noise pollution, for example. "This is noise pollution and something has to be done about it!" "It's not noise pollution!" "It's a nuisance!" "No, this falls well within the scope of what can reasonably be tolerated." What is unfolding in this type of conversation between residents is a conflict over the question of *whether or not* a certain noise is causing damage. In other words, a serious conflict has unfolded before the *reality* of the alleged "suffering" can be ascertained. This suggests that, before analyzing a structure that presupposes the existence of the "damager-damaged" relationship, we must re-conceptualize the very relationship itself (Daimon 2008: 156).

Nor is this the only problem with environmental sociology. If we problematize the loss of the townscape with the word "damage" or "injury," we consign the issue to the register of "rights"—the right to sunlight, for instance. As we have seen in Chap. 2, however, very few rights (apart from the right to sunlight) have been fully authorized or compensated, and ultimately have fallen victim to the two barriers of legality and subjectivity. Following the Great Eastern Japan Earthquake of 2011, the word "damage" has never been more essential to environmental sociology (Horikawa 2012), yet it is hardly sufficient for addressing the issue of townscape preservation. Rather than focus on "damage" or "injury," we should ask how one can verbalize the sense of crisis people experience when the very existence of something is threatened. What I have attempted to depict in this volume is the search for the conditions of an urban community's sustainability, and the contradictions unleashed by the commodification and reification of the urban environment.

This brings us to the theories of "life environmentalism" and "the commons."

Seki describes life environmentalism as "the basis, identified by sociologists in the conflict between governmental authorities and local residents, for building an argument and a theory of persuasion" (Seki et al. 2009: 47). Life environmentalism is not based on the position of the "resident" or "ordinary person," but rather on the ground level of "the experiential world" in which the "resident" lives (Furukawa 1999: 146). The perspective of an individual resident cannot address the diversity of positions within the relevant regional community. That is why this approach does not assume the position of the "inhabitant," or "resident," but rather the position of the inhabitant's or resident's daily life. In other words, this approach enables analysis based on the "logic of preserving people's daily life." This is equally a declaration of methodology (Furukawa 2004).

Based upon the actual site of environmental issues, where the "vital perspectives of society and relevant people are sometimes obscured" (Seki et al. 2009: 47), life environmentalism has clarified the position and legitimacy of everyday citizens as the chief actors in maintaining the environment (Torigoe and Kada [1984] 1991; Torigoe 1989; Kada 1995, 2002; Torigoe 1997; Furukawa 2004). The two "enemies" posited by this approach are "nature environmentalism," which rejects any human interference with the environment, and "modern technicism," which maintains that environmental problems can generally be resolved through technology. Considering the history of Japan's modernization and the enormous impact of national public works projects on local communities, the "life environmentalism" approach has keenly probed such questions as "how have 'residents' and 'inhabitants'" acquired the autonomy to resist or approve these changes?" and "what qualifies them to speak out?" (Kada 2012).

As demonstrated by their exacting and sustained fieldwork in the Lake Biwa region, life environmentalists are chiefly interested in illuminating the methods devised by residents for coexistence with their environment (Kada 1995, 2002; Furukawa 2004). Rather than explicate the nature of "suffering" resulting in grave bodily damage at the site of environmental problems, life environmentalists would prefer to record the local bodily techniques devised by residents to survive in coexistence with the local "environmental carrying capacity." Therein lies the true essence of life environmentalism. Put differently, life environmentalists have sought to relativize development in the form of massive public works projects from the viewpoint of the daily lives that comprise the local "lifestyle."

Yet this "life environmentalism" bears the strong stamp of the rural scenery of the Lake Biwa region, where the theory originated.[3] The theory has described how agricultural villages, or mixed fishing and agricultural villages, have grappled with the environmental destruction wrought by public works projects and other forms of development. This scope does not incorporate the hustle and bustle of the city. Nor does it consider suburban residential districts, located far away from the workplaces of its residents, or factory-lined industrial zones. It would be difficult to find a place for Otaru within the scope of life environmentalism.

We should also note that the theoretical emphasis of life environmentalism centers on the logic of daily life. This is better explained through a symbolic example. If we were to clarify the central claims of life environmentalism through the use of italics, which word would be italicized—life or environmentalism? As the methodology takes "the position of the daily life of residents," then the word to be italicized would be daily "*life*." If we were to italicize a second focus of the theory, theorists' positive assertion of positionality would dictate the following: *life* environmental*ism*. The word "environment" would not be italicized. The communicative relationship between environment and society and the inevitable transformation of the modality of local society as a result of environmental destruction—in other words, the themes of this volume—are pushed to the background by the life environmentalist approach. Furthermore, it is difficult to objectify "the preservation of something with no shape" (Adachi 2010)—in Otaru or any other townscape preservation movement. Thus, while life environmentalism may have broken new ground by expanding its focus to such "shapeless" and intangible objects as culture and tradition, it has also made it more difficult to make the built environment the focus of inquiry. Yet life environmentalism has also performed an enormous service by reminding us that even legends and folk tales comprise a part of our "environment" and, just like aspects

[3]To be sure, my point is *not* to problematize the field's origins. The field has both displayed an excellent resolution power to meet the challenges of decoding a difficult subject, and also bears the strong imprint of its origins, which has led to certain blind spots, and weaknesses. The existence of such blind spots is the point—*not* the origins of the field in and of itself.

of our physical environment, regulate our society and our behavior. However, from the perspective of the present sociological inquiry, which stems from a dual focus on "the physicality of society" and "the sociality of the physical world," the failure of life environmentalism to italicize the word "environment" signals its retreat from the issue of historic environmental preservation. For all these reasons, I cannot count myself among the ranks of the life environmentalists.

The theory of the commons offers yet another methodological perspective. This theory, which has shown enormous vigor in recent years, seeks to clarify the "structure of responsibility" underlying the maintenance and control of the natural environment by local residents.

As Miyauchi Taisuke tells us, the commons theory is rooted in the historic forms of land use in Great Britain; given the relevance of the national trust movement to my research, this is also what drew my attention to the theory (Horikawa 2001). The commons have their origin in the negotiated rights of small farmers to access and use the land of the great landowners. The structure of joint control and joint use, and the physical environment utilized by this structure, are at the heart of the commons theory (Miyauchi 2006). Commons theorists are chiefly concerned with elucidating underlying structures of responsibility in order to identify ways of protecting the physical environment. Understood in this way, the theory surveys the world from a central axis of ownership and control. By examining cases of joint land management from around the world, the theory challenges "ownership" as the only possible relationship between humans and the land, and argues that modern property rights are nothing more than one form of land ownership.

These underlying structures of responsibility do not form automatically (or harmoniously). On the contrary, they emerge only after protracted grappling with questions of publicness and legitimacy—for example, questions regarding whose rights are respected, and the acceptable forms of exercising rights (Inoue and Miyauchi 2001; Inoue 2004). The theory of the commons has thus expanded into debates surrounding publicness, legitimacy, and policy (Miyauchi 2006; Miyauchi 2011).

These characteristics regulate the way in which the theory of the commons deals with the subject of "suffering" or "damage." The theory is defined by an attention to "the structure of responsibility" over "suffering," and "disentangling ownership" over "damage;" this is also its contribution. In terms of the classical schema of "environmental problems" and "coexistence with the environment" (Iijima 2001) the theory's relative proximity to the "coexistence" framework is responsible for much of its distinctiveness. There are indeed many affinities between this approach and my own work. My description of "the Otaru canal as the public sphere" could equally be worded as "the Otaru canal as the commons." Yet my use of the term "the physical foundation of the public sphere" (Horikawa 2011: 54–56) fits less easily within the theory of the commons. For this reason, notwithstanding my debt to the theory of the commons, I cannot identify myself as a "commons theorist."

7.4 The Function of the Townscape: Toward a Sociology of Townscape Preservation

This review has moved from urban sociology to environmental sociology. Where can we turn to next?

As discussed in Chap. 1, Dunlap's HEP-NEP debate is already a thing of the past. What I have discovered in the process of my examination of the Otaru case is the extent of the environment's effect and weight on humans. The disposition of preservation activists, and the course of their preservation efforts, show no indication of being the casual hobby of dilettantes, or a nostalgia-fueled pastime. On the contrary, they reveal a genuine search for a way to control changes to the urban environment and a social circuit through which this control can be exerted. This has not been studied or discussed sufficiently within conventional modes of sociology. How should sociology, which trains its sights on social relationships, discuss the power of the environment to directly influence human beings? How can it address the reciprocal relationship between people and the environment? This is both an old and a new issue. It is also the reason why we must insist upon a sociological approach to the preservation of historic environments.

The work of Barthel (Barthel 1996) represents the first attempt to undertake a sociological examination of preservation efforts targeting the historical environment. Barthel's central concept is that of Staged Symbolic Communities (SSC). Barthel identifies the four characteristics of Staged Symbolic Communities as ahistoricity, a moral order based on social consensus, repetitive activity, and spatial and temporal isolation from other communities (Barthel 1996: 37–48). This definition of the staged symbolic community is exceedingly close to the concept of utopia (Dahrendorf 1968: 107–108). Barthel identifies American open-air museums and historic villages as one form of utopia, and criticizes them as a form of ideology.

This is certainly a significant observation, yet I must admit to some doubt regarding the range of application permitted by Barthel's theoretical approach. Barthel developed her theory with reference to the Amana—a religious commune and agricultural community. Barthel's approach is thus well-suited to the analysis of other agricultural communities, such as that of Hancock Shaker Village. The efficacy of her approach for the analysis of historic preservation movements in urban areas, however, is less obvious. It seems we must develop a different concept and logic with which to approach cities, which do not allow for a historicity or cyclical action—or isolation. The theory of change-rejecting staged symbolic communities (SSCs) cannot be applied to the urban environment.

Indeed, what the Otaru case reveals is the active tolerance of change displayed by city-based historic preservation movements. Urban preservation and its attendant ideology, which can be summarized by the idea that "to preserve is to change," demands a new theoretical framework.

While life environmentalism has diffused the subject of research, it is now time to re-position the environment as the focus. With a critical acceptance of Barthel's work as a starting point, I have set out to elucidate a theory of urban preservation.

Moreover, I have attempted to develop the architectural concept of change and apply it sociologically to detect the gaps between physical reality and social consciousness. Using preservation as a lens, I have clarified residents' ideas of "what makes a city a city." How can urban changes be placed under social control? How can we establish a new forum for government officials and residents to discuss the city as a public good, and implement a system for controlling change? How can we even conceive of such a structure? The answer to this string of questions seems to be the development of a new urban sociology, and a new environmental sociology. Put differently, this theoretical consideration of the interaction between humans and the urban environment is an urban sociological study that also indicates a new direction for environmental sociology, away from the current focus on global environmental issues.

If we employ the language of Matsumoto Yasushi (Matsumoto 2014: 4–8) this conclusion would sound something like this:

What leads to preservation? The drastic and fundamental changes to the urban environment have elicited the reaction of "preservation." Let us recall that it was the destruction of the Arihoro warehouses that inspired the preservation movement in Otaru. As Barthel (1996) observes, preservation is also a response to industrialization. Everyday existence is never abstract—it is always lived in highly specific and physical forms. As we have examined in detail in Chap. 5, the loss of this form is equally the loss of human life.

How does the process of preservation work? This entire volume is a reply to this question. What we have learned is that the act of preservation is far from monolithic. Fueled by nostalgia, it was fought through many different layers, "*in a highly specific time and space*" (Matsumoto 2014: 7. Emphasis in original).

What, then, has preservation itself created? Preservation has created an awareness of the potential for self-determination among city residents—in other words, it has made residents conscious of their desire to exert control over the inevitable changes to the built environment of the city. Their recognition that "to preserve is to change" emerged through the act of preservation. Let us recall the words of Otaru preservation activist Ogawara Tadashi: "change in itself is not the problem." The fundamental question was actually *how* change was brought about. This volume presents a detailed discussion of the structure underlying the perceptions of Otaru's preservation activists and the function of changes to Otaru's built environment. My term for this research is "the sociology of historic environment preservation." I conceive of it as a subfield of sociology that takes into account the interrelationship between residents and the urban environment in which they live.

In this sense, my research on Otaru has identified both the potential and limitations of existing sociological approaches, and suggests opportunities for exchange with new theoretical frameworks being developed across the world. If we describe the architectural data presented in this volume as "hard data," and the life histories of Otaru's preservation activists as "soft data," then this volume is an attempt to simultaneously address the "sociality of the material world" and the "material nature of society" through the combination of these data types. This new experiment has only just begun, and all sorts of theoretical gaps and distortions must be addressed.

Yet one thing is clear: we now have a prototype for the future "sociology of historic environment preservation."

References

Adachi, S. (2010). *Gujō Hachiman Dentō o Ikiru: Chiiki Shakai no Monogatari to Riariti*. Shinyōsha.

Ashiwara, Y. (1979). *Machinami no Bigaku*. Iwanami Shoten.

Ashiwara, Y. (1983). *Zoku, Machinami no Bigaku*. Iwanami Shoten.

Barthel, D. (1996). *Historic preservation: Collective memory and historical identity*. New Brunswick, NJ: Rutgers University Press.

Barthel-Bouchier, D. (2001). Authenticity and identity: Theme-parking the Amanas. *International Sociology, 16–2,* 221–239.

Castells, M. (1983). *The city and the grassroots: A cross-cultural theory of urban social movements*. Berkeley, CA: University of California Press.

Dahrendorf, R. (1968). Out of Utopia: Toward a reorientation of sociological analysis. In Ralph Dahrendorf (Ed.), *Essays in the theory of society* (pp. 107–128). Stanford, CA: Stanford University Press.

Daimon, S. (2008). Sekinin Jissen Toshite no Kinrin Sōon Mondai: 'Higai o Uttaeru Koto no Imi' no Kihan Rironteki Kōsatsu. *Kankyō Shakaigaku Kenkyū, 14,* 155–169.

Furukawa, A. (1999). "Kankyō no Shakaishi Kenkyūno Shiten to Hōhō: Seikatsu Kankyōshugi to Iu Hōhō." Funabashi, Harutoshi and Furukawa, Akira, (Eds.), *Kankyō Shakaigaku Nyūmon: Kankyō Mondai Kenkyū no Riron to Gihō* (Shakaigaku Kenkyū Shirīzu: Riron to Gihō 25, pp. 125–152). Bunka Shobō Hakubunsha.

Furukawa, A. (2004). *Mura no Seikatsu Kankyōshi*. Kyoto: Sekai Shisōsha.

Hara, H. (1987). *Kūkan: Kinō kara Yōsō e*. Iwanami Shoten.

Holleran, M. (1998). *Boston's "Changeful Times:" Origins of preservation and planning in America* (Creating the North American Landscape). Baltimore, MD: The Johns Hopkins University Press.

Horikawa, S. (2001). Keikan to Nashonaru Torasuto: Keikan wa Shoyū Dekiru ka. In Torigoe, H. (ed.), *Shizen Kankyō to Kankyō Bunka* (Kōza Kankyō Shakaigaku 3, pp. 159–189). Yūhikaku.

Horikawa, S. (2005). Toshi Seikatsu to Seikatsu Kankyō Hendō: Rōkaruna Kūkan Seigyo Shisutemu, Saikō. In H. Fujita & M. Urano (Eds.), *Toshi Shakai to Risuku: Yutakana Seikatsu o Motomete* (Shirīzu Shakaigaku no Akuchuariti: Hihan to Sōzō 8, pp. 173–204). Tōshindō.

Horikawa, S. (2009a). Kichō Hōkoku 'Hozon Undō no Keika to Seika,' Otaru Shinpojiumu Jikkōiinkai, ed. Otaru Unga to Sekizō Sōkogun no Hozon Undō Kara Nani o Uketsugu ka: Chiiki ni Iki, Chiiki o Mamoru…Machizukuri Undō no Senkusha Mineyama Fumi-shi ga Tsutaeru Koto Shinpojiumu Kaisai Hōkoku: 9–30. Sapporo: Nihon Kenchikugakkai Hokkaido Shibu (A4, 76 pp.).

Horikawa, S. (2010). Basho to Kūkan no Shakaigaku: Toshi Kūkan no Hozon Undō wa Nani o Imi Suru no ka. *Shakaigaku Hyōron, 60–64,* 517–534.

Horikawa, S. (2011). Kindai Toshi no Mizubeto Kōkyōken: Toshin no Mizube wa Dare no Mono ka. *Kantō Toshi Gakkai Nenpō, 13,* 50–59.

Horikawa, S. (2012). Kankyō Shakaigaku ni totte 'Higai' to wa nani ka: Posuto 3.11 no Kankyō Shakaigaku o Kangaeru Tame no Ichi Sozai Toshite. *Kankyō Shakaigaku Kenkyū, 18,* 5–26.

Horikawa, S. (2017). Jueki-ken to Juku-ken. In T. Tomoeda, H. Hama & M. Yamada (Eds.) *Shakaigaku no Chikara: Saijūyō Gainen, Meidaishū* (pp. 132–135). Yūhikaku.

Iijima, N. (2001). Kankyō Shakaigaku no Seiritsu to Hatten. In N. Iijima, H. Torigoe, K. Hasegawa & H. Funabashi (Eds.), *Kankyō Shakaigaku no Shiten* (Kōza Kankyō Shakaigaku 1). Yūhikaku.

Inoue, M. (2004). *Komonzu no Shisō o Motomete: Kalimantan no Mori de Kangaeru*. Iwanami Shoten.

Inoue, M., & Miyauchi, T. (Eds.). (2001). *Komonzu no Shakaigaku: Mori, Kawa*. Shinyōsha: Umi no Shigen Kyōdō Kanri o Kangaeru.
Jinnai, H. (1985). *Tokyo no Kūkan Jinruigaku*. Chikuma Shobō. [Nishimura, K. trans. (1995). *Tokyo: A Spatial Anthropology*. Berkeley, CA: University of California Press].
Jinnai, H. (1988). *Watashi no Tokyogaku* (Toshi Sōsho). Nihon Keizai Hyōronsha.
Jinnai, H. (1993). *Toshi to Ningen* (Iwanami Shimin Daigaku "Ningen no Rekishi o Kangaeru" 6). Iwanami Shoten.
Kada, Y. (1995). *Seikatsu Sekai no Kankyōgaku: Biwako Kara no Messēji*. Nōbunkyō.
Kada, Y. (2002). *Kankyō Shakaigaku*. Iwanami Shoten.
Kada, Y. (2012). *Chiji ni Nani ga Dekiru ka: 'Nihonbyō' no Chiryō wa Chiiki kara*. Nagoya: Fūbaisha.
Matsubara, H., & Nitagai, K. (Eds.). (1976). *Jūmin Undō no Ronri: Undō no Tenkai Katei, Kadai to Tenbō*. Gakuyō Shobō.
Matsumoto, Y. (Ed.). (2014). *Toshi Shakaigaku/Nyūmon* (Yūhikaku Alma). Yūhikaku.
Miyauchi, T. (Ed.). (2006). *Komonzu o Sasaeru Shikumi: Rejitimashī no Kankyō Shakaigaku*. Shinyōsha.
Miyauchi, T. (2011). *Kaihatsu to Seikatsu Senryaku no Minzokushi: Soromon shotō Anokero-mura no Shizen/Ijū/Funsō*. Shinyōsha.
Nishio, M. (1975). Gyōsei Katei ni Okeru Taikō Undō: Jūmin Undō ni Tsuite no Ichikōsatsu. *Seiji Sanka no Riron to Genjitsu* (Nenpō Seijigaku 1974, pp. 69–95). Iwanami Shoten.
Okuda, M. (1993). *Toshi to Chiiki no Bunmyaku o Motomete: 21 Seiki Shisutemu Toshite no Toshi Shakaigaku*. Yūshindō.
Okuda, M. (2000). *Toshi Shakaigaku no Me*. Hābesutosha.
Satō, K. (1993). Komyuniti no Naka no 'Komyuniti.' In O. Hasumi & M. Okuda (Eds.) *21 Seiki Nihon no Neo Komyuniti* (pp. 153–176). University of Tokyo Press.
Seki, R., Nakazawa, H., Maruyama, Y., & Tanaka, M. (2009). *Kankyō no Shakaigaku* (Yūhikaku Aruma). Yūhikaku.
Torigoe, H. (1997). *Kankyō Shakaigaku no Riron to Jissen: Seikatsu Kankyō Shugi no Tachiba Kara*. Yūhikaku.
Torigoe, H. (Ed.). (1989). *Kankyō Mondai no Shakai Riron: Seikatsu Kankyō Shugi no Tachiba Kara*. Ochanomizu Shobō.
Torigoe, H., & Kada, Y. (Eds.). ([1984], 1991). *Mizu to Hito no Kankyōshi: Biwako Hōkokusho*. Ochanomizu Shobō.
Ueda, K. (2016). *Sonzoku no Kiro ni Tatsu Mura: Damu, Saigai, Genkai Shūraku no Saki ni*. Kyoto: Shōwadō.
Whyte, W. F. (1955). *Street corner society: The social structure of an italian slum*. Chicago, IL: University of Chicago Press.
Yoshiwara, N. (2002). *Toshi to Modaniti no Riron*. University of Tokyo Press.

Timeline of Events: The "Otaru Canal Issue" and Subsequent Community Development Initiatives, 1959–2017

Notes

Abbreviated references are shown in italics. Please see the chapter-end reference sections for complete bibliographic information.

Information relayed or confirmed by Mineyama Fumi during interviews with the author is attributed here to "Mineyama."

1959

August 12	The Otaru Port Commission submits its ten-year plan (1958–1967) to the national port authorities for approval. During the planning process, the reclamation of the Otaru canal and construction of railway facilities on the former site of the canal were incorporated into the ten-year plan. *Asahi Shinbun, Ogasawara(1986a).*

1966

August 25	The Ministry of Construction Bulletin 2912 approves the Dōdō Rinkōsen Highway Plan.

1967

April	Construction of the Rinkōsen begins. *Otaru City.*

1970

January	Otaru's canal and stone warehouses are designated as one of the three most important Meiji-era landscapes in Japan (the other two are in Nagasaki and Kobe) by the Architectural Institute of

S. Horikawa, *Why Place Matters*,
https://doi.org/10.1007/978-3-030-71600-4

Japan's subcommittee on Meiji-era architecture. *Kankōshigen Hogo Zaidan, ed. (1979).*

1971–1973

- As construction of the Rinkōsen continues, warehouses in the Arihoro district of Otaru are demolished to make way for this new road. *Asahi Shinbun, Ogasawara(1986a).*

1973

December 4 The inaugural meeting of the newly formed Association to Protect the Otaru Canal (APOC) is held at the Kaiin Kaikan. The twenty-four founding members of the APOC select Koshizaki Sōichi (chairman) and Fujimori Shigeo (secretary) to lead the organization. *Asahi Shinbun, Ogasawara(1986a), Mineyama.*

1974

May 17 APOC secretary Fujimori Shigeo embarks on a pilgrimage to the sites of earlier townscape preservation campaigns across Japan. Fujimori visited twelve towns in total, including Kurashiki and Tsumago, and met with local preservation campaigns. Fujimori concluded his two weeks of travel in Tokyo, where he petitioned the Agency for Cultural Affairs and the Ministry of Construction to save the Otaru canal. *Asahi Shinbun, Ogasawara(1986a).*

1975

August City officials and APOC representatives hold their first meeting. Six subsequent meetings were convened through September 1976. *Hokkaido Shinbun, Ogasawara(1986a).*

1976

July 9–14 An exhibit of "Plans to Reuse the Otaru Canal," submitted by graduates of the Hokkaido University architecture department, is held on the fifth floor of an Otaru department store. The exhibit was organized by the Toshi Isan Kenkyūjo and supported by the APOC, the Hokkaido University engineering department's engineering and architectural history laboratory, Otaru Bungei (a local arts circle), and the Hokkaido chapter of the Japan Scientists' Association. *Hokkaido Shinbun, Ogasawara(1986a).*

1977

November 1	Thirty-seven economic and neighborhood associations, led by the Otaru Chamber of Commerce and Industry, established the Association for the Promotion of the Otaru Rinkōsen (Otaru Rinkōsen Seibi Sokushin Kiseikai). *Hokkaido Shinbun, Ogasawara(1986a).*

1978

May 25	Mineyama Fumi is the elected chairwoman of the APOC at a general meeting of APOC members. *Hokkaido Shinbun, Ogasawara(1986a).*
June 5	At a standing meeting of the city council's construction committee, the director of the city's civil engineering department asserts that "there is no alternative to the route that appears in the original draft" for the Rinkōsen. *Minutes of the Otaru City Council.*
June 24	In Sapporo, ten academics and prominent members of Hokkaido's cultural world formally launched the Association to Think about the Otaru Canal (Otaru Unga wo Kangaeru Kai). *Ogasawara(1986a).*
July 8	Roughly 100,000 people turn out for the first annual Port Festival in Otaru. *Hokkaido Shinbun, Ogasawara(1986a).*
August 13	The *Yume no Machizukuri* Executive Committee is launched. *Interview with Sasaki Kyōjirō.*
October	Three organizations—the APOC, *Otaru Unga wo Kangaeru Kai*, and the *Otaru Unga wo Aisuru Kai*—petition the Agency for Cultural Affairs, the Environment Agency, and the Ministry of Construction to save the Otaru canal. *Ogasawara(1986a).*

1979

February 24	The mayor of Otaru issues a written response to the APOC and three additional organizations, stating that the alternate proposals submitted by the APOC for the location of the Rinkōsen would be too difficult to implement. *Minutes of the Otaru City Council.*

June 29 The details of the plan devised by Hokkaido University Assistant Professor Iida Katsuyuki(the "Iida Plan") commissioned by the city are announced at a meeting of the city's standing committee on construction. This revised plan called for the Otaru canal to be narrowed but not entirely paved over. *Minutes of the Otaru City Council.*

July The Otaru city government launches its first conservation effort targeting the city's historic buildings and landscapes by establishing the Otaru Historic Buildings Council (Otaru-shi Rekishiteki Kenzōbutsu Taisaku Kaigi). *Otaru City.*

December 4 The Otaru city government submits the revised "Iida Plan" to the Hokkaido prefectural government. *Asahi Shinbun, Otaru City.*

1980

May 24–27 The National Townscape Preservation Alliance (Zenkoku Machinami Hozon Renmei) holds the third annual national townscape seminar in Otaru and Hakodate, where members adopt the Otaru and Hakodate Declaration: "the preservation and restoration of historic environments is an issue that should be positioned at the very heart of city planning in the 1980s." *Asahi Shinbun, Hokkaido Shinbun, Mineyama.*

September 20 The revised Rinkōsen plan is officially approved by the Hokkaido prefectural government (Bulletin 2361). *Otaru City.*

1981

December The Otaru City Council adopts the "Iida Plan." *Minutes of the Otaru City Council.*

1982

December The Otaru City Council votes to approve the reclamation of the Otaru canal. The final 2,350 m of the Rinkōsen leading up to the canal has been completed by month's end. *Otaru City.*

September 30 Seibu Distribution Group's Tsutsumi Seiji announces that Seibu "could not cooperate in the redevelopment of [Otaru's] canal district if the canal is filled in." *Asahi Shinbun.*

1983

March	The Otaru city government acquires the "Otaru warehouses" for preservation and redevelopment (land price: 159 million yen, building price: 70 million yen). *Minutes of the Otaru City Council.*
August 17	The president of the Otaru Chamber of Commerce and Industry issues a statement advocating a "reconsideration" of the plan to pave over the canal. *Asahi Shinbun, Hokkaido Shinbun.*
August 28	The executive committee of the sixth annual Port Festival issues an urgent appeal for the preservation of the canal. *Interview with Ogawara Tadashi, Asahi Shinbun.*
September 12	The newly established Committee of One Hundred for the Otaru Canal (Otaru Unga Hyakunin Iinkai) begins a signature-collection campaign to save the canal and distributes a [written] *Appeal* throughout the city. *Asahi Shinbun.*
November 19	The city government formally replies to the open letter of inquiry submitted by the Committee of One Hundred for the Otaru Canal. The official reply states that "the current plan is the best" and that the canal issued had been "decided by the administration after sufficient discussion." *Minutes of the Otaru City Council.*
December 22	The city enacts an ordinance for the "preservation of historic buildings and scenic districts" (Otaru City Ordinance #25). *Otaru City.*

1984

January 17	Members of the Committee of One Hundred for the Otaru Canal meet with Construction Minister Mizuno Kiyoshi and Minister of Home Affairs Tagawa Sei'ichi and present the ministers with the 98,000 signature petitions supporting the canal's preservation. *Interview with Yamaguchi Tamotsu, Asahi Shinbun.*
January 27	The Committee of One Hundred for the Otaru Canal announces it has formed a subcommittee to prepare for a mayoral recall campaign. *Asahi Shinbun, Ogasawara(1986a).*
January 31	At a press conference, Hokkaido Governor Yokomichi Takahiro hints at a reversal on the canal issue, saying: "the situation is beginning to change, and I would also like to work hard

	to gain a new consensus" on the matter. *Hokkaido Shinbun, Ogasawara(1986a).*
March 26	Construction Minister Mizuno Kiyoshi meets with Governor Yokomichi and announces a "temporary suspension of land reclamation work on the canal during the Otaru Expo." At a press conference on the same day, Otaru City Mayor Shimura Kazuo says: "We are not considering any change to the current plan. We will discuss how to respond once we learn the specifics from the prefectural government." *Hokkaido Shinbun, Ogasawara(1986a).*
May 24	Hokkaido Governor Yokomichi convenes the first session of the five-party talks on the canal's fate, which he mediates. *Asahi Shinbun, Hokkaido Shinbun, Interview with Shimura Kazuo.*
June 28	Several members of the Committee of the One Hundred for the Otaru Canal announce they will initiate a campaign to recall Mayor Shimura. Their announcement aggravates conflict within the preservation movement. *Mineyama, Hokkai Taimusu.*
August 18	After a fourth session of the five-party talks on August 17 fails to produce consensus, Governor Yokomichi announces that "reclamation work will continue in accordance with administrative procedure." The governor presents his proposal to establish an "Otaru RevitalizationCommittee" (Otaru-shi Kasseika Iinkai). *Otaru City documents, Asahi Shinbun, Interview with Shimura Kazuo, Mineyama.*
August 24	APOC Chairwoman Mineyama Fumi steps down. Mineyama's resignation solidifies the divisions within the preservation movement. *Asahi Shinbun, Mineyama.*
September 1	Members of the Committee of One Hundred for the Otaru Canal convene to discuss the group's future in light of a recent member survey: of fifty-nine responses, forty-one members had voted to disband, eight to continue; seven others had abstained (three additional votes were voided). After the passage of a motion to have "those present here today vote on whether or not to disband the committee," seventeen members voted to disband and the rest abstain (no opposing votes were cast). Although some members continued to contest its validity, the September 1 vote resulted in the dissolution of the Committee of One Hundred for the Otaru Canal. *Asahi Shinbun.*
November 20	The Otaru Revitalization Committee is established and subsequently convenes twenty times through 1987. Various subcommittees convene six times over the same period. *Otaru City.*

1985

October 29	Otaru citizens who had supported the canal's preservation form the Otaru Rebirth Forum (Otaru Saisei Forum). *Otaru Saisei Forum.*

1986

March	The portion of the Rinkōsen built on land reclaimed from the canal is completed. Authorization is also granted to complete the reclamation of public waters. *Otaru City.*
April	A portion of the Rinkōsen opens. *Otaru City.*

1987

January	The Otaru Revitalization Committee issues its final report, "Water Frontier 21: A Vision for the Revitalization of the Otaru Canal and Port District." *Otaru City.*

1988

January	As part of the formulation of a general plan, the Otaru city government establishes the Otaru Citizen Colloquium on Community Development (Otaru Machizukuri Shimin Kondankai) and appoints 110 citizens to the committee. *Otaru City.*
July 23	The craft store chain Glass Ship opens in a refurbished stone warehouse along the canal. *Asahi Shinbun.*
October	The final portion of the Rinkōsen is completed. *Otaru City.*

1989

January 27	Approximately 120 guests attend a "Celebration of the Otaru Rinkōsen's Completion." A lecture on "Community Development in the twenty-first Century" is held as a joint event. *Otaru City.*
May 30	The Ministry of Construction designates Otaru as a "model city" for urban landscape creation. *Asahi Shinbun.*

1990

May 26	Twenty-one businesses along the Otaru canal begin to dissolve the old "Kaigan-dōri Liason Council" and establish in its place the "Otaru Kaigan-dōri 21," a new council aimed at promoting tourism in the city's canal district. *Otaru Kaigan-dōri 21.*
December	A new footpath running alongside the northern end of the canal is completed. *Otaru City.*

1991

October	The city government organizes a colloquium with twelve related local organizations, including the Otaru Rebirth Forum, which includes an explanation and opinion exchange regarding the Landscape Ordinance (these meetings continue through February 1992). *Otaru Saisei Forum.*

1992

March 31	The "Landscape Ordinance for Community Development through Otaru's History and Natural Environment" (Otaru no Rekishi to Shizen wo Ikashita Machizukuri Keikan Jōrei) is promulgated. *Otaru City.*

1993

December	The city's Architecture and Urban Design Section publishes the first issue of the newsletter *Machinami: Otaru Landscape News (Otaru Keikan Nyūsu). Otaru City.*

1994

August 6	Otaru citizens organize the seventeenth and final Port Festival. *Asahi Shinbun, Port Festival Executive Committee.*

1995

January 31	The Otaru Rebirth Forum organizes the first in a series of three seminars (through the end of February) on the future redevelopment of the Chikkō railyard. *Otaru Saisei Forum.*

March 24	The ninth volume in the *History of Otaru* is published by the municipal government. In Chapter 8 (pp. 1293–1464), entitled "The Rinkōsen and the Otaru Canal," the city makes the first real reference to the canal preservation campaign. *Otaru-shi Shishi Hensan Iinkai ed. (1995).*
May 22	The Otaru Juku, a community development movement launched by a group of former local preservation activists, hosts the "Otaru Juku Community Development Forum." The lecture is given by former "Hokudai trio" member Yanagida Ryōzō. *Otaru Juku.*
August 20	The Otaru Rebirth Forum and the Otaru Juku organize the community development seminar "A Second Chapter in Otaru Community Development" in Otaru. *Otaru Saisei Forum.*

1996

March 2	The newly established Association to Ban the Redevelopment of the Chikkō Yard opens an office. *Chikkō Saikaihatsu Sashitome Haru o Yobu Kai.*
September 2	The Special Committee on the Redevelopment of the Former Chikkō Yard Site is convened. *Otaru City.*

1997

March 2	The Otaru Community Development Council hosts the first in a series of symposia on "The Temiya Line: Open A Way/Otaru," entitled "A Reflection on Community Development in Otaru." The lecturer is Dr. Nishimura Yukio of the University of Tokyo. *Otaru Machizukuri Kyōgikai.*
November 27	The city's consultative council on "landscape and community development based on Otaru's history and natural surroundings" announces its "Eight Views and Eight Sites of Otaru" at the tenth annual "Cityscape Award" ceremony honoring best practices in local landscape improvement, held at a local hotel. *Otaru City.*

1998

January	The Otaru Tourism Promotion Council publishes a report entitled "A Consideration of Otaru Tourism: The Present and Future." *Otaru Kankō Yūchi Sokushin Kyōgikai, ed. (1998).*

March 1 — The Otaru Community Development Council publishes a 243-page report on the symposia *The Temiya Line: Open A Way/Otaru*, aimed at developing a plan for the old Temiya Railway Line. *Otaru Machizukuri Kyōgikai.*

1999

February 11–21 — Approximately 180,000 visitors arrived in Otaru for the first annual Snow Light Path Festival. *Otaru City.*

March 11 — The Otaru Bay City Development Company opens the shopping complex MyCal Otaru. *Hokkaido Shinbun.*

Otaru welcomes a record-breaking number of visitors (9,720,000). *Otaru City documents.*

2000

February 1–22 — The reminiscences of former APOC chairwoman Mineyama Fumi are serialized in the "My History" column of the evening edition of the *Hokkaido Shinbun*. Mineyama's fifteen-part series recounting the history of the canal preservation campaign is entitled "To Live in the Region" (*Chiiki ni Ikiru*). *Hokkaido Shinbun.*

2001

September 28–30 — Otaru hosts the "24th National Townscape Seminar." *Machinami Zemi Jikōiinkai, Hokkaido Shinbun.*

2002

September 2 — The Otaru Trust, a voluntary citizen's group, receives official NPO status. The Otaru Trust goes on to purchase the historic Otaru Mujin Building, which had previously housed the Otaru branch of the North Pacific Bank (Hokuyō Ginkō). *Cabinet Office of Japan.*

2003

April 28	Yamaguchi Tamotsu, who played a central role in the campaign to save the canal, wins the election to the Otaru City Council. *Hokkaido Shinbun.*

2004

June 30	Otaru City Mayor Yamada Katsumaro holds a "discussion" with Otaru citizens, at which the Otaru Rebirth Forum recommends "the wider dissemination of the historical background to the canal preservation effort." *Otaru Saisei Forum.*
December	The annual number of visitors to Otaru drops to roughly 7.5 million. *Otaru City.*

2005

February 8	The city's consultative council on "landscape and community development based on Otaru's history and natural surroundings" begins considering the enactment of a new "landscape ordinance" that would include penalties for violations. *Asahi Shinbun.*
September 7	The Otaru Rebirth Forum convenes an emergency meeting of local citizens to address the ongoing construction of a high-rise apartment building in the Ironai district of the city. At the meeting, participants issue a formal demand to developers, asking them to respect the local scenery. *Otaru Saisei Forum.*
September 16–18	A special resolution supporting the preservation of the scenery of Otaru's canal district is adopted at the National Townscape Seminar held in the city of Mino, Gifu Prefecture. *Zenkoku Machinami Hozon Renmei.*

2006

May 16	The Otaru Tourism School is established to "cultivate human resources" for promoting authentic experiences of Otaru for tourists. *Otaru Kankō Daigakkō, Hokkaido Shinbun.*
November 1	The city of Otaru becomes a "landscape administrative association" based on Japan's Landscape Act. *Otaru City.*

2007

March 29	The Otaru Tourism Association and the Otaru Tourism Promotion Council merged. *Hokkaido Shinbun.*

2008

May 30	Mineyama Fumi is awarded the Architectural Institute of Japan's 2008 Culture Prize for "her contribution to the revival and restoration of the Otaru city center through the citizens' campaign to preserve the canal and the stone warehouses of Otaru." *Architectural Institute of Japan.*
October 2	The Otaru City Council decides upon the "Otaru City of Tourism Declaration." *Otaru City.*
November 7	The Hokkaido chapters of the Architectural Institute of Japan and the City Planning Institute of Japan sponsor a symposium on "The Legacy of the Movement to Save the Otaru Canal and Stone Warehouses," attended by more than 450 people. *"Otaru Unga to Sekizō Sōkogun no Hozon Undō Kara Nani o Uketsugu ka" Jikkōiinkai.*
November	Business owners in the popular Otaru tourist district of Sakaimachi form the Sakaimachi Prosperity Council. *Interviews with members of the Sakaimachi Nigiwai Kyōgikai.*

2009

October 28	A Teikoku Data Bank survey reveals that one-third of Otaru's stone warehouses have been demolished over the past twenty years. *Asahi Shinbun.*

2010

August 14–15	The Sakaimachi district of Otaru hosts the first annual summer paper lantern festival. *Otaru Sakaimachi Yukata Chōchin Matsuri Jikkōiinkai.*

2011

June 6	An association promoting bringing a casino to Otaru holds a "Casino Forum." *Otaru ni Cajino o Yūchi Suru Kai.*

2012

March	The merchants of the Sakaimachi district form the Sakaimachi Shopping District Promotion Association (which gains corporation status in July). *Sakaimachi Shōtengai Shinkō Kumiai.*
June	The population of Otaru falls below 130,000. *Otaru City.*

2013

September 9	Otaru Mayor Nakamatsu Yoshiharu indicates his support for bringing a casino to Otaru during a city council meeting. *Hokkaido Shinbun.*

2014

May 11	Local citizens form a group to oppose the effort to bring a casino to Otaru. The principal members of this new group are former members of the canal preservation movement. *Interview with Watanabe Shin'ichirō.*
September 5	The Hokkaido District of the Junior Chamber International Japan holds a convention in Otaru. The Otaru canal preservation movement and community development are the subjects of one session. *Junior Chamber International Japan.*

2015

April 30	Former preservation activist Yamaguchi Tamotsu retires from the Otaru City Council after serving three consecutive terms (twelve years) in office. *Yamaguchi Tamotsu Kōenkai.*
September 13	The NPO Otaru Minka Restoration Project hosts a contest and panel exhibition on "the reuse of historic buildings in Otaru." *Otaru Minka Saisei Project.*

2016

July 23	Nitori Co., Ltd. opens the Otaru Art Base, located in the former Otaru branch office of Mitsui Bank and other neighboring historic structures. *Nitori Co., Ltd.*

2017

April 1 The Otaru city government establishes a "Heritage Section" within the Otaru Port and Industry Department and initiates an effort to receive a heritage site designation for the Otaru Port from the Agency for Cultural Affairs. *Otaru City.*

The annual number of tourists to Otaru increases to 8,060,000. This marks the first time in fourteen years that the number of annual visitors exceeds eight million. *Otaru City.*

Maps

An overview of the districts surveyed

S. Horikawa, *Why Place Matters*,
https://doi.org/10.1007/978-3-030-71600-4

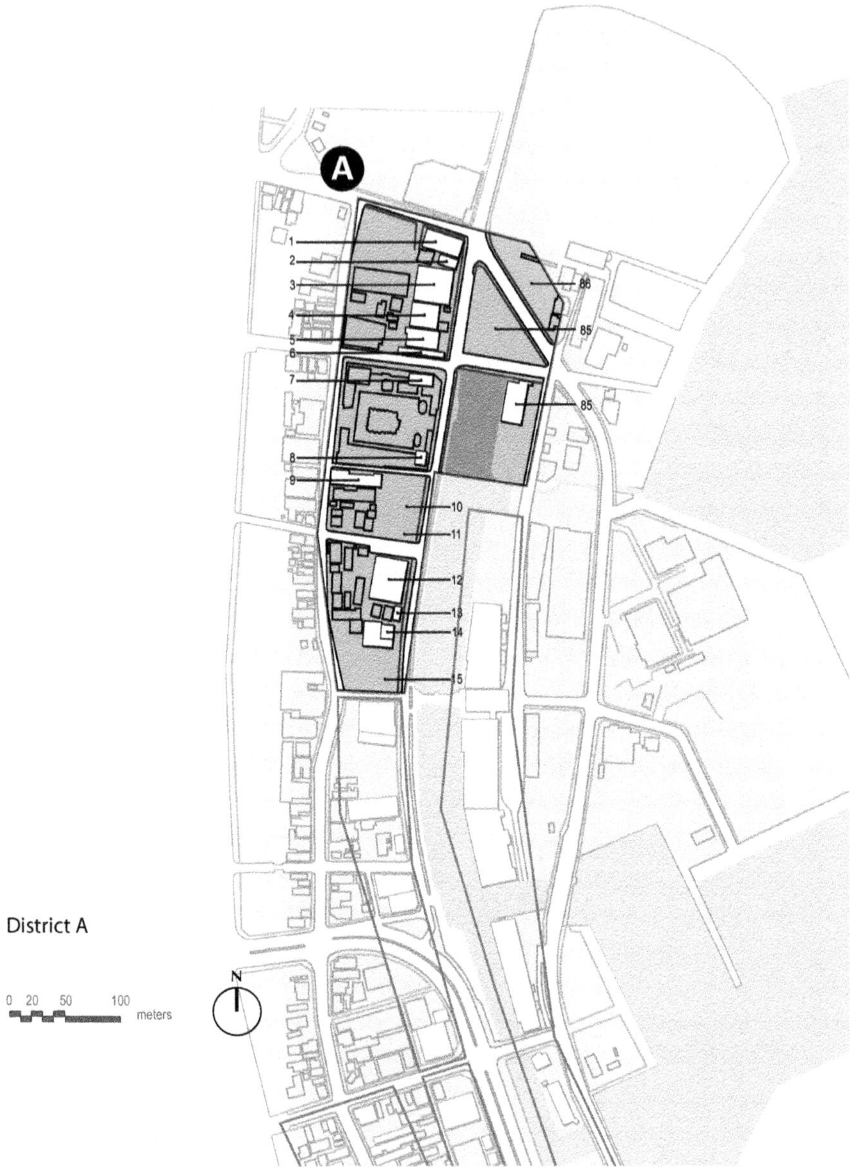
A
1
2
3
4
5
6
7
8
9
10
11
12
13
14
15
85
85
86
District A
0 20 50 100
meters
N

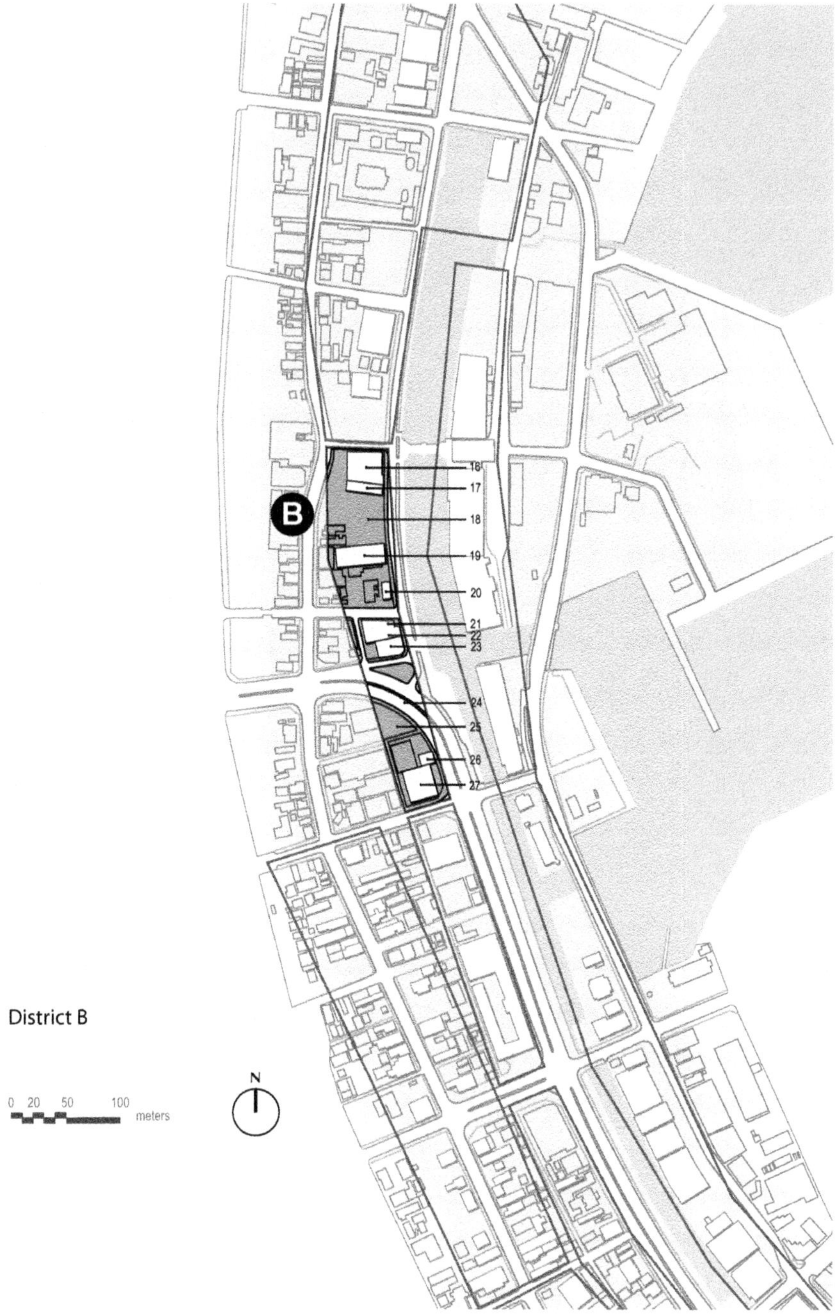

District B

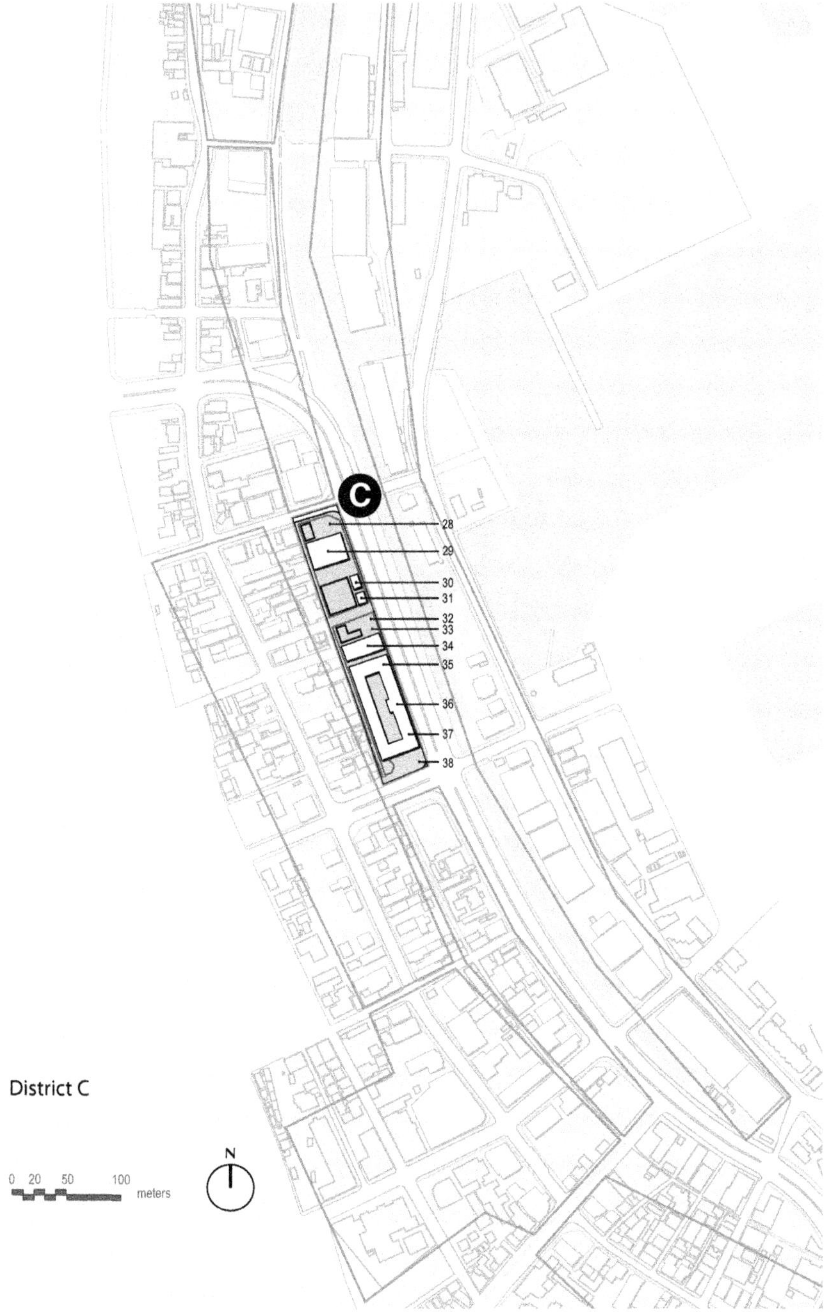
C
28
29
30
31
32
33
34
35
36
37
38
District C
0 20 50 100
meters
N

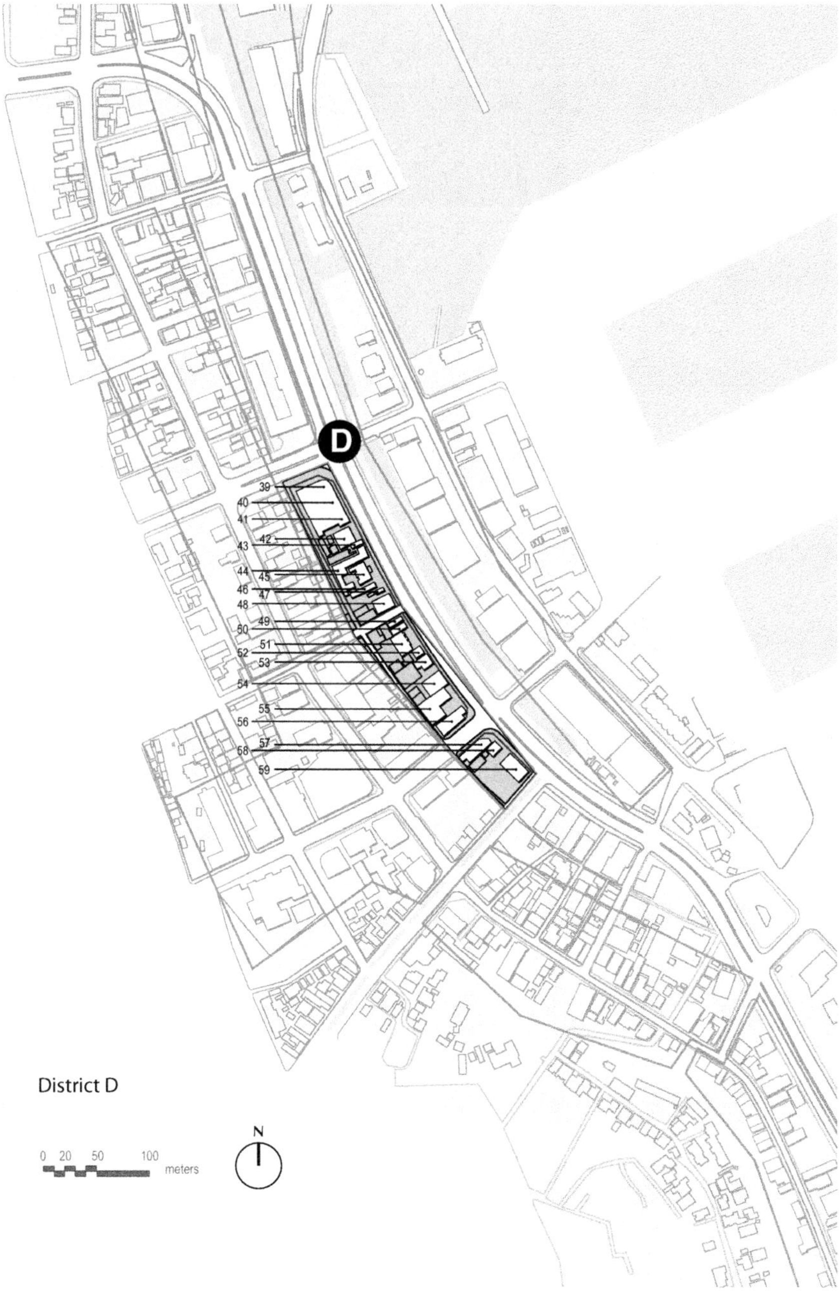
D
39
40
41
42
43
44
45
46
47
48
49
50
51
52
53
54
55
56
57
58
59
0 20 50 100
meters
N

District D

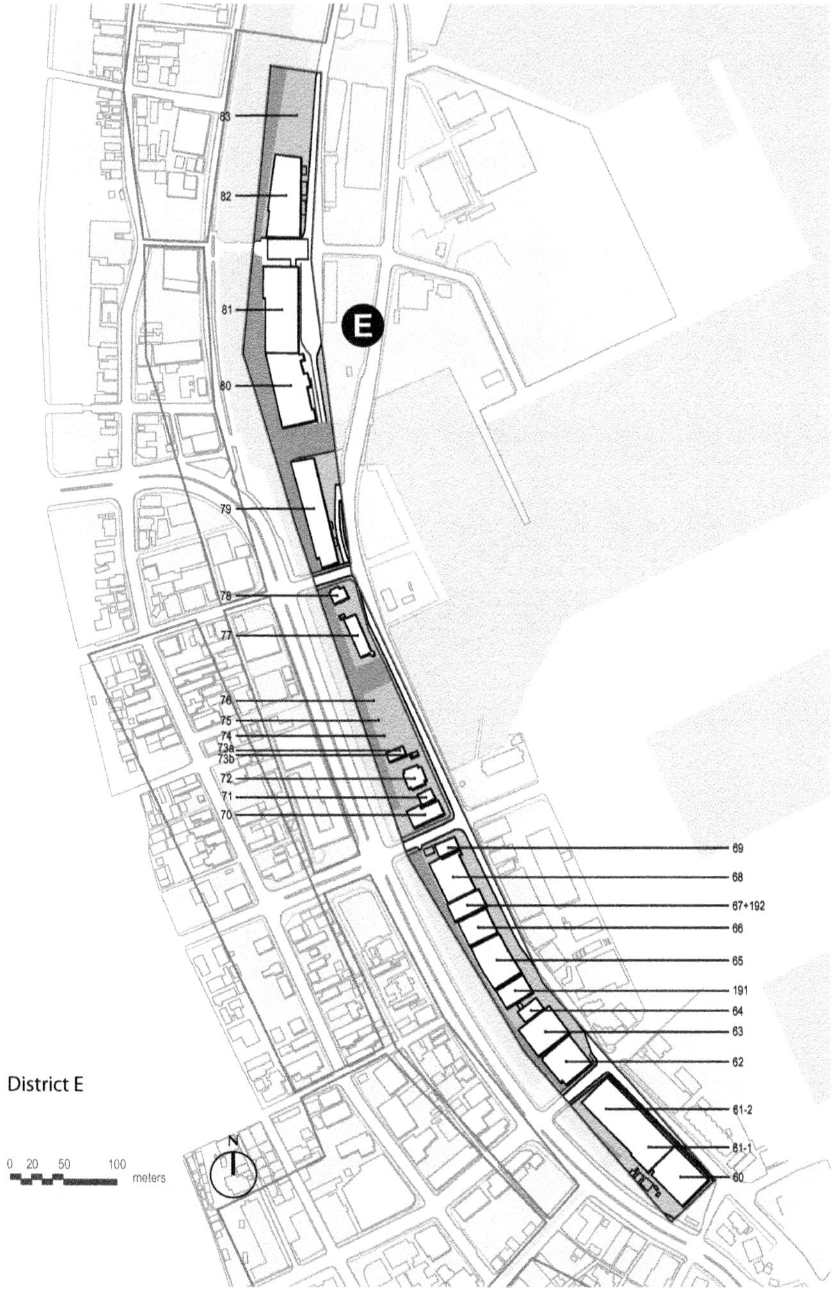
83
82
81
E
80
79
78
77
76
75
74
73a
73b
72
71
70
69
68
67+192
66
65
191
64
63
62
61-2
61-1
60
District E
N
0 20 50 100
meters

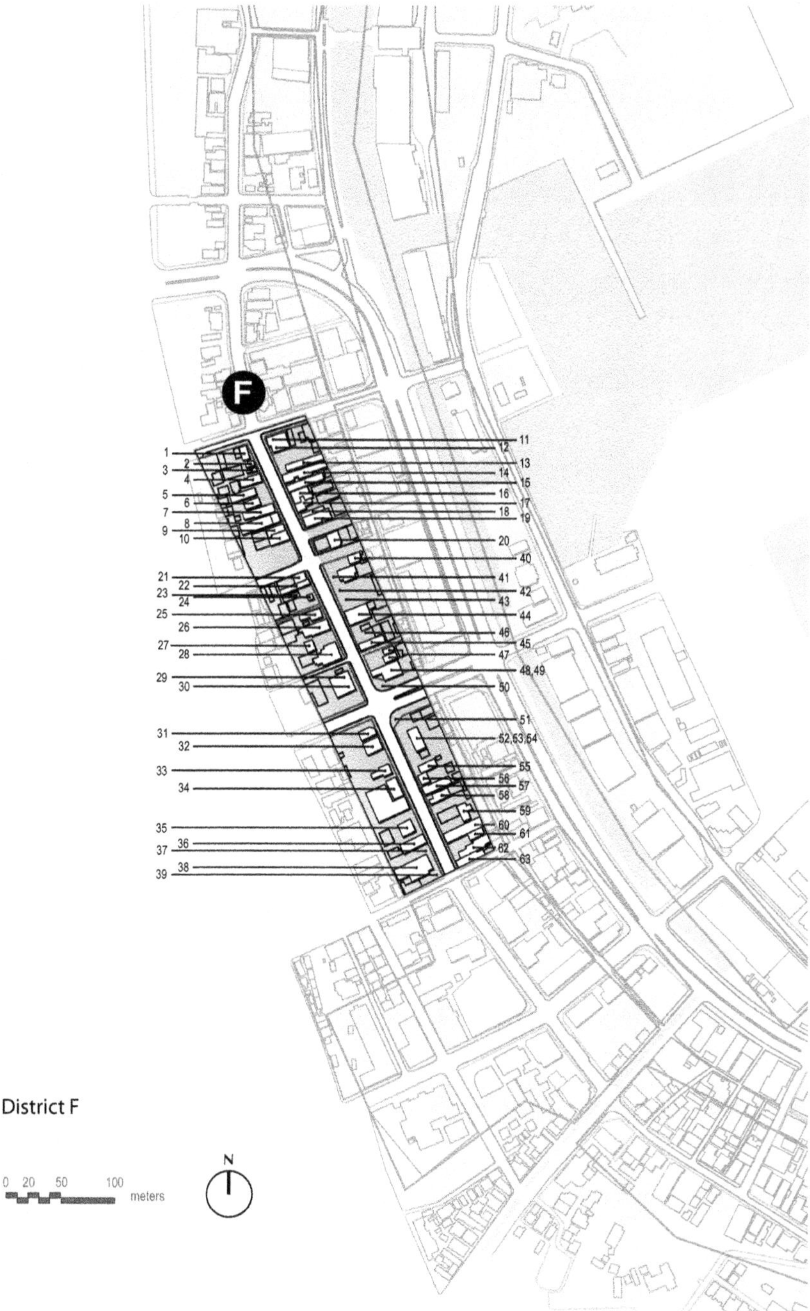
F
1
2
3
4
5
6
7
8
9
10
11
12
13
14
15
16
17
18
19
20
21
22
23
24
25
26
27
28
29
30
31
32
33
34
35
36
37
38
39
40
41
42
43
44
46
45
47
48,49
50
51
52,53,54
55
56
57
58
59
60
61
62
63
0 20 50 100 meters
N

District F

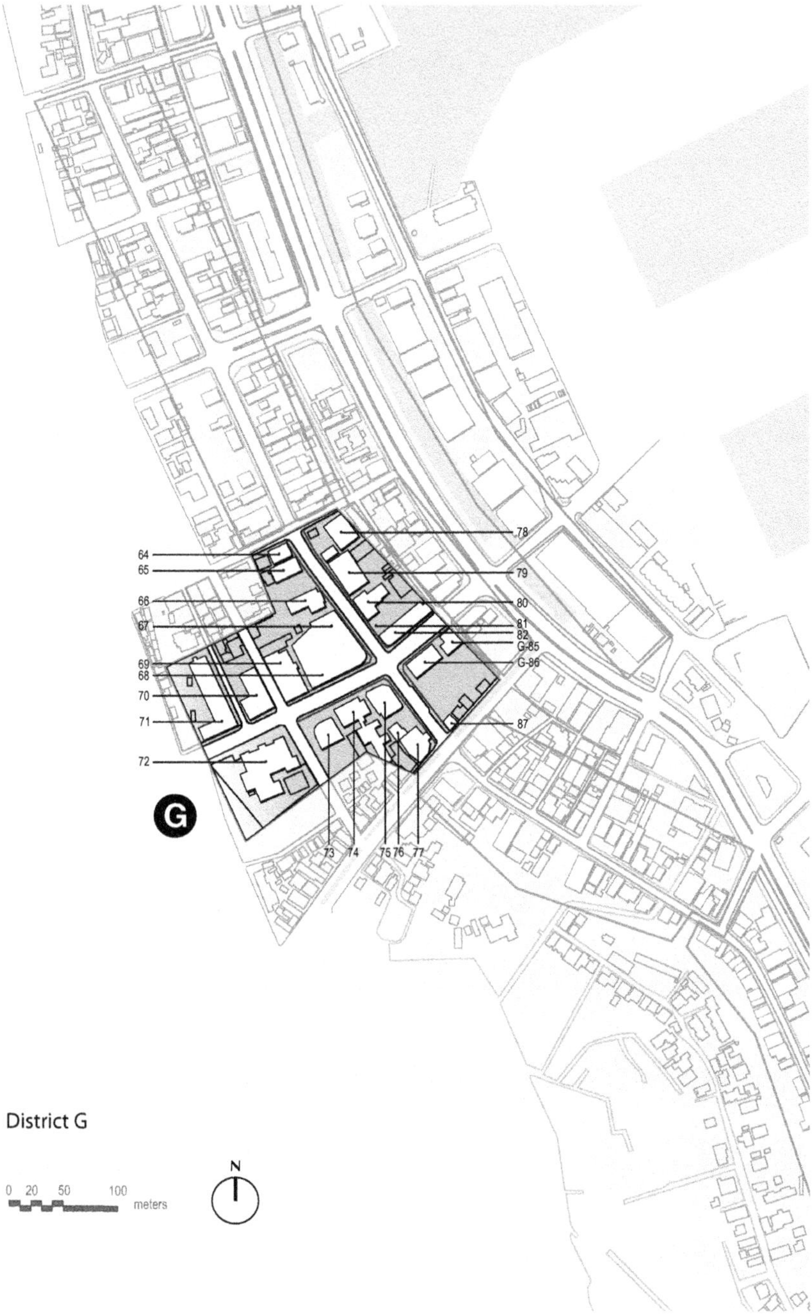
64
65
66
67
69
68
70
71
72
78
79
80
81
82
G-85
G-86
87
73
74
75
76
77
District G
0 20 50 100
meters
N

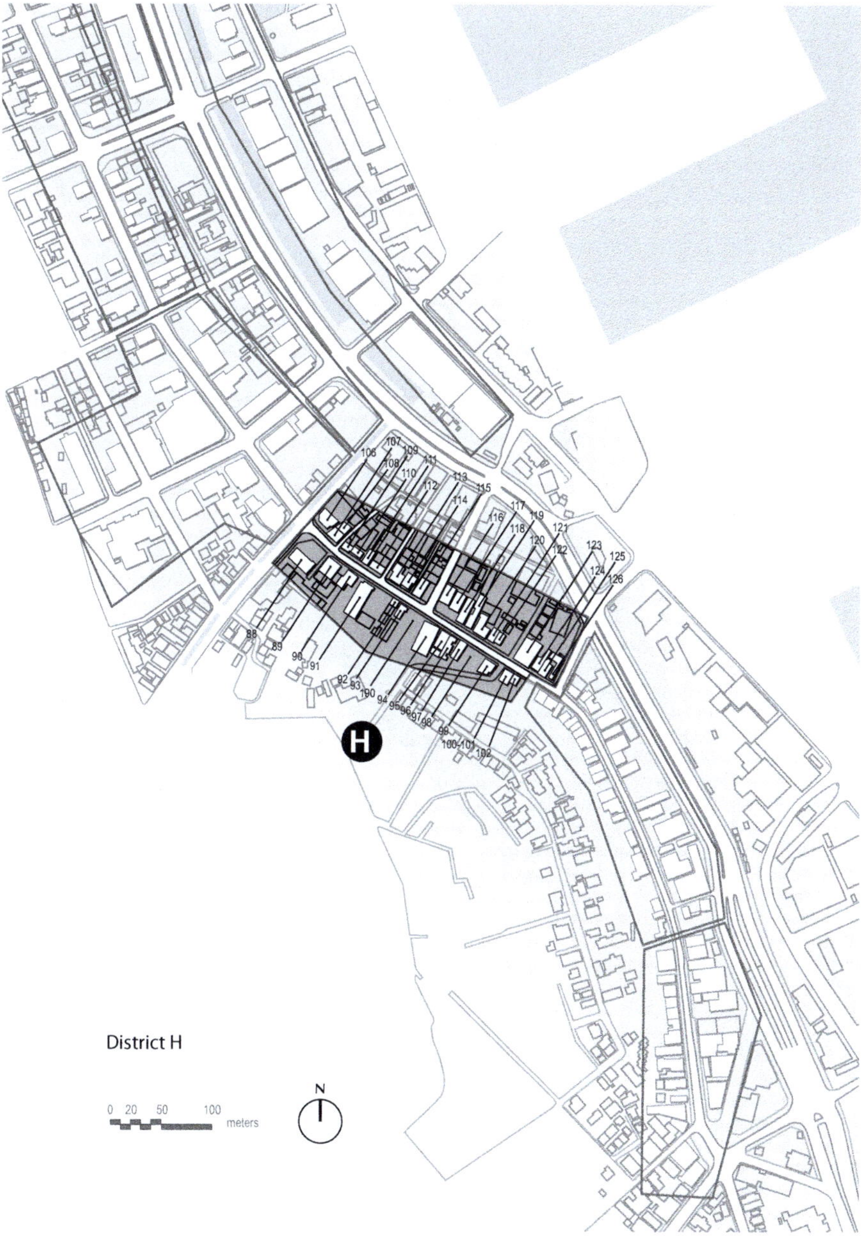
District H
0 20 50 100 meters
N

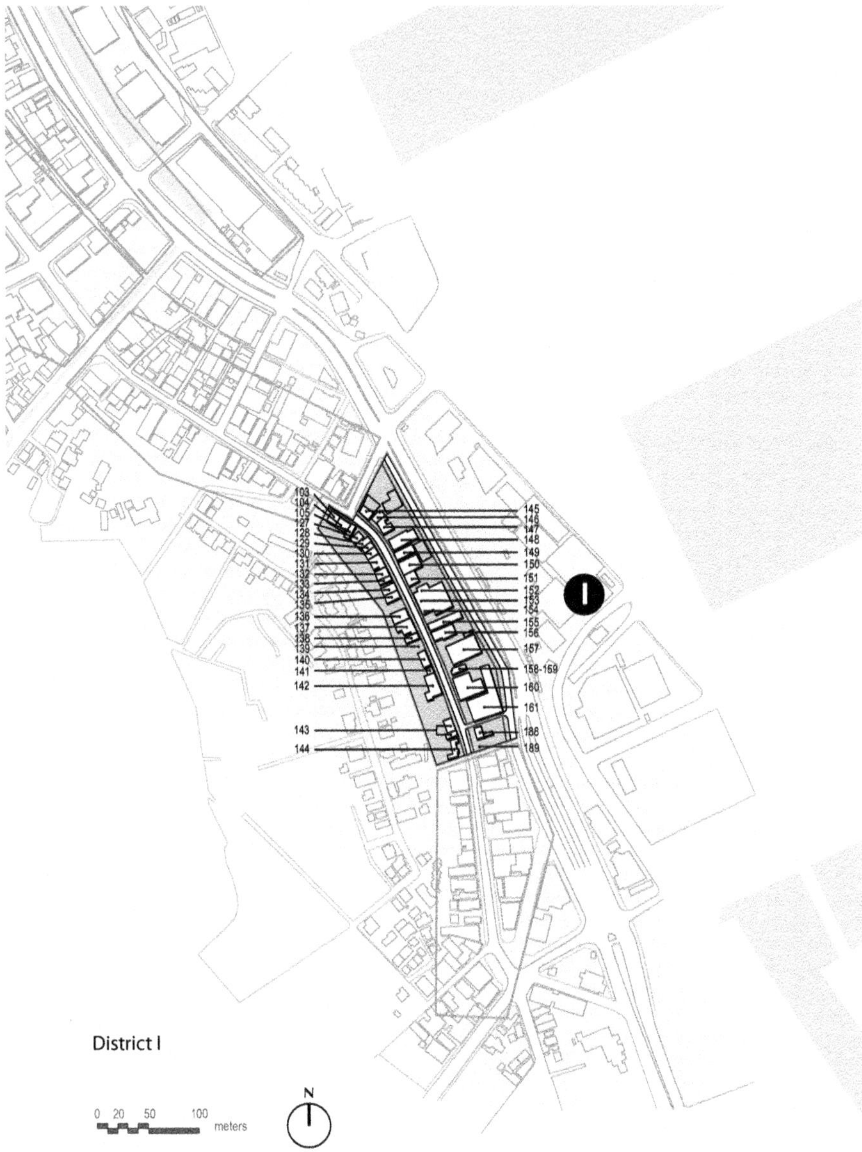
103
104
105
127
128
129
130
131
132
133
134
135
136
137
138
139
140
141
142
143
144
145
146
147
148
149
150
151
152
153
154
155
156
157
158-159
160
161
188
189
I
District I
0 20 50 100
meters
N

178
179
180
181
182
187
183
184
185
176
177
162
163
164
165-166
167
168
169
170
171
172
173
186 (ex-174)
175
J

District J

0 20 50 100
meters
N

List of Buildings Surveyed

District	Nos.	Hokkaido University survey (1986)	Hosei University survey (1998)	Hosei University survey (2006)	Hosei University survey (2016)
A	1	Otaru Rinyū cooperative	Rinyū market	←	Rinyū market
	2	Murai family	←	←	Murai family
	3	Futaba warehouse (former Ukon warehouse)	Kitaichi Glass (inventory warehouse)	←	Kitaichi Glass (inventory warehouse)
	4	Nippon Express (former Hiromi warehouse)	Kitaichi Glass (inventory warehouse)	(left) Kitaichi Glass shōhin sentā (right) Kitaichi inventory warehouse	(left) storeroom (right) Kitaichi glass inventory warehouse
	5	Nichimō (former Masuda warehouse)	Kitaichi Glass (inventory warehouse)	←	Kitaichi Glass (inventory warehouse)
	6	Daidō warehouse	Miyake Body	(front) --- (back) Miyake Body	(front) Sapporo-mura Sweet Potato (back) Miyake Body
	7	Tonbo Haiyā (former Yūsen warehouse)	(planned site of the canal park)	canal park (wooden play structures)	canal park (wooden play structures)
	8	Tonbo Haiyā	(planned site of the canal park)	canal park (public restrooms)	canal park (public restrooms)
	9	Otaru Reitō Seihyō	Sake Ebisuya	←	vacant
	10	Murai Suisan Reitō Seihyō Kōjō	Car Plaza Co., Ltd. World Kikaku	Suzuki Transport Co., Ltd.	Ishihara Transport Co., Ltd.

S. Horikawa, *Why Place Matters*,
https://doi.org/10.1007/978-3-030-71600-4

District	Nos.	Hokkaido University survey (1986)	Hosei University survey (1998)	Hosei University survey (2006)	Hosei University survey (2016)
A	11	Fishery association market	parking lot (association-owned land)	←	parking lot (association-owned land)
	12	Otaru City Fisheries Cooperative frozen ice factory	Otaru City Fisheries Cooperative Local Wholesale Market	←	Otaru City Fisheries Cooperative Local Wholesale Market
	13	Dunlop Kiriyama Tire Co.	←	←	Toyo Tires Kiriyama Tire Shōkai
	14	Nippun Feed Otaru factory	Tonbo Haiyā Co., Ltd.	←	Tonbo Haiyā Co., Ltd.
	15	Nippun Feed Otaru factory (former Shibusawa warehouse)	Vacant land	(Right) construction site/Nihon Seifun (Left) construction site	Vacant land owned by Kuromori Co., Ltd.
	84	Kitahama Jidōsha Kōgyō (former NYK warehouse)	Otaru Canal red brick museum	Kitaichi Glass Product Center	System distribution center
	85	Nichirei Corp. (former Otaru branch office)	………… **(parking lot)**	Unknown	Lawson
	86	Nichirei Corp. Otaru factory	…………**(parking lot)**	monthly parking lot (Esse Co., Ltd)	Hotto Motto

District	Nos.	Hokkaido University survey (1986)	Hosei University survey (1998)	Hosei University survey (2006)	Hosei University survey (2016)
B	16	Futaba Kitahama No. 2 warehouse (former Shibusawa warehouse)	Otaru Rekishi-mura Museum	empty warehouse/Crazy Spice/calm	Gold Stone (Otaru North Canal Live Theater)/Press Café
	17	Kimura Kitahama No. 2 warehouse	←	Unknown	Unknown
	18	Daidō Kitahama No. 2 warehouse (former Satō warehouse)	parking lot (for Pachinko New Otaru)	(left) parking lot (right) unknown	(left) construction vehicle storage space (right) parking lot
	19	Daidō Kitahama No. 1 warehouse (former Satō warehouse)	Kitaichi Glass warehosue	Kitaichi Glass warehouse (under demolition)	Tsūsen Co., Ltd. (parking lot for Blue Cave)
	20	Futaba warehouse (former Kakusan warehouse)	Cosmo Oil service station	←	Kitanihon Nenryō Co., Ltd. Otaru sales office gas stand
	21	Otaru Denkai Kōgyō	←	Hokkai Seikan Co., Ltd. (Otaru scrap factory)	Hokkai Seikan Co., Ltd. (Otaru scrap factory)
	22	Otara Denkai Kōgyō (former Fujiyama warehouse)	←	Hokkai Seikan Co., Ltd. (Otaru scrap factory)	Hokkai Seikan Co., Ltd. Otaru scap press factory
	23	Itaya Merchant Ship Co., Ltd. (former Itaya warehouse)	Kyōei Seimei Otaru Bldg.	Gibraltar Seimei Otaru Bldg.	Gibraltar Seimei Otaru Bldg.

District	Nos.	Hokkaido University survey (1986)	Hosei University survey (1998)	Hosei University survey (2006)	Hosei University survey (2016)
B	24	Itaya Merchant Ship Co., Ltd. (former Itaya warehouse)	………… (Rinkōsen road)	←	(Rinkōsen)
	25	Futaba warehouse (former Shibusawa warehouse)	…………(Rinkōsen road)	Rinkōsen and parking lot	Rinkōsen and parking lot
	26	Shōwa Seiki (former Asahi glass warehouse)	←	←	Hokkai Seikan Co., Ltd.
	27	Shōwa Seiki (Ijiri warehouse)	Shōwa Seiki (new building)	(Shōwa Seiki Co., Ltd.) Duskin Rent-All Recycle Shop Next	Tōto Seikei Co., Ltd. Hokkaido sales office

District	Nos.	Hokkaido University survey (1986)	Hosei University survey (1998)	Hosei University survey (2006)	Hosei University survey (2016)
C	28	Maeno Shōten Co., Ltd.	Otaru Toys (not in use due to bankruptcy)	Unknown	Unknown
	29	Ōya warehouse	Otaru Toys (not used due to bankruptcy)	(Kitaichi-owned warehouse)	Former Ōya warehouse (Kitaichi-owned warehouse not in use)
	30	Kumagaya Tent (former Shioda warehouse)	Kaimeirō Gallery	(not in use)	Otaru City Museum, Ungakan parking lot
	31	Maekawa Kanaami (Shioda warehouse)	under construction (owned by Kaimeirō)	vacant with 2F office	(Left) La Quatrieme (Right) Mitsui Repark parking lot
	32	Takagi Trading (former Okada warehouse)	...........**(Otaru Craft** Museum parking lot)	←	Times Otaru Ironai No. 2 Parking
	33	Takagi Trading (former Okada warehouse)	...........**(Otaru Craft** Museum parking lot)	←	Times Otaru Ironai No. 2 Parking
	34	Komatsu Steel Works (former Saitō warehouse)	Otaru Craft Museum	←	Taiken Kōbō Otaru–Ilponte
	35	Komatsu Steel Works (former Saitō warehouse)	Otaru Craft Museum	←	Taiken Kōbō Otaru–Ilponte

District	Nos.	Hokkaido University survey (1986)	Hosei University survey (1998)	Hosei University survey (2006)	Hosei University survey (2016)
	36	Otaru warehouse office	Otaru Club/Otaru Tourism Association	←	Otaru Club/Otaru Tourism Association
C	37	Otaru warehouse	Otaru City Museum, Otaru City Souvenir Plaza	←	Ungakan/Otaru City Souvenir Plaza
	38	Hokkaido Shingyoren Otaru branch	………… **(open space)**	←	_____(open space)

District	Nos.	Hokkaido University survey (1986)	Hosei University survey (1998)	Hosei University survey (2006)	Hosei University survey (2016)
D	39	Shingyoren Osaka Shōsen Bldg.	Hotel Nord Otaru	←	Hotel Nord Otaru
	40	Yamamoto iron works	Hotel Nord Otaru	←	Hotel Nord Otaru
	41	Japan Cargo Tally Corp.	Hotel Nord Otaru	←	Hotel Nord Otaru
	42	Nichimo Co., Ltd. Otaru factory	Hotel Sonia II	Hotel Sonia II (Kitahama on 1F)	Hotel Sonia II (Otaru Kaisen Sushi Kitahama on 1F)
	43	Otaru Senpaku Yōhin Co., Ltd.	←	←	(Otaru Senyōhin, roofed parking space)
	44	Otaru Senpaku Yōhin Co., Ltd. (former Makino warehouse)	←	←	Otaru Senyōhin
	45	Gōdō Kasei warehouse No. 2	Hotel Sonia I	←	Hotel Sonia
	46	Minamihama police dormitory	←	←	Minamihama police dormitory
	47	Hakodate Seimōsengu (netting) (former Aoki warehouse)	Orix Rental Car	←	Times Otaru Ironai No. 3 parking

District	Nos.	Hokkaido University survey (1986)	Hosei University survey (1998)	Hosei University survey (2006)	Hosei University survey (2016)
D	48	Hisano Kaisōten	Hisano Kaisōten warehouse	Art and Entertainment Gypsy mart／Cafe Restaurant **Bronco's** ／Digital Photo Studio Koohp／Silver Acc. Joker Trap	unoccupied building/Terrazzetta sul Canale
	49	Honma	Fuku Sushi	←	Otaru Eiroku
	50	Kankō Kisen	←	←	Unga no Yado Otaru Furukawa
	51	Otaru Jidōsha Kōgyō	Canal J.B. Inn	Hotel Furukawa	Unga no Yado Otaru Furukawa
	52	Yūbari Transport K.K. Otaru sales office (former Inukami Trading Co., Ltd)	Grandy	Handmade Glass and Glass Bead Store Otaru Canal Lights	Natural Honay shop
	53	Taihei Kōgyō	Taihei Sangyō/Glass Ship/Ijinkan (rental)	Kaimeirō	Nitori Otarugeijutsumura Stained Glass Museum
	54	Tsūshin Densetsu Hama Bldg.	**Kyōwa Hama Bldg. Café** Restaurant Hao	←	**Kyōwa Hama Bldg. Café** Restaurant Hao
	55	Otaru Jūki Center	maraЗИН 928	vacant store	Times Otaru Unga parking
	56	Car Beauty Otaru	Otaru Masa Sushi Unga-ten	Otaru Masa Sushi 2F/Hiyori 1F	Otaru Masa Sushi Zen-an 2F/Hiyori 1F

District	Nos.	Hokkaido University survey (1986)	Hosei University survey (1998)	Hosei University survey (2006)	Hosei University survey (2016)
D	57	Otaru bank association	←	Otaru *denuki kōji*	Otaru *denuki kōji*
	58	Suzuran Shokudō	Machino Sushi Unga-ten	Yōfuku no Aoyama	Yōfuku no Aoyama
	59	Toyota Corolla Dōto Otaru branch	←	Yōfuku no Aoyama	Yōfuku no Aoyama (same as 58)

District	Nos.	Hokkaido University survey (1986)	Hosei University survey (1998)	Hosei University survey (2006)	Hosei University survey (2016)
E	60	Daidō Refrigeration No. 2	←	←	Daidō Refrigeration No. 2
	61	Daidō warehouse (former Naniwa warehouse)	←	Otaru Unga Shokudō (Ramen Kōbō, Asakusabashi Beer Hall)	Otaru Unga Shokudō (Ramen Kōbō, Asakusabashi Beer Hall)
	62	Shimizu warehouse	Italian Kitchen “Pepe Sale”	Hamburger restaurant “Bikkuri Donkey” Otaru Unga-ten/ Shimizu Steel Machinery Co., Ltd. (office)	Hokkaido Gourmet Kaiten Sushi Hakotarō/Laox
	63	Shibusawa warehouse C	Hokkai Aburiyaki “Canal Warehouse,” Café Bar “A Canal”	Hokkai Aburiyaki “Canal Warehouse,” Hanabatake Bokujō	Hokkai Aburiyaki “Canal Warehouse”
	64	Shibusawa warehouse B	Otaru Kaisen Kōbō “Banquina”	Uomaru Otaru Hotori (with adjacent sweet bar)	Indoor parking lot Mitsui Repark
	65	Otaru warehouse	Hamburger restaurant “Bikkuri Donkey” Otaru Unga-ten 1	Beer Pub & Gallery Otaru Sōko No. 1 (Otaru Beer) and Kamendō	Beer Pub & Gallery Otaru Sōko No. 1 (Otaru Beer), No. 2 (“Bikkuri Donkey”)
	66	Daidō warehouse Pier B (former Shinoda warehouse)	Asian seafood restaurant “Otaru Kaisenshō”	vacant shop	Otaru Canal Restaurant “Hikari”

District	Nos.	Hokkaido University survey (1986)	Hosei University survey (1998)	Hosei University survey (2006)	Hosei University survey (2016)
E	67	Daidō warehouse Pier C	Asian seafood restaurant "Otaru Kaisenshō" mechanical room and indoor parking lot for staff	vacant warehouse	Parking lot "Canal Side Parking"
	68	Kitanihon Sōkō Kō-un (former Kita Nihon Kisen warehouse)	Canal shop "Ore no Otaru"	←	vacant store/partial parking lot for rickshaw "Ebisuya"
	69	Yūsen Kairiku Unyu Co., Ltd.	Yūsen Kairiku Unyu Co., Ltd. (rebuilt)	←	North Transport Co., Ltd.
	70	Otaru police, water police deployment station	Shin'gyoren (Hokkaido Fisheries Cooperative Association, Otaru branch)	←	Shin'gyoren (Hokkaido Fisheries Cooperative Association, Otaru branch)
	71	Japan Grain Inspection Association, Otaru office	←	←	Japan Grain Inspection Association, Otaru office and Kokken Corp.
	72	Ministry of Agriculture and Forestry Hokkaido Food Office, Otaru branch	←	Ministry of Agriculture and Forestry Hokkaido Agricultual Administration Office, Otaru	vacant office

District	Nos.	Hokkaido University survey (1986)	Hosei University survey (1998)	Hosei University survey (2006)	Hosei University survey (2016)
E	73a	Residence of Endō Kan'ichi	←	←	vacant land
	73b	(unknown)	Mutō Co., Ltd. Otaru branch	←	Mutō Co., Ltd. Otaru branch
	74	Kita Nihon Kō-un Co., Ltd.	←	parking lot	vacant land
	75	Kyōsei Kōgyō	Kitaichi warehouse	parking lot	vacant land
	76	Kitakura sagyō-ka (warehouse)	Kitaichi warehouse	parking lot	vacant land
	77	Otaru port workers' welfare center	←	Otaru Port Center/Seaside Inn	Otaru Port Center/Seaside Inn
	78	All Japan Dockworker's Union, Otaru branch	←	parking lot (managed by Otaru Port Center)	parking lot (managed by Otaru Port Center)
	79	Hokkai Seikan Co., Ltd. ■	Shōwa Seiki Co., Ltd. ■	←	Shōwa Seiki Co., Ltd. ■ Warehouse No. 3
	80	Hokkai Seikan Co., Ltd.	←	←	Hokkai Seikan Co., Ltd.
	81	Hokkai Seikan Co., Ltd.	←	←	Hokkai Seikan Co., Ltd.
	82	Hokkai Seikan Warehouse No. 1	←	(collapsed during typhoon)	Hokkai Seikan Warehouse No. 1
	83	Shōwa Seiki Co., Ltd. ■	Hokkai Seikan warehouse ■	factory (on premises of Hokkai Seikan)	factory (on premises of Hokkai Seikan)

District	Nos.	Hokkaido University survey (1986)	Hosei University survey (1998)	Hosei University survey (2006)	Hosei University survey (2016)
F *	F-1		Iizaka Fuji Shōten Co., Ltd.	←	Iizaka Fuji Shōten Co., Ltd.
	F-2		Kyōwa Machinery Co., Ltd.	←	vacant lot
	F-3		Araki residence	←	Araki residence
	F-4		Kyōritsu Murakami Fisheries	residence of Murakami Katsutoshi (manufacture and sale of tackle and fishing equipment)	residence of Murakami Katsutoshi (manufacture and sale of tackle and fishing equipment)
	F-5		Hōsan Setsubi Co., Ltd./Nasa Foods sales warehouse	Satō residence	Sanyō Oil Co., Ltd.
	F-6		Marukoshi, Otaru branch	Marukoshi, Otaru branch storefront for rent (vacant)	storefront for rent
	F-7		Fujimatsu fishnet shop	Fujimatsu residence	Fujimatsu residence
	F-8		Orgeldoh Music Box Hall Kaimeirō	Royal Performance BBC	Furudōguya i (1F)/Kumon (2F)
	F-9		Dai-ichi Sewing	←	Dai-ichi Sewing
	F-10		Beauty equipment wholesale "Peace" garage	←	**"Peace" garage**
	F-11		Itō Bicycle Co.	←	Itō Bicycle Co.

District	Nos.	Hokkaido University survey (1986)	Hosei University survey (1998)	Hosei University survey (2006)	Hosei University survey (2016)
F *	F-12		Hokkaido Shima no Ka Bussan	unknown	Saitō Shōkai Ltd.
	F-13		Kyōyū Mansion (apartments)	Daisho Bldg. (3F: Wizard Co., Ltd. 2F: Marudai Daisho store/(joint stock) Umaibeya 1F: parking lot and warehouse)	Daisho Bldg. (3F: Wizard Co., Ltd. 2F: Marudai Daisho store/(joint stock) Umaibeya 1F: parking lot and warehouse)
	F-14		Gift Shop Santa Ship	Daisho Bldg. (3F: Wizard Co., Ltd. 2F: Marudai Daisho store/(joint stock) Umaibeya 1F: parking lot and warehouse)	Daisho Bldg. (3F: Wizard Co., Ltd. 2F: Marudai Daisho store/(joint stock) Umaibeya 1F: parking lot and warehouse)
	F-15		Takimoto Shōji Co., Ltd.	**Apartment building "Germe Otaru Unga"**	**Apartment building "Germe Otaru Unga"**
	F-16		Daishō Shōten	Daishō Shōten (1F)/Aparments	Apartments
	F-17		Chiba fishnet shop/Otaru Kurofuneya	←	OLDECO
	F-18		Kawamata Shōten garage	(not used due to bankruptcy)	Vivre sa vie + mi-yyu

District	Nos.	Hokkaido University survey (1986)	Hosei University survey (1998)	Hosei University survey (2006)	Hosei University survey (2016)
F *	F-19		Kawamata Shōten	(not used due to bankruptcy)	Vivre sa vie + mi-yyu
	F-20		Idemitsu Otaru service station	←	Idemitsu Otaru service station
	F-21		(not displayed)	←	Iizaka warehouse
	F-22		Nukui Co., Ltd.	←	Nukui Co., Ltd.
	F-23		(destroyed by arson)	Hakui no Nozawa-ya	Dia Park Ironai 2 cho-me (parking lot operated by Daiya Electronics Co., Ltd)
	F-24		(destroyed by arson)	Kuji residence	residence of Kuji Nobuko
	F-25		Eiwa Shōkai Co., Ltd.	←	Minagawa Masaharu/Motonori residence
	F-26		Peace Shōkai Ltd.	←	Peace Shōkai Ltd.
	F-27		Munakata Shōten	Orange Sun Shine (1F) / Colony (2F)	orange Otaru/Hair Make Colony (2F)
	F-28		Sanritsu Kiden Co., Ltd.	←	Sanritsu Kiden Co., Ltd. (Hokkaido Hitachi Systems, Hitachi Electronic Services Co., Ltd.)
	F-29		(vacant/city-owned land)	parking lot (for newspaper?)	parking lot, monthly rental parking lot

District	Nos.	Hokkaido University survey (1986)	Hosei University survey (1998)	Hosei University survey (2006)	Hosei University survey (2016)
F *	F-30		Hokkaido Keizai Shinbun (newspaper)	Hokkaido Keizai Shinbun (relocation completed)	Hanagokoro Wa.Bi.Sai
	F-31		vacant land	Unga-don Chiharu/Hair salon Otaru Motoyama/Motoyama parking lot	Glass & sundry shop Yuzu Kōbō/Hair salon Otaru Motoyama/Motoyama residence/Motoyama parking lot
	F-32		Hokkaido Kami Shōji Co., Ltd.	←	vacant shop
	F-33		Alibaba Collection	**Isfahan (café)/Alibaba** Collection (2F)	Otaru glass shop Wakura/2F workshop
	F-34		WALKIE SELECTION	←	**Café Ironai Shokudō**
	F-35		Credit Guarantee Corporation of Hokkaido, Otaru branch	←	Times Ironai 1 cho-me parking
	F-36		Glass Sash Teramoto Co., Ltd.	←	vacant shop
	F-37		monthly parking lot	←	monthly parking lot

District	Nos.	Hokkaido University survey (1986)	Hosei University survey (1998)	Hosei University survey (2006)	Hosei University survey (2016)
F *	F-38		Otaru Chamber of Commerce	←	vacant
	F-39		parking lot	Daishō Dental parking lot	Daishō Dental parking lot
	F-40		Umineko-ya	←	Front: Carbuncle (Zircon Co., Ltd) Rear: Umineko-ya
	F-41		Yoshimura Paint Co., Ltd.	←	Yoshimura Paint Co., Ltd.
	F-42		land for sale	"Green River Finesse Otaru Chūō-dōri Mansion Gallery"	"Green River Finesse Otaru Chūō-dōri" parking lot
	F-43		land for sale	"Green River Finesse Otaru" (high-rise apartment building under construction; same as F-44)	"Green River Finesse Otaru Chūō-dōri" (apartment building; same as F-44)
	F-44		land for sale	"Green River Finesse Otaru" (high-rise apartment building under construction)	"Green River Finesse Otaru Chūō-dōri" (apartment building)
	F-45		Tsuchiya (Tetsuya Kuroganeya) Co., Ltd. workshop	←	Tsuchiya (Tetsuya Kuroganeya) Co., Ltd. workshop
	F-46		land for sale	parking lot for construction	vacant land

District	Nos.	Hokkaido University survey (1986)	Hosei University survey (1998)	Hosei University survey (2006)	Hosei University survey (2016)
F *	F-47		Shōya Kura Sushi	(A) Sushi Kō (B) Kura Sushi	(A) Sushi Kō (B) vacant shop
	F-48		Hokkaido Chūō Shokuryō Co., Ltd., Otaru branch	"Music House Otaru Thriller Karaoke"	"Music House Otaru Thriller Karaoke"
	F-49		Hokkaido Chūō Shokuryō Co., Ltd., Otaru branch	"Music House Otaru Thriller Karaoke"	"Music House Otaru Thriller Karaoke"
	F-50		Sidewalk created by the expansion of Chūō-dōri	←	Sidewalk created by the expansion of Chūō-dōri
	F-51		Sidewalk created by the expansion of Chūō-dōri	sidewalk	sidewalk
	F-52		(1F) "Tax Free Shop Aurora" (2F) Shimizu Hirofumi Dance Studio	COSMO North Japan Oil Co., Otaru Canal service station	COSMO North Japan Oil Co., Otaru Canal service station
	F-53		Sanesu Co., Ltd.	COSMO North Japan Oil Co., Otaru Canal service station	COSMO North Japan Oil Co., Otaru Canal service station
	F-54		Sanesu Co., Ltd.	COSMO North Japan Oil Co., Otaru Canal service station	COSMO North Japan Oil Co., Otaru Canal service station
	F-55		Gōdō Kasei Co., Ltd. Ironai PS	Chiharu Sushi	Chiharu Sushi

District	Nos.	Hokkaido University survey (1986)	Hosei University survey (1998)	Hosei University survey (2006)	Hosei University survey (2016)
F *	F-56		nothing, empty house	Otaru Advance Club	Otaru Advance Club
	F-57		Ushio Bussan, Otaru sales office	empty house	Kameya Fishing Tackle and parking lot
	F-58		Chiba Sushi, canal branch	"Te-uchi Soba Ironai"	"Te-uchi Soba Ironai"
	F-59		JOMO Oil Ironai service station	←	vacant land
	F-60		Takagi Shōji Co., Ltd.	←	La Glasse Otaru Station Square
	F-61		Mazda Rent-A-Car Otaru	←	La Glasse Otaru Station Square
	F-62		Ikeda-tei	←	Mizuho Shōji-managed site
	F-63		Itaya Merchant Ship Co., Ltd. and affiliates	←	parking lot

District	Nos.	Hokkaido University survey (1986)	Hosei University survey (1998)	Hosei University survey (2006)	Hosei University survey (2016)
G *	G-64		Sugino Corp., Otaru sales office	vacant land	Plam Unga
	G-65		Sanko Co., Ltd. Otaru branch	(A) Bankruptcy (B) Esse Otaru Ironai Parking	warehouse (1F)/Sanko Co., Ltd. (1-3F)
	G-66		Otaru Grand Hotel Classic	←	unoccupied building
	G-67		Otaru Post Office (main branch)	←	Otaru Post Office (main branch)
	G-68		Otaru Post Office (main branch; together with G-67)	Otaru Post Office (main branch)	Otaru Post Office (main branch)
	G-69		Otaru Bine	←	Otaru Bine
	G-70		Matsuda Bldg.	←	Matsuda Bldg.
	G-71		Otaru City branch offices (Otaru City Museum of Art, Otaru Literature Museum)	←	Otaru Literature Museum, Otaru City Museum of Art
	G-72		Bank of Japan, Otaru branch	Bank of Japan Otaru Museum	Bank of Japan Otaru Museum
	G-73		Hello Work Otaru public employment security office	←	Hello Work Otaru, Otaru public employment security office

District	Nos.	Hokkaido University survey (1986)	Hosei University survey (1998)	Hosei University survey (2006)	Hosei University survey (2016)
G	G-74		Otaru Ekisaikai Hospital	←	vacant lot
	G-75		TFC Topgent Fashion Core Co., Ltd.	←	TFC Topgent Fashion Core Co., Ltd.
	G-76		KAZU Dyeing Atelier	KAZU Dyeing Atelier (1F: workshop, 2F: shop)	KAZU Dyeing Atelier (1F: workshop, 2F: shop)
	G-77		… (parking lot)	Ironai Ichibankan (1F: Sushi Nakaichi, 2F: Sapporo Beer Garden Hokuyōtei	vacant lot
	G-78		Japan Tobacco, Inc. Otaru sales office	←	Otaru Ironai Ichōme-ten new construction site
	G-79		Sakura Bank, Otaru branch	unoccupied building	Nitori Otaru Art Base Japan Museum of Modern Art
	G-80		Sekiyu Kaiun Co., Ltd.	Magazine 928	Next Step Otaru Ironai Parking
	G-81		Petersburg Museum parking lot	Hotel Vibrnt Otaru and Otaru Music Box Waltz Tearoom	Hotel Vivrant Otaru parking lot
	G-82		Petersburg Museum	Hotel Vibrnt Otaru and Otaru Music Box Waltz Tearoom	Hotel Vivrant Otaru parking lot

District	Nos.	Hokkaido University survey (1986)	Hosei University survey (1998)	Hosei University survey (2006)	Hosei University survey (2016)
G	G-85		Hokkaido Chūō Bus and Building Maintenance	1F: Otaru Canal Terminal, 2F: Hokkaido Chūō Bus and Building Maintenance	Hokkaido Chūō Ironai Office, Chūō Building Maintenance Co. Otaru branch & Chūō Bus Business Service
	G-86		Hokkaido Chūō Bus and Building Maintenance, No. 2 Bldg.	Ironai sales office	1F: Otaru Canal Terminal, 2F: Hokkaido Chūō Bus and Building Maintenance
	G-87		Natori Co., Ltd. Otaru branch	Otaru Taishō Glass Palace	Taishō Glass (1F)/Natori Co., Ltd. (2F)

District	Nos.	Hokkaido University survey (1986)	Hosei University survey (1998)	Hosei University survey (2006)	Hosei University survey (2016)
H	88	Nihon Gas station	Taiheiyō Sekiyu Bar Park Sakaimachi	Taruseki Sakaimachi parking lot	Lawson convenience store, Otaru Sakaimachi shop
	89	Morikawa Bldg.	←	←	Morikawa Bldg. and Taishō Glass Shop Utsuwaya
	90	Taguchi Shōten, Metoki Pharmacy, Morikawa Shōten	(1) Sushi Isshin, (2) Metoki Pharmacy, (3) parking lot	(1)←/(2)←/(3) Sushi Saikō	(1) Ramen Rikyu-tei (2) Metoki Pharmacy (3) Sushi Saikō
	91	Akiyama Aiseikan	parking lot	**"Shussemae Hiroba" and** parking lot	**"Shussemae Hiroba"**
	92	Taguchi Shōten	Craft Shop Mizubashō	←	Craft Shop Mizubashō
	93	Ace Sangyō Ltd.	←	Manjirō	Manjirō
	94	Hokkaido Chūō National Kaden Hanbai Co., Ltd.	Morikawa Sangyō Co., No. 2 distribution center	**(1) K's Blowing (2) Il Ponte**	(1) The Uroko, (2) Uroko-tei, (3) Kaisenshō
	95	Hokkaido National Service Co. Otaru service station	Otaru Ishi no Kura	←	Otaru Ishi no Kura
	96	Kogama Shōten Ltd.	unknown	**"Otaru Music Box Museum Kazariya"**	Glass Gallery

District	Nos.	Hokkaido University survey (1986)	Hosei University survey (1998)	Hosei University survey (2006)	Hosei University survey (2016)
H	97	Otaru Densō Co., Ltd.	←	Rishiriya Minoya (Taishō Coop-kan)	Rishiriya Minoya (Taishō Coop-kan)
	98	Hokkaido Mail Transportation	Monthly parking lot □2	← (parking lot for Rishiriya Minoya customers)	(parking lot for Rishiriya Minoya customers)
	99	Fukaya Denki Co., Ltd.	Otaru Sushi Restaurant Takiji	←	Otaru Take no Sushi, Otaru Ezoya
	100	Fukaya Denki warehouse	Yōrōnotaki	Iso Sushi	Iso Sushi
	101	Fukaya Denki warehouse	Yōrōnotaki	Iso Sushi	Iso Sushi
	102	Ise Shōten, Kojin Ise Taxi	←	"Craft House," Ise Shōten	Ise Shōten
	106	Chiakian	(unoccupied building)	Otaru Romankan/Café Deco	Otaru Romankan/Café Deco
	107	Sanritsu Denki Co. garage	Family Restaurant Raku	unoccupied building	Ruri Kōbō
	108	warehouse	←	unoccupied building	parking lot (Ruri Kōbō)
	109	Shokusan Shōkai	Otaru bean shop Maruji	Craft Kōbō Otaru Club	Taishō Glass Uchū (Sora)
	110	Iwanaga Clock Shop	←	Iwanaga Clock Shop & Café	Otaru Music Box Museum Sakaimachi
	111	Hokkaido Hayashiya Tea Co., Ltd.	Hokkaido Hayashiya Tea Co., Ltd.	Rishiriya Minoya Furōkan/old tool museum	Rishiriya Minoya Furōkan

District	Nos.	Hokkaido University survey (1986)	Hosei University survey (1998)	Hosei University survey (2006)	Hosei University survey (2016)
H	112	Hokkaido Hayashiya Tea Co., Ltd.	Hokkaido Hayashiya Tea Co., Ltd. warehouse	Kagetsudō・Café F Moon	Musix Box Hall Kaimeirō & Café Jimeikin
	113	Bisen Ltd.	Kataoka Eiji residence	Hikari to Kaori Yakata Oatru/residence	Handmade Glass Otaru Aya & Otaru Glass Lights
	114	Tsujichu textile shop	←	**(Left) Italian café Ō**ruri, (Right) Textile shop Tsuji Tadashi	**(Left) Hokkaido Creamy's** (Right) Textile shop Tsuji Tadashi
	115	Fujita Shōten (rented building)	Greengrocer --> Maruyama Liquor--> garage	←	(Left) Shichifuku (Right) monthly parking lot
	116	Nisshin Trading Co., Ltd.	←	Otaru Glass no Akari	(Left) Seafood restaurant Donburi Chaya (Right) Chisō Mono Hikozō
	117	Hoshi Itō Co., Ltd.	←	Takeyama Co., Ltd. Otaru branch	Takeyama Co., Ltd. Otaru branch
	118	Unknown metal structure	(1F) Maekawa Shōten (2F) empty	(1F) Otaru Kani Dōjō (2F) Defy Store	(1F) Ore no Genghis Khan (2F) Ore no Curry Rice
	119	Hokuto Bussan	Ishiya Seiji residence	←	Ishiya Seiji residence (and warehouse)
	120	Okuno Co., Ltd.	←	←	Okuno Co., Ltd.
	121	Okuno Co., Ltd.	←	←	Okuno Co., Ltd.

District	Nos.	Hokkaido University survey (1986)	Hosei University survey (1998)	Hosei University survey (2006)	Hosei University survey (2016)
H	122	Sasson Jidōsha Kōgyō	Asia Club, Otaru branch	**Souvenir store "paw pads"**	**Souvenir store "Kamuy paw pads"**
	123	Hokuren, Otaru branch	Hokuren Fudōkan	←	Hokuren Fudōkan/Otaru Beer
	124	Iwatani Corp.	Otaru Shizen Kōbō	←	Taishō Glass Sake Brewery
	125	Takeya Kōgei, Ltd.	Casa de Pancho Villa	Otaru House Island Sprit/Art Gallery Hokkikō	Otaru House Island Sprit/Art Gallery Hokkikō
	126	Garage	Souvenir store Canal Manjū	Tatsumi Sushi	Otaru Tatsumi Sushi, Sakaimachi shop
	190	—	Hoshi Itō	Hokuyaku Co., Ltd. parking lot **"Next Step Parking"**	Next Step Monthly Parking (Chūō Building Maintenance Co., Ltd.)

District	Nos.	Hokkaido University survey (1986)	Hosei University survey (1998)	Hosei University survey (2006)	Hosei University survey (2016)
I	103	Kamiya Shōji, Ltd.	Kaori Kōbō Phyton/Yūkidō	Kaori Kōbō Phyton/Gas Tō (Gaslight)	Kaori Kōbō Phyton/Gas Tō (Gaslight)
	104	Sakaiya	←	←	Sakaiya
	105	Sakaiya warehouse	←	←	Sun Fresco
	127	Residence	Horafuki Konbukan	←	Horafuki Konbukan
	128	Koyamaya Shōkai	Rishiriya Minoya Otarukan	←	Rishiriya Minoya
	129	Kyōwa Densetsu Co., Ltd.	Rishiriya Minoya, Ltd.	←	Takahashi Suisan
	130	Stone warehouse (unknown)	←	**"Otaru Kiku Sushi"**	**"Otaru Kiku Sushi"**
	131	Thick-walled warehouse	←	←	**"3Ring"**
	132	Miyashita Shōji Co., Ltd.	←	←	Taishō Glass Kanzashiya/3 Stones
	133	Uchida Shōji Sakaimachi warehouse	Otaru Yūyūsen (cruise)	F.O.B. Otaru	Sugi Beer Garden
	134	Hashitani Co., Ltd. Otaru branch office	←	←	Kaisen Donburiya Poseidon **"Aza"**
	135	Residence (Satō Hideyuki)	Satō Shōten	Satō Shōten	Satō Shōten
	136	Tanaka Shōten/Kawamura Pharmacy warehouse	←	Yukijirushi/Snow Brand Parlor Otaru	Snow Brand Parlor Otaru

District	Nos.	Hokkaido University survey (1986)	Hosei University survey (1998)	Hosei University survey (2006)	Hosei University survey (2016)
I	137	Tetsuya *Kizai* warehouse	Otaru Kaisen-tei/Yamabuki Shōten/*Kōshin Sakaimachi* direct sale shop	(1F) Yamabuki Shōten (2F) empty	(1F) Yamabuki Shōten (2F) storeroom
	138	Hokuryu Tosō Kōgyō Ltd.	Otaru Canal Ramen-kan	←	Ramen Misora
	139	Wooden warehouse	…(**Ramen**-kan parking lot)	**Hokkaido gentei "Fancy Shop" (pre**-fab shop)	Maccha Sweets-dokro Chawawa
	140	Kōyō Shōji Co., Ltd.	Kōyō/Channel/Moku	←	Kōyō/Channel/Moku
	141	Residence (Sasaki Yukio)	Otaru Green Farm	**"Uni-ya Shinkichi"** (1-2F)	Taishō Glass Hokkoriya
	142	Dai-ichi Jidōsha Kōgyō Co., Ltd.	←	Unknown, possibly vacant	Island　Spirit
	143	Sapporo Komatsu Forklift	Kaidōya	Under construction (renovation as LeTao Bldg. No. 2)	le chocolat LeTAO
	144	Sakai Shōkai	←	**"Nigiri Sushi Yokochō"**	Handmade chopstick shop Yūzen
	145	Imai Company, Ltd.	(1F) Fine Craft/Live Crab Otaru/Takahashi Suisan (2F) Imai Company	←	Fine Craft／souvenir gallery Otaru

District	Nos.	Hokkaido University survey (1986)	Hosei University survey (1998)	Hosei University survey (2006)	Hosei University survey (2016)
I	146	**Shin'ei Bane/Inazumi** warehouse	Fuji Sushi	Fuji Sushi & prefab store	Fuji Sushi & warehouse
	147	**Shin'ei Bane**	Fuji Sushi	←	Fuji Sushi
	148	**Shin'ya**	←	←	Yanshū Kita no Ryōba, Otaru Canal shop
	149	Nakazawa Fishnet stone warehouse	Ore no Otaru	Kita no Ryōba	Kita no Ryōba, Otaru Canal shop
	150	Warehouse of iron plate gables	Otaru Kaisen Donya seafood store (Abashiri Fisheries)	Kita no Ryōba	Kita no Ryōba, Otaru Canal shop
	151	Shiraishi Kōgyō Co., Ltd.	←	**"Otaru Ruri Kōbō"**	**"Otaru Ruri Kōbō"**
	152	**Café Atelier**	←	Kahisakan entrance/exit	Kahisakan entrance/exit
	153	Arata Shōkai Co., Ltd. delivery dept.	vacant land	Kahisakan entrance/exit	Kahisakan entrance/exit
	154	Miki Shōten	vacant land	**Kahisakan**	**Kahisakan**
	155	Otaru Kyūbin Kumiai Kyōdō Center	Venetian restaurant Marco Polo	←	Souvenir shop Kobushiya, Otaru shop
	156	Yamafuku Bussan Co., Ltd.	Gallery Venini	←	Venini

District	Nos.	Hokkaido University survey (1986)	Hosei University survey (1998)	Hosei University survey (2006)	Hosei University survey (2016)
I	157	Kanemori Otaru sales office	Kitaichi Venetian Art Museum	←	Kitaichi Venetian Art Museum
	158	Residence (Sano Seiichi)	kitaichi Venetian Art Museum	Fuku Sushi	Fuku Sushi, main shop
	159	Fujiyama Manufacturing Co., Ltd.	private house	(1F) Sushi Chōbashi (2) private house	Kaisendon En
	160	Building with unusual painted exterior	Uroko	←	Uroko Shōkai
	161	Sanyo Electric	Kitaichi Glass Building No. 5	←	LeTAO Pathos
	188	—	Okura Galleria	**Sen'en Kōbō** Yume Kurōdo/ (interior) Galleria	Japanese goods studio Yume Kurōdo
	189	—	Kitaichi Glass parking lot	←	Sashi Su Se So 2 & Minatomachi Parking

District	Nos.	Hokkaido University survey (1986)	Hosei University survey (1998)	Hosei University survey (2006)	Hosei University survey (2016)
J	162	Kita Nihon Hōsō	vacant (relocated?)	Gin no Kane Bldg. No. 6	Chirimen accessories **"Fuyūsha"/ Otaru** Bushō-kan
	163	Kita Nihon Hōsō	Otaru Bussan Kurosawa Shōten	Maruyasu Shōji	Maruyasu Corp.
	164	Bōeki Shōkan ONZCO	Maruyasu Shōji	←	Maruyasu Corp.
	165	Shōsan Shōji Co., Ltd., Otaru branch	Kitaichi Glass Crytal Building	←	Kitaichi Glass Crystal Building
	166	Shōsan Shōji Co., Ltd., Otaru branch	Kitaichi Glass Crytal Building	←	Kitaichi Glass Crystal Building
	167	Shōsan Shōji Co., Ltd., Otaru branch	Kitaichi Glass Crytal Building workshop	←	Kitaichi Glass Crystal Building **workshop ("Kitaichi Glass** Kengaku Kōjō")
	168	Hokkaido Salt Co., Ltd., Otaru sales office	←	Ajimasa	Nukumori Kōbō Otaru Fukurō
	169	Hokkaido Salt warehouse	←	Nukumori Kōbō Otaru Fukurō	Nukumori Kōbō Otaru Fukurō
	170	Hokkaido Salt warehouse	←	Hokkai Marumasa Shōten	Nukumori Kōbō Otaru Fukurō

District	Nos.	Hokkaido University survey (1986)	Hosei University survey (1998)	Hosei University survey (2006)	Hosei University survey (2016)
J	171	Yoshikawa Shōji Co., Ltd.	Karakuri Zoo	Karakuri Zoo & Giyaman warehouse	Otaru Taishō Glass Museum & *Tourist Information Center* (Sakaimachi Street Promotion Association)
	172	Zushi Shōji Co., Ltd.	Otaru Bussan Kurosawa Shōten/Daichi no Megumi/Hokkaido no Aji Ichiba (souvenirs)	(Left) glass shop, (Right) Otaru Bussan Kurosawa Shōten/Daichi no Megumi/Hokkaido no Aji Ichiba	(Left) glass shop
	173	Fukukawa Co., Ltd.	Otaru-dō	←	Créme Glacée LeTAO
	174	Carwash Tōshin	Otaru Music Box Hall No. 2 □ 3	#186 Otaru Music Box Hall 2	Otaru Music Box Hall 2
	175	Otaru Sakaimachi Post Office	←	←	Otaru Sakaimachi Post Office
	176	Hokkaido Tsūshin Densetsu Bldg.	Otaru Mesena	Souvenir Otaru-kan	Souvenir Otaru-kan
	177	Toide Bussan Co., Ltd.	Cake Gin no Kane Coffee Bldg. No. 1	←	Cake Gin no Kane Coffee Bldg. No. 1
	178	Kitaichi Glass Country	←	←	Kitaichi Glass Building No. 3 Country Floor

District	Nos.	Hokkaido University survey (1986)	Hosei University survey (1998)	Hosei University survey (2006)	Hosei University survey (2016)
J	179	Kitaichi Glass Gallery	←	←	Kitaichi Glass Building No. 3 Japanese Floor
	180	Kitaichi Glass Bldg. No. 3	←	←	Kitaichi Glass Building No. 3
	181	Kitaichi Glass Building No. 3 office	←	←	Kitaichi Glass Bldg. No. 3/offices
	182	Miyuki Shōji Co., Ltd.	transfer/subsequent use unknown	Kitaichi Glass Lamp Hall satellite shop	Kitaichi Wine Shop Vinoteca
	183	Nittō Corp.	←	(Right) Kitakarō, (left) Rokkatei Kita no Daichi Art Museum	(Right) Kitakarō, (left) Rokkatei Kita no Daichi Art Museum & shoe shop
	184	Kitaichi Glass Products Center	Local liquor shop Kitaichi (right), Kitaichi Glass outlet (left)	←	Local liquor shop Kitaichi (right), Kitaichi Glass outlet (left)
	185	Arata Shōkai Co., Ltd.	LeTAO	←	LeTAO
	186	Fujiya Furniture Center	Otaru Music Box Hall 2	←	Otaru Music Box Hall 2
	187		Kitaichi Glass main office (3F) & Loire (at front)/Kitaichi Glass parking	←	Kitaichi Glass main office (3F)/Kitaichi Plaza (1F)/FM Otaru Broadcasting Station

District	Nos.	Hokkaido University survey (1986)	Hosei University survey (1998)	Hosei University survey (2006)	Hosei University survey (2016)
			lot (1F)/FM Otaru Broadcasting Station (3F)		(3F)
			E 191	Otaru Feliz Church □4	petit wedding
			E 192	empty warehouse □5	Parking lot Canal Side Parking
				F F-34	Germe Otaru Chūō-dōri
				142 (new construction addition)	Otaru Sakaimachi Meiten Nanaho
				142 (new construction addition)	Power Stone ROP/Kinshō croquette & handmade store Sakaimachi
				J 189	Sushi Su Se So

(Notes)

■ The information is known to be incorrect, but cannot be revised until the cause is identified.

□1 **Official name: "Otaru Beer: Hamburger Restaurant Bikkuri Donkey Otaru"**

□2 Now just open space

□3 This appears to be an error in the Hokkaido University surveys. See #186.

□4 Building between E-64 and 65

□5 Building between E-67 and 68

* For 1986, the Hokkaido University survey areas F and G are unknown, as the original survey sheets have been lost.

Index

S. Horikawa, *Why Place Matters*,
https://doi.org/10.1007/978-3-030-71600-4

The manufacturer's authorised representative in the EU is Springer Nature Customer Service Centre GmbH, Europaplatz 3, 69115 Heidelberg, Germany. If you have any concerns regarding our products, please contact ProductSafety@springernature.com

Printed and bound by CPI Group (UK) Ltd, Croydon, CR0 4YY
30/09/2026
02225500-0006